Food chain integrity

© Woodhead Publishing Limited, 2011

Related titles:

Tracing pathogens in the food chain
(ISBN 978-1-84569-496-8)
Understanding the epidemiology of pathogens in production chains and processing environments is highly important for food safety control. Pathogen surveillance is also an essential part of a food-safety programme. This significant collection considers current and emerging techniques to trace pathogens in the food chain and the implications for food safety. Opening sections cover methods to identify and type foodborne pathogens and advances in surveillance techniques. Following chapters concentrate on epidemiological issues in specific food production chains and research situations in which the techniques described in the book are applied.

Foodborne pathogens Second edition
(ISBN 978-1-84569-362-6)
Effective control of pathogens continues to be of great importance to the food industry. The first edition of *Foodborne pathogens* quickly established itself as an essential guide for all those involved in the management of microbiological hazards at any stage in the food production chain. This major edition strengthens that reputation, with extensively revised and expanded coverage, including more than ten new chapters. Part I focuses on risk assessment and management in the food chain. Chapters in this section cover pathogen detection, microbial modelling, the risk assessment procedure, pathogen control in primary production, hygienic design and sanitation, among other topics. Parts II and III then review the management of key bacterial and non-bacterial foodborne pathogens.

Food authenticity and traceability
(ISBN 978-1-85573-526-2)
With recent problems such as the use of genetically modified ingredients and BSE, the need to trace and authenticate the contents of food products has never been more urgent. The first part of this authoritative collection reviews the range of established and new techniques for food authentication. Part II explores how such techniques are applied in particular sectors, whilst Part III reviews the recent developments in traceability systems for differing food products.

Details of these books and a complete list of Woodhead titles can be obtained by:

- visiting our web site at www.woodheadpublishing.com
- contacting Customer Services (e-mail: sales@woodheadpublishing.com; fax: +44 (0) 1223 832819; tel.: +44 (0) 1223 499140 ext. 130; address: Woodhead Publishing Limited, 80 High Street, Sawston, Cambridge CB22 3HJ, UK)

If you would like to receive information on forthcoming titles, please send your address details to: Francis Dodds (address, tel. and fax as above; e-mail: francis.dodds@woodheadpublishing.com). Please confirm which subject areas you are interested in.

© Woodhead Publishing Limited, 2011

Woodhead Publishing Series in Food Science, Technology and Nutrition:
Number 212

Food chain integrity

A holistic approach to food traceability, safety, quality and authenticity

Edited by
J. Hoorfar, K. Jordan, F. Butler and R. Prugger

Oxford Cambridge Philadelphia New Delhi

© Woodhead Publishing Limited, 2011

Published by Woodhead Publishing Limited,
80 High Street, Sawston, Cambridge CB22 3HJ, UK
www.woodheadpublishing.com

Woodhead Publishing, 1518 Walnut Street, Suite 1100, Philadelphia, PA 19102-3406, USA

Woodhead Publishing India Private Limited, G-2, Vardaan House, 7/28 Ansari Road,
Daryaganj, New Delhi – 110002, India
www.woodheadpublishingindia.com

First published 2011, Woodhead Publishing Limited
© Woodhead Publishing Limited, 2011
The authors have asserted their moral rights.

This book contains information obtained from authentic and highly regarded sources. Reprinted material is quoted with permission, and sources are indicated. Reasonable efforts have been made to publish reliable data and information, but the authors and the publisher cannot assume responsibility for the validity of all materials. Neither the authors nor the publisher, nor anyone else associated with this publication, shall be liable for any loss, damage or liability directly or indirectly caused or alleged to be caused by this book.

Neither this book nor any part may be reproduced or transmitted in any form or by any means, electronic or mechanical, including photocopying, microfilming and recording, or by any information storage or retrieval system, without permission in writing from Woodhead Publishing Limited.

The consent of Woodhead Publishing Limited does not extend to copying for general distribution, for promotion, for creating new works, or for resale. Specific permission must be obtained in writing from Woodhead Publishing Limited for such copying.

Trademark notice: Product or corporate names may be trademarks or registered trademarks, and are used only for identification and explanation, without intent to infringe.

British Library Cataloguing in Publication Data
A catalogue record for this book is available from the British Library.

ISBN 978-0-85709-068-3 (print)
ISBN 978-0-85709-251-9 (online)
ISSN 2042-8049 Woodhead Publishing Series in Food Science, Technology and Nutrition (print)
ISSN 2042-8057 Woodhead Publishing Series in Food Science, Technology and Nutrition (online)

The publisher's policy is to use permanent paper from mills that operate a sustainable forestry policy, and which has been manufactured from pulp which is processed using acid-free and elemental chlorine-free practices. Furthermore, the publisher ensures that the text paper and cover board used have met acceptable environmental accreditation standards.

Typeset by Godiva Publishing Services Limited, Coventry, West Midlands, UK

© Woodhead Publishing Limited, 2011

Contents

© Woodhead Publishing Limited, 2011

© Woodhead Publishing Limited, 2011

Part II Food safety and quality

© Woodhead Publishing Limited, 2011

© Woodhead Publishing Limited, 2011

© Woodhead Publishing Limited, 2011

© Woodhead Publishing Limited, 2011

© Woodhead Publishing Limited, 2011

Contributor contact details

(* = main contact)

Editors and Chapter 18

J. Hoorfar*
National Food Institute
Technical University of Denmark
Mørkhøj Bygade 28, Block H
DK-2860 Søborg
Denmark
E-mail: jhoo@food.dtu.dk

K. N. Jordan
Teagasc, Food Research Centre
Moorepark
Fermoy
Co. Cork
Ireland
E-mail: kieran.jordan@teagasc.ie

F. Butler
UCD School of Agriculture, Food Science and Veterinary Medicine
University College Dublin
Belfield
Dublin 4
Ireland
E-mail: f.butler@ucd.ie

R. Prugger
Tecnoalimenti S.C.p.A
Via G. Fara 39
20124 Milano
Italy
E-mail: r.prugger@tecnoalimenti.com

Chapter 1

V. Morreale* and M. Puccio
Intelligent Systems Laboratory
Research & Development Department
Engineering Ingegneria Informatica S.p.A.
Viale Regione Siciliana 7275
90146 Palermo
Italy
E-mail: vito.morreale@eng.it
michele.puccio@eng.it

© Woodhead Publishing Limited, 2011

N. Maiden
Centre for HCI Design
City University London
Northampton Square
London EC1V 0HB
UK
E-mail: n.a.m.maiden@city.ac.uk

J. Molina
Quality, Food Safety & Environment R&D Department
AINIA – Centro Tecnológico
C. Benjamin Franklin, 5–11
E46980 Valencia
Spain
E-mail: jmolina@ainia.es

F. Rosines Garcia
Transport and Manufacturing Sector
Atos Research and Innovation Division
Atos Origin SAE
Av. Diagonal, 200
08018 Barcelona
Spain
E-mail: francesc.rosines@atosorigin.com

Chapter 2

K. N. Jordan*
Teagasc, Food Research Centre
Moorepark
Fermoy
Co. Cork
Ireland
E-mail: Kieran.Jordan@teagasc.ie

M. Wagner
Institute of Milk Hygiene, Milk Technology and Food Science
University of Veterinary Medicine
Veterinärplatz 1
1210 Vienna
Austria
E-mail: martin.wagner@vetmeduni.ac.at

J. Hoorfar
National Food Institute
Technical University of Denmark
Mørkhøj Bygade 28, Block H
DK-2860 Søborg
Denmark
E-mail: jhoo@food.dtu.dk

Chapter 3

M. Garcia Martinez*, A. Silva and J. R. O'Hanley
Kent Business School
University of Kent
Canterbury CT2 7PE
UK
E-mail: m.garcia@kent.ac.uk

Chapter 4

N. Marmiroli
University of Parma
Department of Environmental Sciences
Viale Usberti 11/A
43124 Parma
Italy
E-mail: nelson.marmiroli@unipr.it

Chapter 5

M. Jakobsen*
Department of Food Science
Faculty of Life Sciences, University of Copenhagen
Rolighedsvej 30
1958 Frederiksberg, Copenhagen
Denmark
E-mail: moj@life.ku.dk

© Woodhead Publishing Limited, 2011

Joanna Verran
Department of Biology, Chemistry and Health Science
Manchester Metropolitan University
UK

Liesbeth Jacxsens
Department of Food Safety and Food Quality
University of Ghent
Belgium

Bruno Biavati
Department of Agroenvironmental Sciences and Technologies
AlmaMaterStudiorumBologna University
Italy

Jordi Rovira
Facultad de Ciencias, Tecnología de los Alimentos
University of Burgos
Spain

Luca Cocolin
DIVAPRA
University of Turin
Italy

Chapter 6

J. Hoorfar*
National Food Institute
Technical University of Denmark
Mørkhøj Bygade 28, Block H
DK-2860 Søborg
Denmark
E-mail: jhoo@food.dtu.dk

D. N. Lees
CEFAS Weymouth Laboratory
Barrack Road
The Nothe
Weymouth DT4 8UB
UK
E-mail: david.n.lees@cefas.co.uk

A. Bosch
Virus Entèrics, Dep. Microbiologia
Universitat de Barcelona
Av. Diagonal 645
08028 Barcelona
Spain
E-mail: abosch@ub.edu

Chapter 7

F. Butler
UCD School of Agriculture, Food Science and Veterinary Medicine
University College Dublin
Belfield
Dublin 4
Ireland
E-mail: f.butler@ucd.ie

Chapter 8

M. Eden
TTZ Bremerhaven
Fischkai 1
27572 Bremerhaven
Germany
E-mail: eden@ttz-bremerhaven.de

© Woodhead Publishing Limited, 2011

Chapter 9

J. Hoorfar* and I. Bang-Berthelsen
National Food Institute
Technical University of Denmark
Mørkhøj Bygade 28, Block H
DK-2860 Søborg
Denmark
E-mail: jhoo@food.dtu.dk

F. T. Jones
Performance Poultry Consulting, LLC
PO Box 7107
Springdale, Arkansas
USA
E-mail: ftjones@uark.edu
performpoultry@gmail.com

P. Häggblom
Statens Veterinärmedicinska Anstalt
SE 751 89 Uppsala
Sweden
E-mail: per.haggblom@sva.se

G. Bruggeman
Vitamex N. V.
Booiebos 5
9031 Drongen
Ghent
Belgium
E-mail:
Geert.bruggeman@vitamex.com

J. Zentek
Institut für Tierernährung
Freie Universität Berlin
Brümmerstrasse 34
D 14195 Berlin
Germany
E-mail: zentek.juergen@vetmed.fu-berlin

Chapter 10

C. M. Burgess and G. Duffy*
Teagasc, Food Research Centre
Ashtown
Dublin 15
Ireland
E-mail: geraldine.duffy@teagasc.ie

Chapter 11

L. A. Boyle*
Pig Research Department
Animal & Grassland Research and Innovation Centre
Teagasc
Fermoy
Co. Cork
Ireland
E-mail: laura.boyle@teagasc.ie

K. M. O'Driscoll
Animal and Bioscience Research Department
Animal & Grassland Research and Innovation Centre
Teagasc
Grange
Co. Meath
Ireland
E-mail: keelin.odriscoll@teagasc.ie

Chapter 12

Y. Bertheau
National Institute for Agricultural Research (INRA)
Route de Saint Cyr
78026 Versailles cedex
France
E-mail:
yves.bertheau@versailles.inra.fr

© Woodhead Publishing Limited, 2011

Chapter 13

C. Brera*, B. De Santis, E. Prantera, M. De Giacomo and R. Onori
Italian National Institute of Health (ISS)
GMO and Mycotoxins Unit
Dept of Public Veterinary Health and Food Safety
Viale Regina Elena 299
00161 Roma
Italy
E-mail: carlo.brera@iss.it

Chapter 14

S. Kelly
Food and Environment Research Agency
Sand Hutton
York YO41 1LZ
UK
E-mail: simon.kelly@fera.gsi.gov.uk

Chapter 15

J. T. Martinsohn*
European Commission, DG Joint Research Centre
Institute for the Protection and Security of the Citizen
Maritime Affairs Unit
Via Enrico Fermi 2749 (TP 051)
I-21027 Ispra (VA)
Italy
E-mail: jann.martinsohn@jrc.ec.europa.eu

A. J. Geffen
Department of Biology
University of Bergen
PO Box 7800
5020 Bergen
Norway
E-mail: Audrey.Geffen@bio.uib.no

G. E. Maes
Katholieke Universiteit Leuven
Laboratory of Animal Diversity and Systematics
Charles Debériotstraat 32
3000 Leuven
Belgium
E-mail: Gregory.Maes@bio.kuleuven.be

E. E. Nielsen
Technical University of Denmark
National Institute of Aquatic Resources
DK-8600 Silkeborg
Denmark
E-mail: een@aqua.dtu.dk

R. Ogden
TRACE Wildlife Forensics Network
Royal Zoological Society of Scotland
Edinburgh EH12 6TS
UK
E-mail: rob.ogden@tracenetwork.org

R. S. Waples
NOAA Fisheries
Northwest Fisheries Science Center
2725 Montlake Blvd East
Seattle, WA 98112
USA
E-mail: Robin.Waples@noaa.gov

G. R. Carvalho
Molecular Ecology and Fisheries Genetics Laboratory
School of Biological Sciences, Environment Centre Wales
Bangor University
Bangor LL57 2UW
UK
E-mail: g.r.carvalho@bangor.ac.uk

© Woodhead Publishing Limited, 2011

Chapter 16

W. Verbeke
Ghent University
Department of Agricultural Economics
Coupure links 653
B-9000 Ghent
Belgium
E-mail: wim.verbeke@ugent.be

Chapter 17

M. Garcia Martinez* and F. M. Brofman Epelbaum
Kent Business School
University of Kent
Canterbury CT2 7PE
UK
E-mail: m.garcia@kent.ac.uk
fmb7@kent.ac.uk

© Woodhead Publishing Limited, 2011

Woodhead Publishing Series in Food Science, Technology and Nutrition

1 **Chilled foods: a comprehensive guide** *Edited by C. Dennis and M. Stringer*
2 **Yoghurt: science and technology** *A. Y. Tamime and R. K. Robinson*
3 **Food processing technology: principles and practice** *P. J. Fellows*
4 **Bender's dictionary of nutrition and food technology Sixth edition** *D. A. Bender*
5 **Determination of veterinary residues in food** *Edited by N. T. Crosby*
6 **Food contaminants: sources and surveillance** *Edited by C. Creaser and R. Purchase*
7 **Nitrates and nitrites in food and water** *Edited by M. J. Hill*
8 **Pesticide chemistry and bioscience: the food–environment challenge** *Edited by G. T. Brooks and T. Roberts*
9 **Pesticides: developments, impacts and controls** *Edited by G. A. Best and A. D. Ruthven*
10 **Dietary fibre: chemical and biological aspects** *Edited by D. A. T. Southgate, K. W. Waldron, I. T. Johnson and G. R. Fenwick*
11 **Vitamins and minerals in health and nutrition** *M. Tolonen*
12 **Technology of biscuits, crackers and cookies Second edition** *D. Manley*
13 **Instrumentation and sensors for the food industry** *Edited by E. Kress-Rogers*
14 **Food and cancer prevention: chemical and biological aspects** *Edited by K. W. Waldron, I. T. Johnson and G. R. Fenwick*
15 **Food colloids: proteins, lipids and polysaccharides** *Edited by E. Dickinson and B. Bergenstahl*
16 **Food emulsions and foams** *Edited by E. Dickinson*
17 **Maillard reactions in chemistry, food and health** *Edited by T. P. Labuza, V. Monnier, J. Baynes and J. O'Brien*
18 **The Maillard reaction in foods and medicine** *Edited by J. O'Brien, H. E. Nursten, M. J. Crabbe and J. M. Ames*
19 **Encapsulation and controlled release** *Edited by D. R. Karsa and R. A. Stephenson*
20 **Flavours and fragrances** *Edited by A. D. Swift*
21 **Feta and related cheeses** *Edited by A. Y. Tamime and R. K. Robinson*
22 **Biochemistry of milk products** *Edited by A. T. Andrews and J. R. Varley*
23 **Physical properties of foods and food processing systems** *M. J. Lewis*

© Woodhead Publishing Limited, 2011

24 **Food irradiation: a reference guide** *V. M. Wilkinson and G. Gould*
25 **Kent's technology of cereals: an introduction for students of food science and agriculture Fourth edition** *N. L. Kent and A. D. Evers*
26 **Biosensors for food analysis** *Edited by A. O. Scott*
27 **Separation processes in the food and biotechnology industries: principles and applications** *Edited by A. S. Grandison and M. J. Lewis*
28 **Handbook of indices of food quality and authenticity** *R. S. Singhal, P. K. Kulkarni and D. V. Rege*
29 **Principles and practices for the safe processing of foods** *D. A. Shapton and N. F. Shapton*
30 **Biscuit, cookie and cracker manufacturing manuals Volume 1: ingredients** *D. Manley*
31 **Biscuit, cookie and cracker manufacturing manuals Volume 2: biscuit doughs** *D. Manley*
32 **Biscuit, cookie and cracker manufacturing manuals Volume 3: biscuit dough piece forming** *D. Manley*
33 **Biscuit, cookie and cracker manufacturing manuals Volume 4: baking and cooling of biscuits** *D. Manley*
34 **Biscuit, cookie and cracker manufacturing manuals Volume 5: secondary processing in biscuit manufacturing** *D. Manley*
35 **Biscuit, cookie and cracker manufacturing manuals Volume 6: biscuit packaging and storage** *D. Manley*
36 **Practical dehydration Second edition** *M. Greensmith*
37 **Lawrie's meat science Sixth edition** *R. A. Lawrie*
38 **Yoghurt: science and technology Second edition** *A. Y. Tamime and R. K. Robinson*
39 **New ingredients in food processing: biochemistry and agriculture** *G. Linden and D. Lorient*
40 **Benders' dictionary of nutrition and food technology Seventh edition** *D. A. Bender and A. E. Bender*
41 **Technology of biscuits, crackers and cookies Third edition** *D. Manley*
42 **Food processing technology: principles and practice Second edition** *P. J. Fellows*
43 **Managing frozen foods** *Edited by C. J. Kennedy*
44 **Handbook of hydrocolloids** *Edited by G. O. Phillips and P. A. Williams*
45 **Food labelling** *Edited by J. R. Blanchfield*
46 **Cereal biotechnology** *Edited by P. C. Morris and J. H. Bryce*
47 **Food intolerance and the food industry** *Edited by T. Dean*
48 **The stability and shelf life of food** *Edited by D. Kilcast and P. Subramaniam*
49 **Functional foods: concept to product** *Edited by G. R. Gibson and C. M. Williams*
50 **Chilled foods: a comprehensive guide Second edition** *Edited by M. Stringer and C. Dennis*
51 **HACCP in the meat industry** *Edited by M. Brown*
52 **Biscuit, cracker and cookie recipes for the food industry** *D. Manley*
53 **Cereals processing technology** *Edited by G. Owens*
54 **Baking problems solved** *S. P. Cauvain and L. S. Young*
55 **Thermal technologies in food processing** *Edited by P. Richardson*
56 **Frying: improving quality** *Edited by J. B. Rossell*
57 **Food chemical safety Volume 1: contaminants** *Edited by D. Watson*
58 **Making the most of HACCP: learning from others' experience** *Edited by T. Mayes and S. Mortimore*
59 **Food process modelling** *Edited by L. M. M. Tijskens, M. L. A. T. M. Hertog and B. M. Nicolaï*

© Woodhead Publishing Limited, 2011

60 **EU food law: a practical guide** *Edited by K. Goodburn*
61 **Extrusion cooking: technologies and applications** *Edited by R. Guy*
62 **Auditing in the food industry: from safety and quality to environmental and other audits** *Edited by M. Dillon and C. Griffith*
63 **Handbook of herbs and spices Volume 1** *Edited by K. V. Peter*
64 **Food product development: maximising success** *M. Earle, R. Earle and A. Anderson*
65 **Instrumentation and sensors for the food industry Second edition** *Edited by E. Kress-Rogers and C. J. B. Brimelow*
66 **Food chemical safety Volume 2: additives** *Edited by D. Watson*
67 **Fruit and vegetable biotechnology** *Edited by V. Valpuesta*
68 **Foodborne pathogens: hazards, risk analysis and control** *Edited by C. de W. Blackburn and P. J. McClure*
69 **Meat refrigeration** *S. J. James and C. James*
70 **Lockhart and Wiseman's crop husbandry Eighth edition** *H. J. S. Finch, A. M. Samuel and G. P. F. Lane*
71 **Safety and quality issues in fish processing** *Edited by H. A. Bremner*
72 **Minimal processing technologies in the food industries** *Edited by T. Ohlsson and N. Bengtsson*
73 **Fruit and vegetable processing: improving quality** *Edited by W. Jongen*
74 **The nutrition handbook for food processors** *Edited by C. J. K. Henry and C. Chapman*
75 **Colour in food: improving quality** *Edited by D. MacDougall*
76 **Meat processing: improving quality** *Edited by J. P. Kerry, J. F. Kerry and D. A. Ledward*
77 **Microbiological risk assessment in food processing** *Edited by M. Brown and M. Stringer*
78 **Performance functional foods** *Edited by D. Watson*
79 **Functional dairy products Volume 1** *Edited by T. Mattila-Sandholm and M. Saarela*
80 **Taints and off-flavours in foods** *Edited by B. Baigrie*
81 **Yeasts in food** *Edited by T. Boekhout and V. Robert*
82 **Phytochemical functional foods** *Edited by I. T. Johnson and G. Williamson*
83 **Novel food packaging techniques** *Edited by R. Ahvenainen*
84 **Detecting pathogens in food** *Edited by T. A. McMeekin*
85 **Natural antimicrobials for the minimal processing of foods** *Edited by S. Roller*
86 **Texture in food Volume 1: semi-solid foods** *Edited by B. M. McKenna*
87 **Dairy processing: improving quality** *Edited by G. Smit*
88 **Hygiene in food processing: principles and practice** *Edited by H. L. M. Lelieveld, M. A. Mostert, B. White and J. Holah*
89 **Rapid and on-line instrumentation for food quality assurance** *Edited by I. Tothill*
90 **Sausage manufacture: principles and practice** *E. Essien*
91 **Environmentally friendly food processing** *Edited by B. Mattsson and U. Sonesson*
92 **Bread making: improving quality** *Edited by S. P. Cauvain*
93 **Food preservation techniques** *Edited by P. Zeuthen and L. Bøgh-Sørensen*
94 **Food authenticity and traceability** *Edited by M. Lees*
95 **Analytical methods for food additives** *R. Wood, L. Foster, A. Damant and P. Key*
96 **Handbook of herbs and spices Volume 2** *Edited by K. V. Peter*
97 **Texture in food Volume 2: solid foods** *Edited by D. Kilcast*
98 **Proteins in food processing** *Edited by R. Yada*
99 **Detecting foreign bodies in food** *Edited by M. Edwards*

© Woodhead Publishing Limited, 2011

100 **Understanding and measuring the shelf-life of food** *Edited by R. Steele*
101 **Poultry meat processing and quality** *Edited by G. Mead*
102 **Functional foods, ageing and degenerative disease** *Edited by C. Remacle and B. Reusens*
103 **Mycotoxins in food: detection and control** *Edited by N. Magan and M. Olsen*
104 **Improving the thermal processing of foods** *Edited by P. Richardson*
105 **Pesticide, veterinary and other residues in food** *Edited by D. Watson*
106 **Starch in food: structure, functions and applications** *Edited by A.-C. Eliasson*
107 **Functional foods, cardiovascular disease and diabetes** *Edited by A. Arnoldi*
108 **Brewing: science and practice** *D. E. Briggs, P. A. Brookes, R. Stevens and C. A. Boulton*
109 **Using cereal science and technology for the benefit of consumers: proceedings of the 12th International ICC Cereal and Bread Congress, 24–26th May, 2004, Harrogate, UK** *Edited by S. P. Cauvain, L. S. Young and S. Salmon*
110 **Improving the safety of fresh meat** *Edited by J. Sofos*
111 **Understanding pathogen behaviour in food: virulence, stress response and resistance** *Edited by M. Griffiths*
112 **The microwave processing of foods** *Edited by H. Schubert and M. Regier*
113 **Food safety control in the poultry industry** *Edited by G. Mead*
114 **Improving the safety of fresh fruit and vegetables** *Edited by W. Jongen*
115 **Food, diet and obesity** *Edited by D. Mela*
116 **Handbook of hygiene control in the food industry** *Edited by H. L. M. Lelieveld, M. A. Mostert and J. Holah*
117 **Detecting allergens in food** *Edited by S. Koppelman and S. Hefle*
118 **Improving the fat content of foods** *Edited by C. Williams and J. Buttriss*
119 **Improving traceability in food processing and distribution** *Edited by I. Smith and A. Furness*
120 **Flavour in food** *Edited by A. Voilley and P. Etievant*
121 **The Chorleywood bread process** *S. P. Cauvain and L. S. Young*
122 **Food spoilage microorganisms** *Edited by C. de W. Blackburn*
123 **Emerging foodborne pathogens** *Edited by Y. Motarjemi and M. Adams*
124 **Benders' dictionary of nutrition and food technology Eighth edition** *D. A. Bender*
125 **Optimising sweet taste in foods** *Edited by W. J. Spillane*
126 **Brewing: new technologies** *Edited by C. Bamforth*
127 **Handbook of herbs and spices Volume 3** *Edited by K. V. Peter*
128 **Lawrie's meat science Seventh edition** *R. A. Lawrie in collaboration with D. A. Ledward*
129 **Modifying lipids for use in food** *Edited by F. Gunstone*
130 **Meat products handbook: practical science and technology** *G. Feiner*
131 **Food consumption and disease risk: consumer–pathogen interactions** *Edited by M. Potter*
132 **Acrylamide and other hazardous compounds in heat-treated foods** *Edited by K. Skog and J. Alexander*
133 **Managing allergens in food** *Edited by C. Mills, H. Wichers and K. Hoffman-Sommergruber*
134 **Microbiological analysis of red meat, poultry and eggs** *Edited by G. Mead*
135 **Maximising the value of marine by-products** *Edited by F. Shahidi*
136 **Chemical migration and food contact materials** *Edited by K. Barnes, R. Sinclair and D. Watson*
137 **Understanding consumers of food products** *Edited by L. Frewer and H. van Trijp*
138 **Reducing salt in foods: practical strategies** *Edited by D. Kilcast and F. Angus*
139 **Modelling microorganisms in food** *Edited by S. Brul, S. Van Gerwen and M. Zwietering*

© Woodhead Publishing Limited, 2011

140 **Tamime and Robinson's Yoghurt: science and technology Third edition** *A. Y. Tamime and R. K. Robinson*
141 **Handbook of waste management and co-product recovery in food processing Volume 1** *Edited by K. W. Waldron*
142 **Improving the flavour of cheese** *Edited by B. Weimer*
143 **Novel food ingredients for weight control** *Edited by C. J. K. Henry*
144 **Consumer-led food product development** *Edited by H. MacFie*
145 **Functional dairy products Volume 2** *Edited by M. Saarela*
146 **Modifying flavour in food** *Edited by A. J. Taylor and J. Hort*
147 **Cheese problems solved** *Edited by P. L. H. McSweeney*
148 **Handbook of organic food safety and quality** *Edited by J. Cooper, C. Leifert and U. Niggli*
149 **Understanding and controlling the microstructure of complex foods** *Edited by D. J. McClements*
150 **Novel enzyme technology for food applications** *Edited by R. Rastall*
151 **Food preservation by pulsed electric fields: from research to application** *Edited by H. L. M. Lelieveld and S. W. H. de Haan*
152 **Technology of functional cereal products** *Edited by B. R. Hamaker*
153 **Case studies in food product development** *Edited by M. Earle and R. Earle*
154 **Delivery and controlled release of bioactives in foods and nutraceuticals** *Edited by N. Garti*
155 **Fruit and vegetable flavour: recent advances and future prospects** *Edited by B. Brückner and S. G. Wyllie*
156 **Food fortification and supplementation: technological, safety and regulatory aspects** *Edited by P. Berry Ottaway*
157 **Improving the health-promoting properties of fruit and vegetable products** *Edited by F. A. Tomás-Barberán and M. I. Gil*
158 **Improving seafood products for the consumer** *Edited by T. Børresen*
159 **In-pack processed foods: improving quality** *Edited by P. Richardson*
160 **Handbook of water and energy management in food processing** *Edited by J. Klemeš, R. Smith and J.-K. Kim*
161 **Environmentally compatible food packaging** *Edited by E. Chiellini*
162 **Improving farmed fish quality and safety** *Edited by Ø. Lie*
163 **Carbohydrate-active enzymes** *Edited by K.-H. Park*
164 **Chilled foods: a comprehensive guide Third edition** *Edited by M. Brown*
165 **Food for the ageing population** *Edited by M. M. Raats, C. P. G. M. de Groot and W. A. Van Staveren*
166 **Improving the sensory and nutritional quality of fresh meat** *Edited by J. P. Kerry and D. A. Ledward*
167 **Shellfish safety and quality** *Edited by S. E. Shumway and G. E. Rodrick*
168 **Functional and speciality beverage technology** *Edited by P. Paquin*
169 **Functional foods: principles and technology** *M. Guo*
170 **Endocrine-disrupting chemicals in food** *Edited by I. Shaw*
171 **Meals in science and practice: interdisciplinary research and business applications** *Edited by H. L. Meiselman*
172 **Food constituents and oral health: current status and future prospects** *Edited by M. Wilson*
173 **Handbook of hydrocolloids Second edition** *Edited by G. O. Phillips and P. A. Williams*
174 **Food processing technology: principles and practice Third edition** *P. J. Fellows*
175 **Science and technology of enrobed and filled chocolate, confectionery and bakery products** *Edited by G. Talbot*

© Woodhead Publishing Limited, 2011

176 **Foodborne pathogens: hazards, risk analysis and control Second edition** *Edited by C. de W. Blackburn and P. J. McClure*
177 **Designing functional foods: measuring and controlling food structure breakdown and absorption** *Edited by D. J. McClements and E. A. Decker*
178 **New technologies in aquaculture: improving production efficiency, quality and environmental management** *Edited by G. Burnell and G. Allan*
179 **More baking problems solved** *S. P. Cauvain and L. S. Young*
180 **Soft drink and fruit juice problems solved** *P. Ashurst and R. Hargitt*
181 **Biofilms in the food and beverage industries** *Edited by P. M. Fratamico, B. A. Annous and N. W. Gunther*
182 **Dairy-derived ingredients: food and neutraceutical uses** *Edited by M. Corredig*
183 **Handbook of waste management and co-product recovery in food processing Volume 2** *Edited by K. W. Waldron*
184 **Innovations in food labelling** *Edited by J. Albert*
185 **Delivering performance in food supply chains** *Edited by C. Mena and G. Stevens*
186 **Chemical deterioration and physical instability of food and beverages** *Edited by L. Skibsted, J. Risbo and M. Andersen*
187 **Managing wine quality Volume 1: viticulture and wine quality** *Edited by A. G. Reynolds*
188 **Improving the safety and quality of milk Volume 1: milk production and processing** *Edited by M. Griffiths*
189 **Improving the safety and quality of milk Volume 2: improving quality in milk products** *Edited by M. Griffiths*
190 **Cereal grains: assessing and managing quality** *Edited by C. Wrigley and I. Batey*
191 **Sensory analysis for food and beverage control: a practical guide** *Edited by D. Kilcast*
192 **Managing wine quality Volume 2: oenology and wine quality** *Edited by A. G. Reynolds*
193 **Winemaking problems solved** *Edited by C. E. Butzke*
194 **Environmental assessment and management in the food industry** *Edited by U. Sonesson, J. Berlin and F. Ziegler*
195 **Consumer-driven innovation in food and personal care products** *Edited by S. R. Jaeger and H. MacFie*
196 **Tracing pathogens in the food chain** *Edited by S. Brul, P. M. Fratamico and T. A. McMeekin*
197 **Case studies in novel food processing technologies: innovations in processing, packing and predictive modelling** *Edited by C. J. Doona, K. Kustin and F. E. Feeherry*
198 **Freeze-drying of pharmaceutical and food products** *T.-C. Hua, B.-L. Liu and H. Zhang*
199 **Oxidation in foods and beverages and antioxidant applications Volume 1: understanding mechanisms of oxidation and antioxidant activity** *Edited by E. A. Decker, R. J. Elias and D. J. McClements*
200 **Oxidation in foods and beverages and antioxidant applications Volume 2: management in different industry sectors** *Edited by E. A. Decker, R. J. Elias and D. J. McClements*
201 **Protective cultures, antimicrobial metabolites and bacteriophages of food and beverage biopreservation** *Edited by C. Lacroix*
202 **Separation, extraction and concentration processes in the food, beverage and nutraceutical industries** *Edited by S. S. H. Rizvi*
203 **Determining mycotoxins and mycotoxigenic fungi in food and feed** *Edited by S. De Saeger*
204 **Developing children's food products** *Edited by D. Kilcast and F. Angus*

© Woodhead Publishing Limited, 2011

205 **Functional foods: concept to product Second edition** *Edited by M. Saarela*
206 **Postharvest biology and technology of tropical and subtropical fruits Volume 1** *Edited by E. M. Yahia*
207 **Postharvest biology and technology of tropical and subtropical fruits Volume 2** *Edited by E. M. Yahia*
208 **Postharvest biology and technology of tropical and subtropical fruits Volume 3** *Edited by E. M. Yahia*
209 **Postharvest biology and technology of tropical and subtropical fruits Volume 4** *Edited by E. M. Yahia*
210 **Food and beverage stability and shelf-life** *Edited by D. Kilcast and P. Subramaniam*
211 **Processed meats: improving safety, nutrition and quality** *Edited by J. P. Kerry and J. F. Kerry*
212 **Food chain integrity: a holistic approach to food traceability, safety, quality and authenticity** *Edited by J. Hoorfar, K. Jordan, F. Butler and R. Prugger*
213 **Improving the safety and quality of eggs and egg products Volume 1** *Edited by Y. Nys, M. Bain and F. Van Immerseel*
214 **Improving the safety and quality of eggs and egg products Volume 2** *Edited by Y. Nys, M. Bain and F. Van Immerseel*
215 **Feed and fodder contamination: effects on livestock and food safety** *Edited by J. Fink-Gremmels*
216 **Hygiene in the design, construction and renovation of food processing factories** *Edited by H. L. M. Lelieveld and J. Holah*
217 **Technology of biscuits, crackers and cookies Fourth edition** *Edited by D. Manley*
218 **Nanotechnology in the food, beverage and nutraceutical industries** *Edited by Q. Huang*
219 **Rice quality** *K. R. Bhattacharya*
220 **Meat, poultry and seafood packaging** *Edited by J. P. Kerry*
221 **Reducing saturated fats in foods** *Edited by G. Talbot*
222 **Handbook of food proteins** *Edited by G. O. Phillips and P. A. Williams*
223 **Lifetime nutritional influences on cognition, behaviour and psychiatric illness** *Edited by D. Benton*

© Woodhead Publishing Limited, 2011

Foreword

Why focus on food chain integrity? Throughout the world, consumers expect safe food. However, on an ongoing basis, food safety issues arise, many of which have an international dimension, for example the 2010 listeriosis outbreak in Austria/Germany/Czech Republic, and the 2009 Salmonella outbreak in Ireland/England/Denmark. Increasingly, food chains are becoming global in their scope; and food chain integrity, which not only includes safety concerns but also origin fraud and quality, has become a central concern to the food industry and regulatory agencies everywhere. After the BSE scare in England in the late 1990s, the EU came under pressure to fund research into complete traceability systems along the food and feed chain with the objective of increasing consumer confidence. This called for ways to ensure that products can be linked to their source, while also protecting products of declared origin; and ways to trace genetically modified organisms, and other products or contaminants, based on recent biotechnology developments, from raw material origin to purchased food product. As a result, the EU, through the Sixth EC Framework Programme (2002–2006), has funded a significant amount of cooperative research in food chain integrity and traceability under Priority Area 5 of the programme. Traceability was one of eight distinct areas of Priority 5, resulting in over 14 different research projects, and a total research budget of more than €140 million. The projects involved up to 400 different participating institutes across 30 countries, and for EU trade reasons had significant international partnerships from around the world, including South America, Australia/New Zealand, the Far East, Russia and South Africa. In addition, there was a high industrial and SME participation from both within and from outside the EU. With many of the projects completed or nearing completion, much of this work is now coming to fruition. The background for this book is a joint effort by the coordinators of these research projects in the area of food chain

© Woodhead Publishing Limited, 2011

integrity to distil the main outcomes of the projects into a single collective publication. As part of a need to improve the dissemination of finished EU research projects, the European Commission facilitated this initiative through several meetings in Brussels with the project coordinators between 2009 and 2010 and decided to support a conference in 2011. The present book is to be published in connection with the conference and is intended to give a broad overview of the different technical and procedural aspects of food chain integrity.

The overall objective of the funded food chain integrity research was to increase consumer confidence in the food supply chain by strengthening technologies needed to ensure complete traceability, transparency and safety along the entire food and animal feed chains. The technologies investigated link products to their source, or declared origin, and facilitate detection of trace contamination or substances present, such as bacteria, toxins, or genetically modified organisms, from the raw material origin to the consumed food product.

A wide number of food sectors were covered, from beef, fish, water, olive oil, to different food formats: fresh, chilled, or farm livestock. Accordingly, a wide range of technologies have been investigated from sophisticated electronic sensors, navigation systems and software, to an extensive range of genomic molecular diagnostics and complex mathematical models. New products such as decision support tools, time–temperature indicators, tracking devices, pathogen detection methods, etc., have been developed. These are now at various stages of implementation, through the participation of industry and SMEs.

Scale of production is an important consideration in food chain integrity, as in many cases the technological solutions for large-scale processes may not be transferable to small- or medium-scale production. The research actions funded by the EU are now generally near completion and have generated a huge amount of data, results, and actual technologies. A single collective publication, as is represented in this volume, summarises the true scale of the advances that have been made in this area, and the excellent possibilities for technology transfer that exist. This publication has extensive cross referencing, where appropriate, to parallel publications and outcomes that have been generated by the individual funded projects. The book is, by its cross-disciplinary approach, a highly valuable resource for food industry practitioners, food regulators, academics and students active in the application and development of food chain integrity solutions.

As Commissioner responsible for European Research and Innovation, I am particularly happy to see that such a fruitful collaboration has resulted from such extensive cross project framework programme collaboration. This book is a timely and useful distillation of the achievements that have been made in protecting our food over the past decade. In addition it represents an important benchmark for both the food industry and for the consumer towards regaining full confidence in the food chain.

Máire Geoghegan-Quinn
European Commission
Directorate E – DG R&I

© Woodhead Publishing Limited, 2011

Part I

Tracing and tracking in the food chain

© Woodhead Publishing Limited, 2011

1

The role of service orientation in future web-based food traceability systems

V. Morreale and M. Puccio, Engineering Ingegneria Informatica S.p.A, Italy, N. Maiden, City University London, UK, J. Molina, AINIA Technological Centre, Spain and F. Rosines Garcia, Atos Origin SAE, Spain

Abstract: This chapter discusses the influences of the next wave of Internet and service orientation on the development of food traceability systems. In particular, it is argued that future food traceability systems should be implemented on the Web as an integration of both existing and new services. An enabler for such a development is a service-oriented reference architecture, which provides a specification for a future traceability management system. The adoption and integration of devices for identification, food analysis, etc., are also discussed in the context of the Internet of 'Things'. Finally two case histories of the successful implementation of the approach discussed in the chapter are presented.

Key words: future Internet, service-orientation, reference architecture, Internet of things, traceability systems.

1.1 Introduction

Traceability in food chains is 'the ability to trace and follow a food, feed, food-producing animal or substance intended to be, or expected to be incorporated into a food or feed, through all stages of production, processing and distribution' (Regulation EC no. 178/2002). It also entails the 'ability to trace the history, application or location of an entity, by means of recorded identification' (Moe, 1998). Traceability systems support processes through which information on food origins, attributes and processing technologies is gathered and made

© Woodhead Publishing Limited, 2011

available throughout a supply chain. Generally speaking, they are data recording and management systems designed to track forwards as well as trace backwards, following not only product flows (or 'items' as an abstract concept), but also product 'attributes'. Traceability contributes to the demonstration of the transparency of the supply chain through the use of verifiable records and labelling.

The purpose of a traceability system is to prevent the introduction of an unwanted agent (e.g. a pathogen) to the food chain. Should this occur, though, a system can reduce the magnitude and impact of such an incident by facilitating the identification of the product(s) and/or batches affected and by allowing the parties involved to specify what has occurred, determine when and where the lapse in food integrity took place in the supply chain, and indicate who is responsible. Traceability operations can range in scale from internal systems organized by individual companies (in-house traceability) to systems enabling interrelated and complete traceability in the whole supply chain (chain traceability). The speed of response and the reliability of information are key aspects of any solution to traceability. Efforts to improve these lead to a move from paper-based traceability to ICT-enabled and ICT-supported traceability.

From the ICT point of view, the traceability problem is simple and complex at the same time. Indeed, while the issues to resolve are generally understood – products must be tracked and traced during a specific part of their life cycle – normally this goal has to be pursued in very complex and heterogeneous situations and contexts. In other words, users can have different, party-specific, and sometimes conflicting requirements, needs and goals and different circumstances present distinct constraints, for example in the use of particular technologies. Companies within the same supply chain manage traceability in different ways according to sector-specific legal requirements, company quality management policies, predefined target markets and so on. Moreover the level of process automation and adoption of ICT can be very different within the same supply chain: from paper-based systems to very advanced ICT solutions integrating traceability with production and logistics processes. As a consequence, a system for supply-chain-level traceability should be appropriate for a very complex environment, in which the ICT systems, among other components, are heterogeneous.

New technologies are invented and become popular in the software industry each year, and sometimes even faster. Therefore, people have to switch to these new technologies quite quickly. As a consequence, previous technologies lose value and may even become worthless. Yet, the situation is even more complex than this, because the technologies themselves change and are available in different versions, without a guarantee that they are backwards compatible. Moreover, software systems rarely exist in isolation: they need to interoperate with new systems built using innovative technologies. Even when systems are completely built from scratch, they often span multiple technologies, sometimes both old and new. Also, the Internet and the Web offer so many opportunities, services, and information that traceability processes and operations could benefit from and should exploit.

© Woodhead Publishing Limited, 2011

Thus, in this era of the Internet, and due to all the above-mentioned technological issues, a modern approach to traceability should be adopted. Stand-alone or client-server traceability systems must be considered the past. Instead, novel operational and business models based on the Web, service-orientation, social networks, knowledge sharing and creation should be considered the foundations for the traceability systems of the present as well as those of the near future. As individuals, we are users of the evolving Internet; why should we not seek to bring our businesses with us?

1.1.1 Future Internet

The Internet is nowadays the most common 'highway' for data, information, knowledge, and experience exchanges all over the world. These exchanges enable a wide range of applications, services, and scenarios. The concept of 'computer networks', dating from the 1970s, only took into account fixed networked computing machines. The modern concept of 'networks' involves all devices endowed with (autonomous) information processing capabilities. Common examples of such devices are personal mobile devices (smart phones, tablets, etc.), but potentially any addressable device or sensor can play a part. Societies and cultures are becoming digital, and enterprises and businesses should be moving towards more internetworked approaches.

Since its beginning, the Internet has evolved from 'sharing' in (Web or Web 1.0) to 'contribution' in Web 2.0 (user-generated content). The next development is Web 3.0, the age of 'co-creation' (collaborative production of information and services). The evolution of the Internet and the move towards Web 3.0 has been supported by the convergence of telephony, video/TV and other multimedia content, all now delivered via the Internet (Papadimitriou, 2009). One thing is sure: the Internet is going to grow, evolve and impact all kinds of human activity. It will become the one common, global and 'ubiquitously' accessible infrastructure for exchanging (and creating) human knowledge, culture, and services, supporting mobility, businesses, economies, processes and our social lives. This evolution will be supported by innovative business models that are emerging (and may emerge in the future), which enable entities (including businesses and individuals) to provide and exploit services on the Future Internet (FI).

The main pillars of this FI are:

- *Internet by and for people*. The FI will support and facilitate everyday life. It will continuously empower through free exchange of ideas, virtual communities, and social networks.
- *Internet of contents and knowledge*. The FI will not just enable access to information and content. It will 'intelligently' support intellectual activities such as thinking, learning, reasoning and recollection.
- *Internet of things*. Today, devices such as computers, printers, actuators and mobile phones are connected to the Internet, but in the FI, any object around us, anywhere, at any time, will connect, creating a 'universally addressable continuum'.

© Woodhead Publishing Limited, 2011

- *Internet of services.* Loosely coupled entities will provide services such as resources, data or software that could be exploited and composed.

Therefore, what is the (near) future of traceability systems? Will current approaches still be appropriate? Of course, they will not.

1.2 The need for a novel approach to food traceability

Universal traceability systems for *any* product, *any* food chain and *any* company cannot exist, due to the great heterogeneity of specific situations, supply chains, products, food industry players and their needs. Instead, specialized traceability information systems and services have to exist and usually need to be customized to meet specific requirements and to provide the necessary level of flexibility (e.g. to cope with changes in processes and procedures). Most current systems do not support and integrate objective and reliable acquisition and analysis of information about product origin and features. Moreover most of them are for single companies, while it is well understood that food traceability should be managed at the level of entire supply chains. Finally many companies are still managing traceability issues and processes manually, without the support of ICT.

The above considerations lead to the following needs and/or opportunities:

- to manage traceability *at the level of supply chains*;
- to make existing (and heterogeneous) traceability information systems *interoperable* between one another and with new services;
- *to set up ICT-based traceability* in many food companies that still manage traceability by paper-based approaches;
- *flexibility and adaptability* in managing traceability (and other processes) within food companies: changes should not always lead to new software development;
- to provide *low-cost and open* solutions, which are easy to deploy, use, configure and update;
- to exploit *innovative (micro-)devices* and technologies for objective analysis of food (and potentially exploiting any ‘thing’ that provides (traceability) information);
- (valued-added) *services that use traceability information as a foundation* (e.g. management of food quality and safety, statistics, monitoring, certificates, etc.);
- to *preserve existing assets* (i.e. tools, procedures, processes, systems, etc.) as much as possible;
- to exploit the Internet and the Web as resources providing relevant (and updated) services, information and knowledge, etc.

It should be understood that traceability systems are not only ICT systems. They are socio-technical systems which involve a reciprocal interrelationship between humans and machines. Both parties are engaged in goal-directed

© Woodhead Publishing Limited, 2011

behaviour with the aim of creating the conditions for successful performance of an organization (i.e. a company, supply chain, society, etc.). Therefore, technologies alone are insufficient to bring about the desired conditions for integrated and reliable traceability: other changes are needed in the social conditions of the food chain. The reciprocal interrelationships between humans and machines in that environment are complex: the food chain needs to be modelled and analysed as a socio-technical system.

The list of needs and opportunities for traceability systems above was further analysed and elaborated in the TRACEBACK project. TRACEBACK is an integrated project supported by the European FP6 with the aim to create an integrated set of tools to develop European traceability systems for entire food supply chains, from the field to the shelf (www.traceback-ip.eu). The aim was to define a coherent set of more specific requirements and goals for a reliable and flexible approach to food traceability (Lockerbie *et al.*, 2008).

The wide range of stakeholder groups in the food chain have different priorities. Clearly, all players in the production of food have a major stake in improved traceability. The other key stakeholder groups are consumers, authorities, third-party certifiers and owners of related legacy systems, which are 'computer systems or application programs that continue to be used because of the prohibitive cost of replacing or redesigning them, despite their poor competitiveness and compatibility with modern equivalents' (Dennis Howe, Free Online dictionary of computing, http://foldoc.org). These are typically invaluable assets with embedded business logic representing years of developments, enhancements, and modifications. Consumers want to obtain information about a product easily and they want the information they procure to be relevant. Authorities need to see information relating to a product when a food crisis occurs, or when suspicion is raised about a product. Third-party certifiers provide certification for standards compliance and need to check information about a product. Owners of related legacy systems want to provide information to and receive information from more modern traceability solutions.

More specifically, players in the food chain need to be able to:

- obtain information about the history and quality of the product, its origin and journey through the food chain. This could include, for example details of tests undertaken, storage temperatures, processing conditions, ingredients and additives used;
- obtain information about the current status of a product, such as its current location and current temperature;
- process product information to obtain the information needed to make informed business decisions;
- generate alerts warning of problems with products, such as contamination, temperature abuse and exceeded limits, as soon as they are discovered;
- generate and manage action plans that respond to serious alerts. These would include the information needed to recall the product, identify the cause, handle the problem and notify other partners in the food chain.

© Woodhead Publishing Limited, 2011

All the above requirements and needs cannot be met only by 'configuring' or 'customizing' a general-purpose (or universal) solution to food traceability. What is really needed is a *paradigm shift*, i.e. a move from software development, configuration, and customization to **service integration** (SI) and **service-oriented architectures** (SOAs). Many systems and services for SI and SOAs are already available, though some of them are generic and not specifically for traceability. Could Google Maps be useful for traceability? Could reliable and certified data storage services (external to food companies) be more appropriate for the management of traceability information? Could business intelligence benefit traceability? Of course, they could, although most of these services are not natively interoperable. They need to evolve to be part of the Internet of services, where functionalities, systems, facilities, information and processes are available as web services in the context of SOAs.

1.3 Service-oriented architectures (SOAs) for traceability

Web services are a new breed of web application that users access via the Internet through well-defined interfaces independent of where the service is executed. They are self-contained, self-describing, modular applications that can be published, discovered, and invoked across the Internet by multiple service consumers. At its simplest, a web service is a piece of software that performs one or more operations, such as locate an address or compute the time to arrive. According to the World Wide Web Consortium (W3C, 2004), each web service needs to have an interface described in a machine-processable format. Other systems interact with the service in a manner prescribed by its description using messages structured according to different web standards. Furthermore services that are heterogeneous and run on different architectures and platforms can still communicate independent of language or platform. The result is that businesses can invoke services directly over the Internet to reuse successful software without needing to purchase and install it locally.

Service-centric systems integrate web services from different providers seamlessly into applications that discover, compose and monitor these services. Each implemented configuration of web services is called a SOA. A typical SOA consists of three different entities – the service provider, the service consumer and the service broker. The service provider offers services that consumers can invoke when they need to. To enable this to happen it develops descriptions of these services and publishes them via a service broker that is most often a registry of all available services. The service consumer wants to consume services to complete processes and tasks with the help of one or more services. To discover and select the right services, the service consumer makes requests to a service broker. When an adequate service is discovered, the service consumer uses the information from the service broker to invoke the services.

SOAs offer a number of major advantages over more traditional techniques with which to implement software applications. The most obvious advantage is

© Woodhead Publishing Limited, 2011

that services facilitate the reuse of popular and useful software more effectively – an organization can invoke a web service offered by a third-party service provider much more simply and easily than integrated software components into their local systems architectures. As a result service-based applications can be simpler. SOAs that are based on industry standards can also reduce the complexity of software when compared with integrating systems on a solution-by-solution basis. This standardization also leads to lower integration costs. They enable future applications to mesh seamlessly with existing standards-based services. Because of the standard interfaces that exist, applications are potentially easier and hence cheaper to maintain, which in turn frees up valuable technical staff for other more strategic work. The resulting applications can also be more flexible: the standardization of interfaces means that new services can, in principle, be discovered, invoked and replaced without much need to change the application code.

Solutions for the integration of heterogeneous systems based on different technical architectures and platforms are especially necessary when interaction with legacy systems (both within businesses and between different businesses) is required. In the context of traceability processes, existing IT legacy systems are a core asset for many organizations, as they usually manage the necessary traceability information. Therefore these systems need to be able to interact with other applications, systems, and services, regardless of their underlying technologies and data formats. In order to simplify this interaction and avoid implementing *ad-hoc* system interfaces, the 'legacy wrapping/service enablement' approach (i.e. exposing existing functionality through the use of wrapping and web services interfaces) can be adopted as part of the SOA development/deployment process. This is a non-invasive technique, which can reduce integration costs and incrementally re-engineer and transfer legacy components to a new platform without any replacement of the system (Erradi *et al.*, 2006).

Developing a reference model of services for traceability (which includes conditions such as what functions/operations the service has to provide or what data it has to share) could provide a common starting point for the development and deployment of service-oriented traceability systems. These systems would be able to exploit new and existing services with the aim of meeting the above-mentioned requirements as well as supply chain and company-specific needs. Going further than this, though, what is really needed is a specification for a traceability 'service network', together with a model of a set of tools that can easily and effectively exploit it. The service-oriented reference architecture for traceability described and discussed in the following sections aims to fulfil this need.

1.4 A service-oriented reference architecture for traceability

A novel approach to flexible traceability in the context of the Future Internet and service networks has been developed recently (http://is.eng.it/ratis). It relies on service-oriented **Reference Architecture for Traceability Information**

© Woodhead Publishing Limited, 2011

Systems (RATIS). A reference architecture is a coordinated set of specifications that defines 'the infrastructure common to several end systems and the interfaces of components that will be included in the end systems' (from W3C glossary, http://www.w3.org/TR/ws-gloss). A set of software tools and libraries for the development of systems according to the specifications of the RATIS has also been developed. This is called the RATIS Framework. Additional technical guidelines and design patterns to develop traceability information systems according to the RATIS and fully exploit the facilities offered by the RATIS Framework have also been developed.

The RATIS is defined according to a traceability reference model (TRM) and a set of recommended standards. The TRM defines the traceability problem domain, considering what elements/ideas are meaningful, relevant, appropriate, useful when solving that problem. It is an abstract model for understanding significant relationships among entities in the traceability environment; a minimal set of unifying concepts, axioms, constraints and relationships within a particular problem domain. The reference architecture is an abstract solution to the problems identified in the reference model. It defines the basis of classes of solutions, according to the reference model and general requirements (MacKenzie *et al.*, 2006). While the TRM is independent of specific technologies, implementations, and other concrete issues, the RATIS relies on and supports some technologies. Concrete (i.e. actual, real-world) architectures of actual traceability systems are specific solutions and describe how things actually work. Therefore, they solve the problems defined in the reference model, starting from and using the reference architectures, by adapting, specializing, and (if needed) extending it for the needs and purposes of specific contexts and situations. The following subsections provide a brief description of the key elements of the discussed approach, i.e. TRM, RATIS Specifications and Framework.

1.4.1 Traceability reference model (TRM)

A reference model can be used to organize concepts in a coherent framework. The TRM defines all the factors that contribute to the goal of guaranteeing traceability along food chains according to some agreed objectives. It also supports possibilities for wider application of the traceability concept, through the ability to trace new data dimensions or 'data levels', mainly relating to food quality and food safety. It was built on the basis of extensive research carried out in several food chains stretching from Turkey to Italy and from Spain to Germany, Poland, the United Kingdom and Finland. Fresh and processed tomatoes and dairy products (from animal feed to final product) were the sectors studied.

The TRM establishes the structure and necessary rules to adequately manage the generation of data that is produced or that should be produced in the interaction of the binomial 'Item-Process', the gathering of such data and their record to assure traceability along the food chain.

© Woodhead Publishing Limited, 2011

1.4.2 The Reference Architecture for Traceability Information Systems (RATIS)

The RATIS is a *coordinated set of* specifications on:

- *external services*: what services a company traceability information system is supposed to both *expose* and *provide* so as to implement traceability at the level of the entire supply chain;
- *internal services*: what services a company traceability information system is supposed to *require* from other internal information systems already owned by food players and *exposed* to them;
- *third-party services*: what new services (logically external to food players) are needed to manage and implement traceability at the level of the entire supply chain. Examples of such services are those associated with security, data storage, data analysis and processing. Within a service-oriented economy, these can be offered by third-party service providers, preventing the need for food companies to implement them, and saving resources (the costs of using these services should be less than the costs of developing and managing them).

The abstraction of 'service' is used here to define the basic building blocks of traceability information systems both at the level of single party and at the level of entire supply chains. It is worth mentioning that the notion of service focuses on *what* benefits, added-value, functionalities are offered, rather than *how* this is internally achieved and implemented. Therefore the RATIS, as a service-oriented reference architecture, complies with this approach, by defining what is expected from the services, rather than how they must be implemented. Then specifications of services need to be implemented in order to be 'really' available and exploitable. Several implementations of the same service specification might be possible. Indeed this is largely desired in a service economy: the aim is to create a service ecosystem in which new implementations of services with better performance, quality, or lower price might replace old ones.

A key element of the RATIS is that the physical flow of goods is complemented and supported by a parallel flow of interactions among several services distributed over the Web. As a result, information relevant for traceability might be distributed along the supply chain (i.e. the current and complete record of information about traceability of a given set of traceable units could consist of several pieces), even though the model also supports a central repository common to one or more supply chains. Thus, traceability processes and procedures involve distributed systems providing and requesting RATIS-based services and actions built and performed by people from different companies. According to this approach and model, services might be physically hosted and provided by traceability systems internal to food companies or by some external service providers that also guarantee management, integrity of data, security, backup and all qualities specified by given service level agreements (SLAs). With the aim of supporting the concept that 'the Web is the reference operating system for traceability information systems and services', on

© Woodhead Publishing Limited, 2011

top of such a network of services, several web applications, mainly fulfilling specific requirements, can be developed and made available to end users (e.g. food players, end consumers, other stakeholders, etc.).

The RATIS' services are specified by Unified Modelling Language (UML) models and the model elements (e.g. entities, services, concepts, interactions, processes) are further documented by additional descriptions in both formal and informal/textual languages. Thus the RATIS' specifications are provided at the three model-driven architecture (MDA) (Kleppe *et al.*, 2003) levels, i.e. Computation-Independent Model (CIM), Platform-Independent Model (PIM), and Platform-Specific Model (PSM), and focus on both services providing functionalities and information *used* by services (see Fig. 1.1).

The relationship between services and information is represented by a UML dependency, which means that 'the services are dependent on the information, as they are supposed to manage, to create, to exchange, to destroy, to store it'. The information models and their entities might be reused by several services. Moreover, while information models are mainly representations of static and structural entities, service models are supposed to represent the behavioural facets of systems and their components.

The RATIS specifications are organized as follows:

- *RATIS Business Service Specifications (BSS)* – the CIM of the RATIS. These include **business services**, identified and defined on the basis of the users' traceability system requirements (in terms of families of functionalities). They represent the solutions users desire to fulfil their requirements, goals, and needs. In the RATIS, business services are represented by UML package diagrams, in order to highlight the dependencies between them. They are described in more details using the RATIS Business Service Description. Business services are supposed to rely on and use the entities described in the RATIS **domain model**, a visual glossary of concepts, objects, elements that are relevant for traceability and the business services under consideration. They are represented by class diagrams, where the classes represent the domain entities.
- *RATIS Logical Service Specifications (LSS)* are the core of the RATIS, as they are supposed to represent services independent from the adopted platform and technologies. The LSS are supposed to be the RATIS' PIM, where services result from a mapping/transformation of the business services defined in BSS. Logical services are represented by classes in UML class diagrams, where classes are the services and operations are the functionalities offered by those services. The services are further described by the Logical Service Description, in which their semantics and functionalities are specified and documented in detail. Logical services are supposed to rely on and use entities described in the **logical information model**, which represents the detailed specification of the classes of objects that services are supposed to exchange.
- *RATIS Web Service Specifications (WSS)*, which are the platform specific models of services, in which a technology has been chosen, i.e. web services.

© Woodhead Publishing Limited, 2011

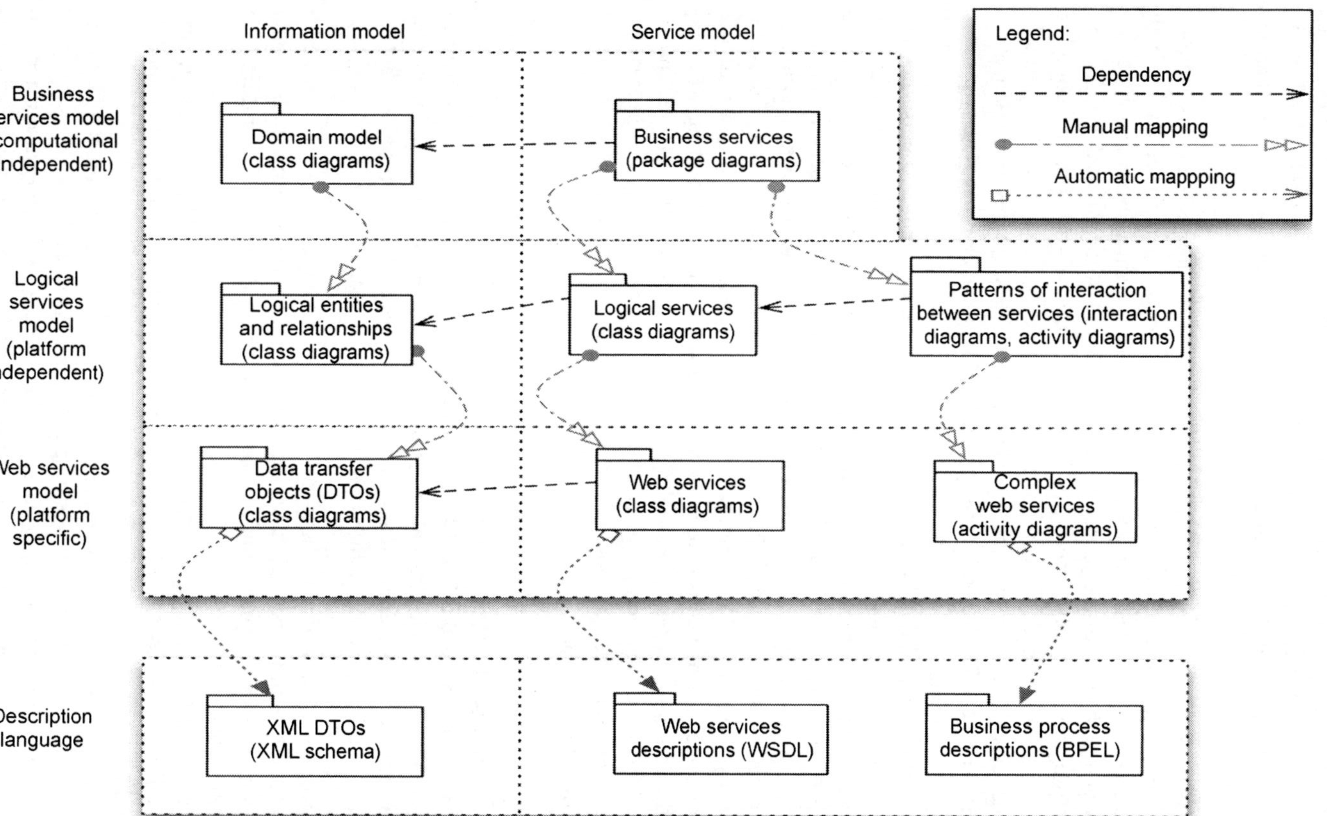

Fig. 1.1 RATIS service models.

Thus, in this layer, logical services are mapped/transformed into web services and represented as such. WSS includes an information model, which is the actual structure of data exchanged by web services during their interactions (i.e. data transfer objects – DTO).

RATIS WSS can be *automatically* transformed into service descriptions (specific for the adopted technology, i.e. web services) represented by Web Service Description Language (WSDL) and XML Schemas, from which can be easily generated part of the source code to implement or invoke such services (in many programming languages, e.g. Java, C#, and so forth). Finally, complex web services defined as compositions of web services can be partially automatically transformed into business process descriptions (represented in Business Process Execution Language – BPEL) that can be processed by process enactment systems.

It should be noted that business services are more than requirements: they are a set of abstract solutions that are supposed to fulfil requirements. Some business services are mapped/transformed into more than one logical service, while more than one business service can also be mapped into one logical service.

This reference architecture approach supports change and/or adaptation of the configuration and the behaviour of services developed in the following ways: by replacing old services with new ones (e.g. those have higher performance, are of better quality or are cheaper, etc.), by indicating where to install and deploy services and which is the best way to access them, and so forth. Moreover, the composition of services is the best way to create new value-added services, providing higher-level functionalities, with high quality and performance, etc.

1.4.3 The RATIS Framework

The idea underlying the RATIS Framework is to help software developers of *concrete* (i.e. real-world) traceability architectures to easily develop timely new implementations by using and reusing the facilities and ideas provided by the RATIS. The Framework includes the following elements:

- a *runtime infrastructure* where services can be deployed and executed. Some parts of this infrastructure have been developed by third parties under some Open Source licenses and were included and integrated within the Framework;
- *implementation of some services* defined within the RATIS, such as storage of information about external traceability, supply chain configurations, and so forth;
- *implementation of some general services* defined within the RATIS (e.g. authentication, authorization, storage, data analysis, etc.);
- *implementation of some web-based user interfaces and applications* that can be reused in many contexts and situations, such as reception or dispatch of goods;

© Woodhead Publishing Limited, 2011

- *Application Programming Interfaces (APIs)* supporting, making easier and quickening the development of new implementations of RATIS services and the supported applications.

The RATIS framework supports the development of traceability information systems as a *service integration task*. It aims to enable users to exploit and reuse new or existing (and useful) services to meet specific (and emerging in the future) needs, constraints, opportunities, configurations and goals. If a given component, system, subsystem, or device can provide one or more services defined in the RATIS, it can be part of the developed traceability systems as long as it is able to fulfil what the RATIS specifies, in terms of interfaces, protocols, manageability, and quality. The Framework helps bridge this gap.

1.5 The Internet of 'things' for traceability

Automatic identification and data capture (AIDC) refers to methods to automatically identify objects, collect data about them, and enter that data directly into computer systems without human involvement. Technologies typically considered part of AIDC include barcodes, radio frequency identification (RFID), optical character recognition (OCR), smart cards, and voice recognition. AIDC is normally introduced in traceability systems to collect and exchange information in processing chains faster than manual tools. AIDC allows fast, direct and secure exchange of information between information systems and the food processing chain, thus reducing the errors and costs. If identification standards are followed, items can be traced along the complete food chain. Up to this point, the most popular AIDC technologies have been based on barcodes, but other technologies are gaining in importance due to their innovative characteristics. RFID is the most widespread of these new technologies. Other important technologies are artificial vision, OCR, voice recognition and iris. Niche technologies include smart-labels, smartcards, biotags, Bluetooth and ZigBee. All these technologies need to be standardized to facilitate the common understanding of the descriptors used in the selected technology. The efforts of GS1 should be mentioned in this respect (www.gs1.org).

In the context of the Future Internet, the 'Internet of things' (IoT) refers to a world-wide network of interconnected objects, which are uniquely addressable, based on standard communication protocols. It is generally viewed as a self-configuring wireless network of sensors whose purpose would be to interconnect 'all' things (Conner, 2010). The IoT will create a dynamic network of billions or trillions of wireless identifiable 'things' communicating with one another, creating an addressable continuum of 'things' such as computers, sensors, actuators, TVs, vehicles, mobile phones, clothes, food, medicines, books, passports, luggage, etc. (Guillemin and Friess, 2009): all these 'things' will be able to exchange information. The IoT, together with AIDC technologies can assist in the implementation of food traceability and supply chain management processes. For instance, if radio frequency identification (RFID) tags are

© Woodhead Publishing Limited, 2011

attached to items, then traceability information can be stored and updated on the items themselves.

It is a known fact that two different devices might not be interoperable, even if they use the same standard. This is a major barrier to wide adoption of IoT technologies, which will remain an obstacle until global and well-defined standards have been adopted (Basi and Horn, 2008). In general, the integration of devices to the IoT is based on a three-tier approach:

- A first tier composed of the 'things' themselves. In the case of a food chain these are measurement devices and AIDC technologies in general, such as RFID.
- A second tier, which is a hardware abstraction layer facilitating object connectivity and mediating in data transmission and adaptation, as well as information modelling.
- A third tier composed of platform enablers, which contribute to data collection and analysis, remote management of objects, secure data, event communications management, and B2B interaction.

Machine-to-machine interaction is one of the main aspects of the IoT and is also very relevant to the food chain. The complete range of machines are involved: from large pieces of equipment to microcontrollers. Lab-on-a-chip devices are particularly worth mentioning in this context.

1.6 Developing traceability systems and services

The ultimate goal of the RATIS is to support, guide, and facilitate the development of service-oriented traceability information systems. Therefore the users of the RATIS are considered 'system and services integrators', who are supposed to provide the 'right solution' according to:

- *the real and specific needs* of the food companies and supply chains under consideration (further analysis by traceability experts may aid with that);
- *technological analysis* of the existing systems and services available within the companies under consideration (if any exist, they need to be integrated, not replaced);
- *parameters to be measured, data* to be captured and *devices* available;
- *services* made *available* on the Web by third-party service-providers;
- *new services* (specific for traceability) developed (or to be developed) by software and service developers.

The result of this service integration task should be company or supply chain specific traceability information systems, which fulfil specific requirements. Indeed, as argued by Golan *et al.* (2004), traceability systems can have different degrees of *breadth* (how many different types of information are included), *depth* (how many layers of the supply chain are included) and *precision* (the level of detail at which information is gathered). The selection of a traceability

© Woodhead Publishing Limited, 2011

system depends on trade-offs between costs and benefits, which in turn vary across industries and through time, reflecting market dynamics, changes in technologies and consumer preferences. Such a tailoring process requires a high degree of flexibility as well as rapid development, which are both enabled and supported generally by a service-oriented approach and specifically by the RATIS-based tools.

Every specific traceability information system developed according to the RATIS will be a distributed, consistent, relevant, and coordinated set of services provided and/or used by food players and other stakeholders (e.g. logistic providers, software service providers, analysts, etc.) through their (existing or new) software systems. Services are logically *distributed* between several players and their corresponding locations. However, logically distributed services can be provided physically by the same system. Services must be *consistent* in order to be properly selected, composed, and used by any party who requires them. This implies that they must be explicitly and clearly described. Services should be *relevant* to traceability issues, providing and requesting only relevant information about traceability processes. Finally, services should be *coordinated* in order to perform traceability-related processes and to create new complex services on the basis of existing services. Therefore, as shown in Fig. 1.2, each party/actor (represented by A_w, A_e, A_r, A_t, etc.) provides some services (lollipop notation) and needs some other services (socket notation) provided by someone else (services are represented with S_a, S_d, S_g, S_h, etc.). Thus, each party is responsible for providing and requesting some services

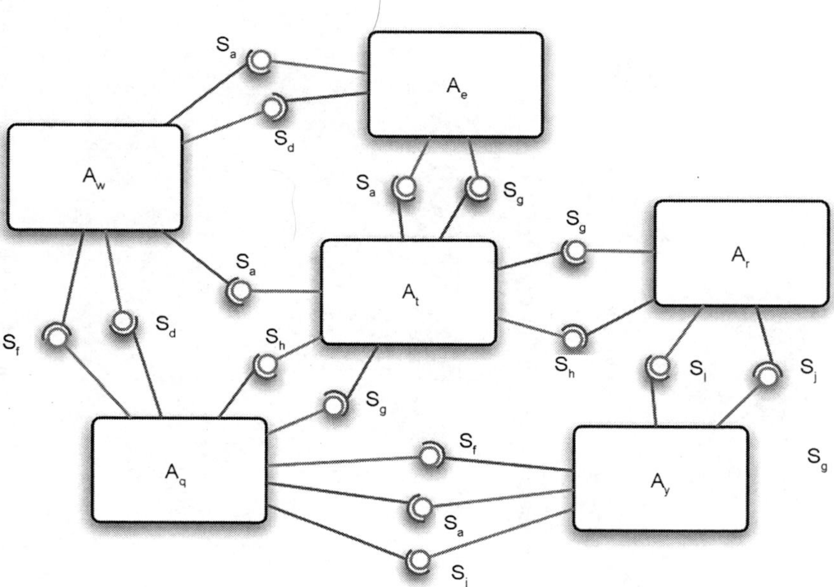

Fig. 1.2 Traceability information systems as a coordinated set of services.

© Woodhead Publishing Limited, 2011

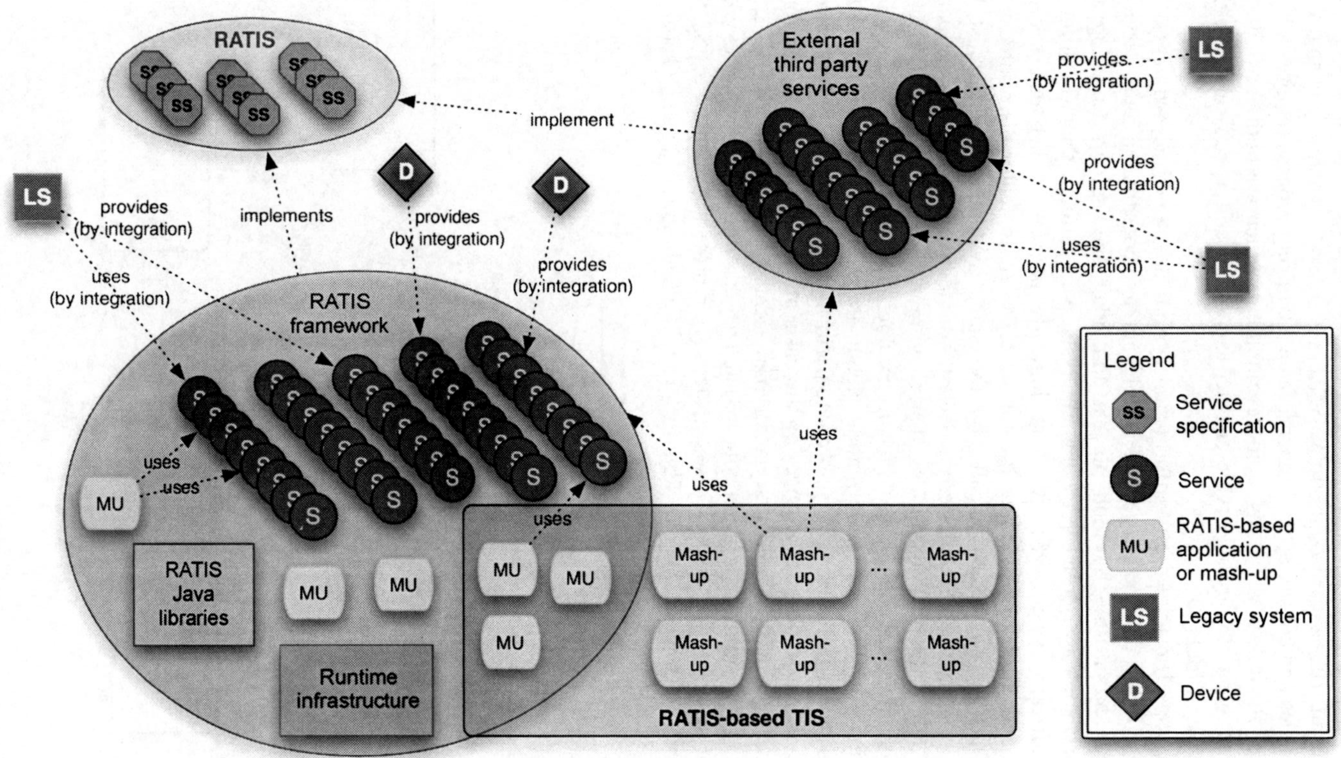

Fig. 1.3 TISs, RATIS framework, RATIS, third-party services, legacy systems, and devices.

© Woodhead Publishing Limited, 2011

according to the RATIS specifications and its role in the specific configuration of supply chain.

Traceability information systems developed according to the RATIS approach are intended to be an 'integrated' set of (web-based) applications or mash-ups (Fig. 1.3), which will let users access and use the system through a rich web-based interface (i.e. web browser). A mash-up is a web page or application that uses, integrates and combines data, services, presentation and/or functionality from two or more sources to create new services. These applications exploit the potential of new web-based interface technologies (i.e. Ajax, Google Web Toolkit, etc.) to provide users with many facilities normally provided by desktop applications (e.g. drag-and-drop, notifications, complex widgets, etc.). RATIS-based mash-ups will rely on and use RATIS service implementations. Some such services will be included within the RATIS framework, which also includes some built-in mash-ups that can be reused in several traceability information systems (TISs). Thus, in order to develop a new TIS, mash-ups that have been developed already and made available within the framework could be used. It should be noted that services have to be implemented according to the RATIS service specification. Some services could be developed by third parties (e.g. Google Maps, news, data storage, etc.) and made available to the RATIS service network to be integrated in RATIS-based TISs.

1.6.1 Real applications of RATIS

The RATIS-based approach has been applied to two real contexts, i.e. a tomato chain in Spain and a milk chain in Italy. The first application focused on a Spanish fresh tomato 'farm to fork' food chain (La Sala, Pasqual Marketing, Consum), tracing the 'kumato' variety. This chain traces the tomato from seed to supermarket distribution. The following roles were identified:

- plant nursery, developing small plants from seeds, and transport to tomato producer;
- tomato production in farms;
- selection and packaging of produced tomatoes;
- transport to logistic platform of product pallets;
- redistribution of pallets at the logistic platform and transport to supermarkets;
- reception in supermarkets and delivery to consumers.

Following the service-oriented approach, the participating companies' existing systems were connected together (previously they were not connected as each used different technologies) allowing information to flow between the companies with minimum changes to existing systems.

Safety and quality parameters are now measured, using proper devices according to the IoT vision. Several of these devices are new or are improvements of existing devices. Several of the parameters measured in different nodes of the chain include: shape, colour, defects, specific metals, moulds, origin,

© Woodhead Publishing Limited, 2011

gaseous emissions, temperature and humidity. Devices and techniques used include the electronic nose, spectrophotometer, machine vision and temperature and humidity active RFID. Additionally, two identification technologies have been included: barcodes and RFID. As part of this system, several scenarios are supported, such as dispatch, maturity prediction, product withdrawal and traceability certification. There is also support for product recall along the food chain, in case of a food safety crisis.

Many of the functionalities described above in relation to the tomato chain in Spain are common to the case of traceability of Parmalat fresh milk 'Blu Premium'. This chain stretches from the cattle feed producer to the supermarket and includes the milk producers, processors, warehouses and all transporters of goods in their various forms (i.e. feed, raw milk, pallets of processed milk, smallest boxes of milk, etc.). Eight food players have been linked to share information on traceability (such as movements of goods, processing and transformation, etc.) and quality and safety issues and problems. This allows them to properly manage traceability processes and crises both efficiently (i.e. in a timely fashion) and effectively (i.e. without information gaps and breaks).

One important feature of the systems developed is that the companies access them as web services and applications, by only using browsers and an Internet connection. Some of the services are also available on and accessible by personal mobile devices (e.g. smart phones, PDA). This ensures that the traceability system supports people's mobility and doesn't impact their normal operations.

1.7 Conclusions

The RATIS should be considered a tool that supports a new paradigm in traceability systems development. It facilitates the move to a service-oriented approach. Instead of producing traceability systems from scratch or reusing existing software, components, and frameworks, the RATIS encourages developers to move towards exploiting the (potentially huge and for sure evolving) network of available and relevant services. The underlying assumption is that many services already exist, so why shouldn't they be exploited? If something new needs to be developed, why not do it using a service-oriented approach following an existing specification?

The mains users of the RATIS are service and system integrators and this is who it is designed for. Systems designed using the RATIS benefit from the adoption of a service-oriented approach, based on the integration of available (web) services, which should be well managed, tested, widespread, and conform to an open specification. The benefits are evident throughout the whole life cycle of traceability systems, from development and integration, though deployment to maintenance and management. Anyone can develop services compliant to the RATIS and make them available within the network. They can build their business around such an activity, by selling software they develop or, a better

© Woodhead Publishing Limited, 2011

method, by selling use of those services. It should be noted that service developers cannot also be service providers.

What is the benefit for food companies if they ask for a RATIS-based traceability system, compared with alternative solutions? RATIS-based systems are supposed to be based on open specifications (i.e. the RATIS) and open, well-known and tested technologies and to be flexible, dynamic and adaptable thanks to the service-oriented approach. Services integrated into RATIS-based systems are provided by specialized operators and are reliable implementations. It should (hopefully) be cheaper to use those services than develop them from scratch. Users do not need to buy, install and manage expensive and complex components (such as database management systems): their services are provided remotely. Thus, the overall cost, quality and performance of such traceability systems should be improved by the adoption of RATIS approach and service-oriented computing.

1.8 Acknowledgement

This work is financially supported by the European Commission's TRACEBACK Project (FOOD-CT-036300).

1.9 References

BASI, A., HORN, G. (2008) *Internet of things in 2020.* Report on joint European Commission/EPoSS expert group workshop 11–12 Feb 2008. http://www.smart-systems-integration.org/public/internet-of-things.

CONNER, M. (2010) Sensors empower the 'Internet of Things'. *Electronics Design, Strategy, News*, 27 May, 31–37.

ERRADI, A. ANAND, S. KULKARNI, N. (2006) *Evaluation of Strategies for Integrating Legacy Applications as Services in a Service Oriented Architecture.* Proc. of IEEE International Conference on Services Computing, 2006, pp. 257–260.

GOLAN, E., KRISSOFF, B., KUCHLER, F., CALVIN, L., NELSON, K., PRICE, G. (2004) *Traceability in the U.S. Food Supply: Economic Theory and Industry Studies*, AER-830, USDA/ERS, March 2004.

GUILLEMIN, P., FRIESS, P. (2009) *Internet of Things – Strategic Research Roadmap.* http://ec.europa.eu/information_society/policy/rfid/documents/in_cerp.pdf.

KLEPPE, A., WARMER, J., BAST, W. (2003) *MDA Explained: The Model Driven Architecture™: Practice and Promise*, Addison Wesley.

LOCKERBIE, J., KARLSEN, K., PUCCIO, M., MORREALE, V., BONURA, S. (2008) Using Requirements to Define Services for Service-Centric Food Traceability Information Systems. In Proceedings of the 2008 International Workshop on Service-Oriented Computing Consequences for Engineering Requirements. IEEE Computer Society, pp. 15–23.

MACKENZIE, C., LASKEY, K., MCCABE, F., BROWN, P., METZ, R., HAMILTON, B. (2006) *Reference Model for Service Oriented Architecture 1.0.* OASIS Committee Specification.

© Woodhead Publishing Limited, 2011

MOE, T. (1998) Perspectives on traceability in food manufacture. *Trends in Food Science and Technology*, 9, 211–214.
PAPADIMITRIOU, D. (2009) *Future Internet – The Cross-ETP Vision Document.*
W3C (2004) *Web services architecture.* W3C Working Group Note 11 2004, http://www.w3.org/TR/2004/NOTE-ws-arch-20040211/.

© Woodhead Publishing Limited, 2011

2

Biotracing: a new integrated concept in food safety

K. N. Jordan, Teagasc, Ireland, M. Wagner, University of Veterinary Medicine, Vienna, Austria and J. Hoorfar, Technical University of Denmark, Denmark

Abstract: Traceability is a common concept in the food chain today. As generally understood, the concept refers to tracing the source of food from the dinner table back to the farm. The concept of bio-traceability is the tracing and tracking of bacteria (and their toxins) along the food chain. Tracing bacteria along the food chain is a new and complementary concept to quantitative microbial risk assessment (QMRA) and hazard analysis of critical control points (HACCP), two approaches that are already widely used. However, QMRA and HACCP are generally concerned with public health impacts and with hazard control at food production enterprises. A comprehensive concept for tracing hazards along a chain, up- and downstream, is lacking. Biotracing strives for that goal and attempts to identify the source of contamination by integrating food microbiology, food chain information and mathematical modelling. This chapter introduces the concept of biotracing and addresses some of the important questions relating to it.

Key words: foodborne pathogens, modelling, detection, tracing, tracking, typing, food safety.

2.1 Introduction to biotracing

Biotracing is tracing (backward) and tracking (forward) biological contamination in the food/feed chain. Advances in detection technologies, improvements in molecular marker identification, clearer understanding of pathogenicity markers, improved modelling methodologies and, more importantly, the

© Woodhead Publishing Limited, 2011

integration of these disciplines will lead to better capability in full-chain tracing and tracking of biological contaminations (Barker *et al.* 2009). This will lead to faster intervention in contamination events, limited product recalls and more targeted remedial action, resulting in a more rapid resolution of contamination incidents.

At a practical level, biotracing will combine finite and limited amounts of sampling with improved detection methods and models. It will bring about the ability to use downstream information to indicate materials, processes or actions within a particular food chain that may be sources of undesirable agents. Biotracing is a driver for improved food safety rather than a mechanism for establishing liability for a particular contamination event.

2.1.1 Biotracing is different from risk assessment and hazard analysis of critical control points (HACCP)

Biotracing is a concept that is fully complementary to already existing strategies to ensure food safety in processing. It facilitates food enterprises to make scientifically sound normative decisions on the origin of contamination and on the impact of a contamination scenario on posterior processing steps. The relationships between biotracing, HACCP and QMRA are shown in Fig. 2.1.

A unique aspect of biotracing, compared to HACCP or risk assessment, is virtual contamination scenarios; i.e. in biotracing, food producers work out for each food chain a contamination scenario that may or may not happen sometime in the future. The contamination can be inadvertent (accidental) or be caused on purpose as part of an attack on a food supply chain. For instance, an ice cream factory has a HACCP plan for the production site. Sampling and testing are carried out as part of HACCP. The ingredients such as pasteurized milk, egg and additives used in the production plant are provided by sub-contractors and it is the responsibility of the sub-contractor to guarantee that they are safe to use. This is normally done by documentation. Biotracing goes beyond this docu-

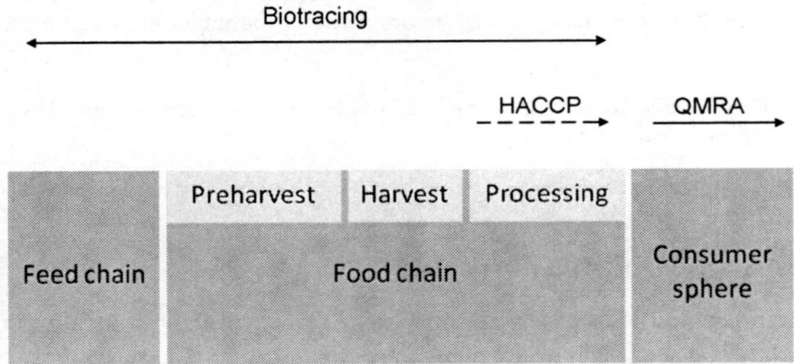

Fig. 2.1 Diagrammatic representation of the relationship between the novel concept of biotracing and the existing concepts of HACCP and QMRA.

© Woodhead Publishing Limited, 2011

mentation to include a sub-contractor's testing and data in the contamination scenario. Without biotracing, it can be difficult to find the source of contamination at the ingredient plant. In addition, does the ice cream factory have a plan for making a targeted product recall in case of problems after the product has left the factory? Is there a preparedness plan for targeted product recall?

Biotracing combines food microbiology knowledge and technology with complex systems approaches and modelling. From a food microbiology point of view it is used to answer questions such as what strain of pathogen can, or has, contaminated the chain, can detection capabilities be improved and what are the growth or no-growth boundaries of the pathogen in the various processes or production points in the chain.

The mathematical modelling aspect will identify a probable time and place of contamination using information about the chain and data from inline monitoring and, in addition, domain models can support improved predictions. All of this facilitates informed decision making in relation to remedial actions, interventions and cost-efficient recalls. While tracking and tracing are sometimes achieved through other systems, biotracing aims to bring a new level of integration, incorporating decision support tools based on modern techniques for pathogen detection and modelling with a more cost-effective sampling.

As biotracing is a new concept in data collection, improved detection and modelling, it goes beyond traditional modelling concepts such as predictive modelling and source attribution and it aims to complement established food safety assessment schemes by combining data flows with mathematical models to give quantitative information about the likely sources of particular hazards or potential problems. Biotracing relies on integration of valid analytical chains, from sampling to data interpretation, into modelling of actual feed and food chains and aims to operate within food chain systems in such a way that responses to safety signals are increasingly targeted at particular sources, and so are less intrusive and are ultimately more efficient. The combination of laboratory experiments and chain-related information flow leads to the all-embracing nature of biotracing; for example, a biotrace might include microbiological data, but also transport data and process failure data into a consistent modelling scheme that provides quantitative information about source level propositions.

2.2 Tools required for biotracing

Biotracing requires collective integration of laboratory methods and mathematical models to estimate and identify where the critical points are across the whole food chain and which contamination scenarios are the most relevant. The likely impacts on food safety can then be quantified using food chain models (Fig. 2.2).

Specifically, biotracing requires advances in the following areas.

© Woodhead Publishing Limited, 2011

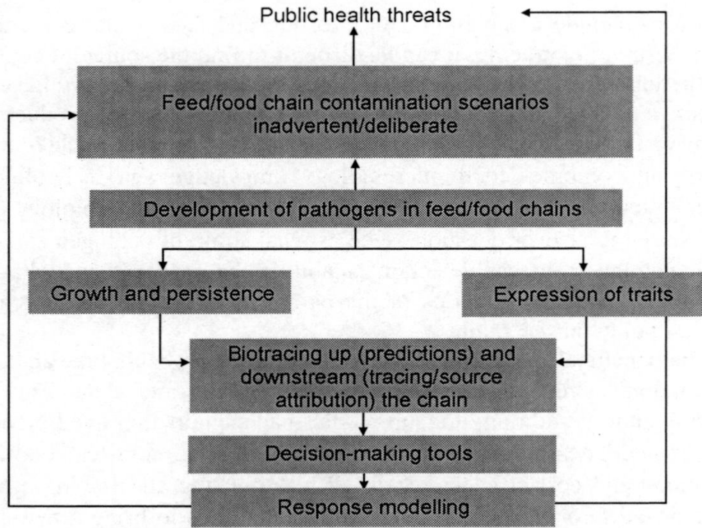

Fig. 2.2 The conceptual food chain modelling of biotracing.

2.2.1 Feed and food chain analysis

Traditional methods for microbial analysis of food borne pathogens are time consuming, have limited sample throughput, are not quantitative, and do not provide all the information needed to assess virulence potential. Even molecular genotyping, which usually involves some form of electrophoresis following culture, is labour intensive, time consuming and not easily automated. Biotracing requires reliable standardized test systems for fast and precise characterization of pathogenic microorganisms, i.e. viable pathogens that are above the known infectious dose for humans. Here, methods to be able to distinguish between viable and dead cells are crucial (Josefsen *et al.* 2010).

2.2.2 The impact of modelling

Biotraceability requires deeper knowledge of microorganisms than is currently available in order to estimate their persistence and virulence and to predict their behaviour in a particular food type at a particular stage of production (Dorrell *et al.* 2005). Currently, most microbial models for predicting the behaviour of microorganisms in food environments are based on laboratory studies (e.g. Combase and Micromodel). As microorganisms will behave differently in food, there are obvious limitations to this and food-based models are required (Schvartzman *et al.* 2010a).

2.2.3 Contamination control technologies

The traditional approach to traceability in the food industry is to bring about the ability to identify the source, and ideally the full history, of a product and its

© Woodhead Publishing Limited, 2011

ingredients. The term 'track and trace' is often used, with 'track' meaning to follow a product through its life and 'trace' meaning to find its origins. Current legislation obliges each company to provide their own records to food safety authorities on request, but there are no standards on the delivery media or format of the records. Biotracing requires standardized formats for data in order that information is comparable across different facilities.

2.2.4 Genome sequencing

Biotracing requires the application of molecular tools to analysis of bacteria so that maximum information can be obtained from a minimum amount of analysis. The sequencing of microbial genomes (http://www.ncbi.nlm.nih.gov/sites/entrez?db=genomeprj) has made it possible to construct whole genome DNA microarrays and to construct tailor-made PCR (polymerase chain reaction) typing methods that can be used to give useful information about the genetic composition of bacterial strains. However, to date, these advanced technologies, which can yield valuable information in a single step, have not been applied to analysis of pathogens in the food chain (Champion *et al.* 2005).

2.3.5 European feed and food law

The feed hygiene regulation (EC No. 183/2005, which became effective 1 January 2006) lays down a set of measures on hygiene, traceability and conditions and arrangements for establishments. The focus of the regulation is on the safety of animal feed, which is the primary responsibility of the operators in the feed chain. One of the most important aspects of this regulation is the registration of establishments and rules for record keeping and improved traceability throughout the entire feed chain. Furthermore, by making European food business operators (FBO) responsible for safe food production, the EU Directive No. 178/2002 has put a huge burden on their shoulders. Since the size of food business operators varies from small family businesses to large globalized companies, the management capabilities with regards to competing contamination scenarios varies dramatically (e.g. in the UK the smallest dairy producers typically process ~300 thousand litres of milk per annum, whereas the largest operations handle ~300 million litres a year). Biotracing is a concept that supports FBOs in taking the right, or at least justifiable managerial decisions.

2.3 Novel aspects of biotracing

In order to develop the concept of biotracing, the EU funded a 6th Framework project, BIOTRACER (www.biotracer.org). The project focused on the development of novel frontier technologies for exploitation in tracking and tracing microorganisms in selected models of the food and feed chains, and modelling the behaviour of pathogens in these environments. In biotracing,

© Woodhead Publishing Limited, 2011

modelling the flow of microorganisms is essentially coupled to decision making that is still done in many crisis situations on an heuristic as opposed to a scientific basis. It is this sound decision-making process that will allow improved responses and make them transparent and targeted.

The work has facilitated inter-disciplinary cooperation and added novel aspects to research that enhance the concept of biotracing:

- There has been integration of modelling studies of the different parts of the food/feed chain (domain modelling) resulting in increased awareness of food safety risk factors and what measures can be used to reduce the identified risks in the entire food chain (Pin *et al.* 2010).
- The integration of global expression data into models has been advanced. This allows a holistic view of the biological systems studied, as physiological data is integrated with information on virulence, gene expression and metabolite composition under particular conditions.
- Sophisticated data-mining and inference schemes have been developed (Barker *et al.* 2009 and http://bbn.ifrn.bbsrc.ac.uk/btmodeller). These can be used to analyse food-chain data relating to when, where and how products are transported within food chain systems and where and when contamination occurred, in addition to information on primary loads. In some cases these have been converted to user-friendly tools (software, expert systems).
- Risk assessments based on the tracing of metabolic markers and the physiological status of whole cells *in situ* have been undertaken.
- Food chain and mathematical models have been linked to developments in detection and tracing technologies. The matrix of data generated is the basis for predicting the behaviour of pathogens under given conditions in relevant scenarios (Manios *et al.* 2009, Schvartzman *et al.* 2010b).
- There have been developments in novel sampling and sample preparation technologies to overcome existing bottlenecks in end-use analytical laboratories (Olsen *et al.* 2009).
- More reliable and faster test systems based on molecular methods have been developed (Josefsen *et al.* 2007, Löfström *et al.* 2009a, 2009b, Fach *et al.* 2009, Walcher *et al.* 2010).
- Methodologies allowing direct tracing of protein-based substances (e.g. enterotoxins) through the food chain have been developed.

2.4 Strategic impacts of biotracing

Strategically, biotracing will foster the continuous traceability of micro-organisms along the food and feed chain, thus establishing a scientific basis for a traceability guideline applying to all food, animal feed, food-producing animals. It will result in an integrated understanding of the growth and spread of pathogenic microorganisms and their toxins in the food chain structures. Both novel data supplies and models will promote the identification of the most vulnerable chain structures, and their elements. This is in keeping with the aims

© Woodhead Publishing Limited, 2011

of the White Paper on Food Safety (Comm 199, 719 final), which calls for 'a comprehensive and integrated approach' to 'ensure a high level of human health and consumer protection'.

To date, selected food-chain/pathogen combinations have been studied, the results of which can be broadly applied in other areas. These include:

- mycotoxin and *Salmonella* traceability in the feed chain;
- *Campylobacter* spp. traceability in chickens;
- *Salmonella* spp. traceability in pigs;
- *Listeria* spp. and *S. aureus* traceability in dairy products;
- traceability of toxin and spore forming bacteria (*C. botulinum* and *B. cereus/anthracis* (bioterror);
- viruses in bottled water.

Most food and feed chains are complex trans-national structures involving non-European countries. The production of food and the threat of its contamination by microorganisms is not only a national issue, but must be handled on a European basis to develop common strategies, rules and policies. Because biotracing is an integrated, borderless concept, it will promote the production of food and feed that will have a positive health impact on European citizens. Thus, it will:

- enhance well-controlled food chains and improve the cost/yield ratio in food control and safety;
- reduce economic, health and social risks in the production of food and feed;
- contribute to implementation of the European Food Safety and Quality strategy;
- provide recommendations and tools for a more reliable decision process, as well as policy process;
- contribute to optimizing the rapid alert system and targeted response systems;
- create a basis for decision making to the European food industry and other institutions;
- integrate the results (completed to date) of biotracing into the development of the food law enforcement practitioners.

Biotracing will provide authorities and FBOs with guidelines and scientific advice on how to respond in different crises, how to facilitate laboratory-based testing excellence and appropriate diagnostic tools, and how to improve overall preparedness for hazard response actions.

The virtual situations will highlight bottlenecks and vulnerabilities associated with the tracing and tracking of etiological agents along the food chain. The proposed response to these scenarios will be a framework for general prohibitive efforts needed to detect risk accumulation and to strengthen response capabilities. Risk assessment must be based on solid scientific evidence, which helps decision making, including drafting of food safety legislation.

The contribution of biotracing to risk assessments is mainly in three areas. First, provision of new methods for fast and sensitive enumeration of pathogens,

© Woodhead Publishing Limited, 2011

since quantitative methods are necessary for quantitative risk assessments (Malorny *et al.* 2008, Krämer *et al.* 2010). Second, domain modelling to assess vulnerability of food production chains and to fill in the data gaps. Finally, development of new genetic markers that can be used as risk determinants (Champion *et al.* 2005).

2.5 Significance of biotracing for production chains

Food safety science is being driven by an expanding and increasingly accessible data supply and by increasingly sophisticated computational science. Fragmented data supplies, e.g. public health statistics, food surveillance studies, microbial research and process risk assessments, and mathematical models can be collected, integrating different elements of the food and feed chains, and different groups of stakeholders.

In particular, physiological data on survival, growth, virulence, gene expression and metabolite composition *in food/feed* has been translated into model systems such that the influence of the food matrix on physiology and detection can be investigated throughout the food chain. As these models encompass all steps in the food chain, from primary production to the consumer, biotracing will be able to trace microbiological contaminations for multiple combinations of food/pathogen at any point in the chain.

Contamination with a biological agent raises many unanswered questions and scientific hypotheses, which will be answered by biotracing. For example: is it possible to estimate the initial source of contamination? What actions can be suggested to prevent this contamination in the future? How should a a strategy to better inform producers, consumers and regulators be developed? The outcome of biotracing will facilitate:

- better understanding of the fundamental characteristics, particularly the basic processing conditions and agent dynamics, for each selected food and feed chain;
- definitions of food and feed chain structures, including time/space ordering, that embrace both the food processing and microbiological dynamics information to support biotraceability;
- provision of a systematic approach for quantitative food chain analysis, which can be applied to analyse the relevant food chains and to encompass the current status of food safety decision making;
- predicting the impact (endpoints of concern) and inferring the origins of (un)intentional agents that occur within the relevant food chain structures, thus leading to food chain intervention decision making, and to construction of a reliable communication structure with various stakeholders;
- provision of an organizational framework for existing information (i.e. a database), and a platform that supports improved decisions surrounding food safety.

© Woodhead Publishing Limited, 2011

2.5.1 Feed chain

The feed chain is an essential element of a safety concept for the production of high quality and safe human food from animal origin. It is well known that animal feed is frequently contaminated with *Salmonella,* causing infection of food animals, although tracing contamination in the food chain to its ultimate source is often difficult.

Biotracing covers relevant aspects in the whole feed chain from validation of sampling, analytical method development and validation of test methods for mycotoxins and *Salmonella*, and finally model generation. Advances have recently been made in reducing existing uncertainties in different areas, specifically in the knowledge of uncertainties in sampling of feedstuffs. The integration of sampling activities at different locations, comparison and cross-validation of analytical methods, and establishment of conceptual models for the introduction and spread of *Salmonella* in the complex feed chain have been made. Understanding of the complexity of the chain has facilitated new ideas for traceability tools that will be developed in the future.

In the feed chain, biotracing has been used in close collaboration between scientists and industry and has resulted in:

- validated cultural and rapid molecular methods for detection of *Salmonella* contamination in feed (Koyuncu and Häggblom 2009);
- rapid and economic methods for mycotoxin clean-up in feed (Reiter *et al.* 2009);
- knowledge on sampling uncertainty for feed contaminants (Andersson and Häggblom 2009).

2.5.2 Meat chain

Within the last decades, the pig and poultry industry has changed markedly. With the introduction of large, centralized breeding farms and subsequent transportation to centralized slaughterhouses, traceability is required for modern animal breeding and for effective generalized prophylaxis and disease control. The meat production and processing chain, including control measures, is directly affected by the desire of consumers to know more about the origin and health status of the poultry meat and pork they eat. Animals are the reservoir for a variety of organisms and can be carriers of human pathogenic bacteria. For this reason, strict controls are required along the food chain to prevent contamination. While these controls are successful in most cases, frequent failures occur, as evidenced, for example, by the numbers of *Salmonella* spp. related food-borne disease incidences and outbreaks (EFSA 2010). When these occur, tracking and tracing methodologies need to be used in order to minimize the associated risks to the consumer.

Campylobacter is one of the most important causes of bacterial gastro-intestinal disease worldwide (EFSA 2010). It is estimated that the cost is €500/incident. Given an estimate of 60 cases per 100 000 (the reported incidence varies between countries, probably due to differences in surveillance systems as

© Woodhead Publishing Limited, 2011

well as real differences in incidence) resulting in an annual cost to Europe of about €30 million (Duff *et al.* 2003). While the disease is usually self-limiting, it can result in severe and serious complications in approximately 1/1000 cases.

In the pork chain, biotracing has addressed issues of improving tracing methods for *Salmonella* spp., providing quantitative data based on genomic tools and led to the study of models that have the potential to incorporate the effects of environmental factors (Löfström *et al.* 2009a). This has resulted in the ability to predict the impact of *Salmonella* spp. persistence and behaviour in pork by adjusting/extending static predictive models for use in dynamic food chain structures and linked the adaptation of *Salmonella* Typhimurium in the pork production facilities with the feed chain. In addition, there are now improved and validated detection methods that contribute to source contamination identification (Josefsen *et al.* 2009, 2010).

In the poultry chain, biotracing has addressed the point of entry of *Campylobacter jejuni* in the poultry production chain, the impact of processing conditions on the physiology of the organism and modelling the behaviour of *Campylobacter* along the production chain for the purpose of tracing and tracking contamination (Olsen *et al.* 2009, González *et al.* 2009). This has resulted in advanced user-friendly tools for source attribution and demonstrated that contamination from the slaughterhouse environment is a significant factor in contaminating chicken meat.

2.5.3 Dairy chain

Listeria monocytogenes is a food-borne pathogen and although the incidence is not very high, the mortality rate of 20–30% is of great concern. Those most at risk include the young, the old, the pregnant and the immunocompromised. With an ageing population in certain regions, an increase in life expectancy, increasing occurrence (and treatment) of cancer and the threat of AIDS, the proportion of the population at risk is increasing.

There are a number of predictive models available for the behaviour of *L. monocytogenes*, e.g. Combase and Micromodel. There is limited and fragmented knowledge on microbial behaviour (survival, proliferation, virulence) in association with food- and feed-related matrix structures, limited food-based experimental model systems and fragmented information on the responses of pathogenic microorganisms to adverse conditions they experience in food products and in experimental food model systems at the level of transcriptome. The current models are primarily based on data obtained from growth of pure cultures in liquid laboratory media, or on data from a variety of different sources. Stress resistance, which is likely to be a factor in the persistence of *L. monocytogenes* in food, and possible changes in virulence due to food composition, are rarely accounted for (Schvartzman *et al.* 2010a).

Staphylococcus aureus is among the most common causes of bacterial food poisoning worldwide. An outbreak in Japan (Asao *et al.* 2003) demonstrated how a significant number of people could be affected by a single contamination

© Woodhead Publishing Limited, 2011

event. *S. aureus* produces a number of enterotoxins which can cause illness, even in the absence of the organism itself.

The concept of biotracing in the dairy chain will be enhanced by recent advances in knowledge of the physiology of *L. monocytogenes* and *S. aureus* and the inclusion of additional parameters into mathematical models (Manios *et al.* 2009, Schvartzman *et al.* 2010b, Le Marc *et al.* 2010, Reichert-Schwillinsky *et al.* 2009). Development and validation of an ELISA method for *S. aureus* enterotoxins (Wallin-Carlquist *et al.* 2010), knowledge of the factors controlling enterotoxin production in conditions corresponding to the dairy chain from farm to the consumer and transcriptional changes in *S. aureus* enterotoxin under different conditions all contribute to biotracing. In addition, identification of a genetic marker for identification of *S. aureus* of bovine origin (Jose Penades, personal communication) will facilitate the incorporation of molecular data into mathematical models, leading to improved biotracing.

Similarly, the development of a domain model for *L. monocytogenes* in the cheese chain is a major contribution to biotracing in the dairy chain. This biotracing development was facilitated by study of occurrence and persistence of *Listeria monocytogenes* in dairy environments and cheeses (O'Brien *et al.* 2009), resistance of *L. monocytogenes* to sequential environmental stresses (Belessi *et al.* 2010) encountered in foods (e.g., during fermentation and ripening) and on food contact surfaces (e.g., exposure to sanitizers) and assessment and prediction of the probability and rate of *L. monocytogenes* growth during cheese-making (Schvartzman *et al.* 2010b). It has resulted in enrichment of publicly available databases, such as COMBASE and PULSENET with phenotypic and genotypic data.

2.6 Potential bioterror agents and accidental contaminants in the food and feed supply

Tracing and tracking is the backbone of outbreak emergency management of an intentional contamination event. Therefore, tracing tools need to be integrated within existing general plans for emergency response to maximize the effectiveness of the plans and to minimize the environmental, social, and economic costs associated with the contamination. Current tracing methods lack flexibility, rapidity, and reliability. They are optimized for known pathogens in certain feed and food samples and most of them are based on culture methodology and consequently not useful in a contamination event. In order to optimize biotracing, it is necessary to operate with a number of virtual contamination scenarios (VCS). In situations where food has been adulterated, a rapid response with the best possible methodology is essential to contain the event and reduce the risk to the wider population.

New tools are needed to improve laboratory preparedness for detection of *B. anthracis* and *Clostridium botulinum* in food and feed. In biotracing, *B. cereus* can be used as a simulation model for *B. anthracis* to reduce the risk to staff and

© Woodhead Publishing Limited, 2011

avoid the need to work in contained laboratories during development of new sampling and sample treatment strategies (Fach *et al.* 2009).

Noroviruses are a group of related, single-stranded RNA viruses that cause acute gastroenteritis in humans. Also known as caliciviruses or small round structured viruses (SRSV), they are the cause of what is commonly termed 'Winter vomiting' disease. Because methods for the detection of Noroviruses in water samples are still under development (Schultz *et al.* 2010), the magnitude of contamination in drinking water and the effects of such contamination on public health are unknown. In effect, the disease burden of Norovirus is unknown, but it is estimated to be considerable (see Chapter 6).

Data on the survival and infectivity of Noroviruses in bottled water is scarce and scattered. Simple sensitive diagnostic assays are not available for Noroviruses. There is need to develop methods that are rapid, sensitive, and selective for virus detection and outbreak investigations (Anonymous 2002). This would assist in tracing the source of contamination, understanding the scale of the problem, thus reducing the associated risks to the public.

Biotracing has been facilitated by the introduction of harmonized, robust and validated tracing systems that are crucial in case of an emergency response related to a bioterrorism event or an accidental contamination event. Ring-trials using advanced tracing tools for *B. anthracis* and *C. botulinum* have been undertaken. Sampling is a key step in biotracing and a model of a biotraceability sampling plan has been developed for spore-forming bacteria. DNA extraction efficiency of *Bacillus* spores in different feed and food products has been studied (Wielinga *et al.* 2010). A bioinformatic platform for analysing whole genome characteristics of *B. anthracis* and *C. botulinum* isolates (sequence reads as well as assemblies) has been established.

2.7 Conclusions and future directions

Dynamism in the emergence of 'new' pathogens, for example emerging viruses, and different pathogen/matrix combinations will continue. Thus, the need for biotracing and developments in biotracing will remain. Future developments in biotracing will include:

- improved detection methods, particularly those molecular methods, resulting in more quantification, a pre-requisite for modelling in biotracing;
- improved identification of pathogen markers will improve biotracing;
- incorporation of molecular data into modelling, for example pathogenicity, will continue for the future and the issues to be addressed will change;
- bioinformatics and functional genomics;
- the ability to study entire bacterial communities, including non-culturable organisms will be studied using DNA technology;
- all of this will result in better decision-making tools, better communication, risk-informed decision making.

© Woodhead Publishing Limited, 2011

2.8 Acknowledgements

BIOTRACER is funded by the EU 6th Framework, Contract #036272.The contribution of Gary Barker, IFR, Norwich, for a critical review of the manuscript, is appreciated.

2.9 References

ANDERSSON G. and P. HÄGGBLOM (2009) Review: Sampling for contaminants in feed – estimating the concentration of mycotoxins found in bulk samples can be a complicated process. *Feed International* 3: 16–19.

ANONYMOUS (2002) Opinion of the Scientific Committee on Veterinary measures relating to Public Health on Norwalk-Like Viruses, European Commission Health and Consumer Protection Directorate-General.

ASAO T., Y. KUMEDA, T. KAWAI, T. SHIBATA, H. ODA, K. HARUKI, H. NAKAZAWA and S. KOZAKI (2003) An extensive outbreak of staphylococcal food poisoning due to low-fat milk in Japan: estimation of enterotoxin A in the incriminated milk and powdered skim milk. *Epidemiol. Infect.* 130: 33–40.

BARKER G. C., N. GOMEZ and J. SMID (2009) An introduction to biotracing in food chain systems. *Trends Food Sci. Technol.* 20: 220–226.

BELESSI C.-I. A., Y. LE MARC, S. I. MERKOURI, A. S. GOUNADAKI, S. SCHVARTZMAN, K. JORDAN, E.H. DROSINOS and P. N. SKANDAMIS (2010) Investigating the adaptive growth responses of *Listeria monocytogenes* to acid and osmotic shifts above and across the growth boundaries. *J. Food Protect.* doi:10.1016/j.ifoodmicro.2010.10.020.

COM (1999) '719 final, 12 January 2000', *E.C.* White Paper on Food Safety. Available at: http://ec.europa.eu/dgs/health_consumer/library/pub/pub06_en.pdf (accessed 23 July 2010).

Commission Regulation (EC) No 183/2005 of the European Parliament and of the Council for feed business establishments manufacturing or placing on the market feed additives of the category coccidiostats and histomonostats, OJ L35, 08.02.2005, 1–22.

CHAMPION O. L., M. W. GAUNT, O. GUNDOGDU, A. ELMI, A. A. WITNEY, J. HINDS, N. DORRELL and B. W. WREN (2005) Comparative phylogenomics of the food-borne pathogen *Campylobacter jejuni* reveals genetic markers predictive of infection source. *Proc. Natl. Acad. Sci.* 102: 16043–16048.

DORRELL N, S. J. HINCHLIFFE and B.W. WREN (2005) Comparative phylogenomics of pathogenic bacteria by microarray analysis, *Curr. Opin. Microbiol.* 8: 620–626.

DUFF S. B., E. A. SCOTT, M. S. MAFILIOS, E. C. TODD, L. R. KRILOV, A. M. GEDDES and S. J. ACKERMAN (2003) Cost-effectiveness of a targeted disinfection program in household kitchens to prevent foodborne illnesses in the United States, Canada, and the United Kingdom. *J. Food Prot.* 66: 2103–2115.

EUROPEAN FOOD SAFETY AUTHORITY (2010) The Community Summary Report on trends and sources of zoonoses, zoonotic agents and foodborne outbreaks in the European Union in 2008. Available at: http://www.efsa.europa.eu/en/scdocs/scdoc/1496.htm (accessed July 2010).

FACH P., P. MICHEAU, C. MAZUET, S. PERELLE and M. POPOFF (2009) Development of Real-time PCR tests for detecting botulinum neurotoxins A, B, E, F producing

© Woodhead Publishing Limited, 2011

Clostridium botulinum, Clostridium baratii and *Clostridium butyricum*. *J. Appl. Microbiol.* 107: 465–473.

GONZÁLEZ M., P. N. SKANDAMIS and M.-L. HÄNNINEN (2009) A modified Weibull model for describing the survival of *Campylobacter jejuni* in minced chicken meat. *Int. J. Food Microbiol.* 136: 52–58.

JOSEFSEN M. H., M. KRAUSE, F. HANSEN and J. HOORFAR (2007) Optimization of a 12-hour TaqMan PCR-based method for detection of Salmonella in meat. *Appl. Environ. Microbiol.* 73: 3040–3048.

JOSEFSEN M. H., C. LÖFSTRÖM, H. M. SOMMER and J. HOORFAR (2009) Diagnostic PCR: Comparative sensitivity of four probe chemistries. *Mol. Cell Probes.* 23: 201–203.

JOSEFSEN M.H., C. LÖFSTRÖM, T. B. HANSEN, L. S. CHRISTENSEN, J. E. OLSEN and J. HOORFAR (2010) Rapid quantification of viable *Campylobacter* bacteria on chicken carcasses, using real-time PCR and propidium monoazide treatment, as a tool for quantitative risk assessment. *Appl. Environ. Microbiol.* 76: 5097–5104.

KOYUNCU S. and P. HÄGGBLOM (2009) A comparative study of cultural methods for the detection of salmonella in feed and feed ingredients. BMC *Vet. Res.* 5: 6

KRÄMER N., C. LÖFSTRÖM, H. VIGRE, J. HOORFAR, C. BUNGE and B. MALORNY (2010) A novel strategy to obtain quantitative data for modelling: Combined enrichment and real-time PCR for enumeration of salmonellae from pig carcasses. *Int. J. Food Microbiol.* doi:10.1016/j.ifoodmicro.2010.08.026.

LE MARC Y., P. N. SKANDAMIS, C. I. A. BELESSI, S. I. MERKOURI, S. M. GEORGE, A. S. GOUNADAKI, S. SCHVARTZMAN, K. JORDAN, E. H. DROSINOS and J. BARANYI (2010) Modelling the effect of abrupt acid and osmotic shifts within the growth region and across the growth boundaries on the adaptation and growth of *Listeria monocytogenes. Appl. Environ. Microbiol.* 76(19): 6555–6563.

LÖFSTRÖM C, J. SCHELIN, B. MALORNY, P. RÅDSTRÖM and J. HOORFAR (2009a) In: *Molecular detection of foodborne pathogens*. Ed. Liu D. Taylor & Francis Group, London.

LÖFSTRÖM C., M. KRAUSE, M. H. JOSEFSEN, F. HANSEN and J. HOORFAR (2009b). Validation of a same-day real-time PCR method for screening of meat and carcass swabs for Salmonella, *BMC Microbiology* 9: 85.

MALORNY B., C. LÖFSTRÖM, M. WAGNER, N. KRAMER and J. HOORFAR (2008) Enumeration of *Salmonella* in food and feed samples by real-time PCR for quantitative microbial risk assessments. *Appl. Environ. Microbiol.* 74: 1299–1304.

MANIOS S. G., A. G. SKIADARESIS, K. KARAVASILIS, E. H. DROSINOS and P. N. SKANDAMIS (2009) Field validation of predictive models for the growth of lactic acid bacteria in acidic cheese-based Greek appetizers. *J. Food Prot.* 72: 101–110.

O'BRIEN M., K. HUNT, S. McSWEENEY and K. JORDAN (2009) Occurrence of foodborne pathogens in Irish farmhouse cheese. *Food Microbiology* 26: 910–914.

OLSEN K. N., M. LUND, J. SKOV, L. S. CHRISTENSEN and J. HOORFAR (2009) Towards realtime monitoring of broiler flocks: detection of Campylobacter in air samples for continuous monitoring of Campylobacter colonization in broiler flocks, *Appl. Environ. Microbiol.* 75: 2074–2078.

PIN C., G. AVENDAÑO-PEREZ, E. COSCIANI-CUNICO, N. GÓMEZ-TOMÉ, A. GOUNADAKI, G.-J. NYCHAS, P. SKANDAMIS and G. BARKER (2010) Modelling Salmonella concentration throughout the pork chain by considering growth and survival in fluctuating conditions of temperature, pH and Aw. *Int. J. Food Microbiol.* doi:10.1016/j.ifoodmicro.2010.09.025.

REICHERT-SCHWILLINSKY F., C. PIN, M. DZIECIOL, M. WAGNER and I. HEIN (2009) Stress and growth-rate-related differences between plate count and real-time PCR data during

© Woodhead Publishing Limited, 2011

growth of *Listeria monocytogenes*, *Appl. Environ. Microbiol.* 75: 2132–2138.

REITER E., J. ZENTEK and E. RAZZAZI-FAZELI (2009) Review on sample preparation strategies and methods used for the analysis of aflatoxins in food and feed. *Mol. Nut. Food Res.* 53: 508–524.

SCHULTZ A. C., S. PERELLE, S. DI PASQUALE, K. KOVAC, D. DE MEDICI, P. FACH, H. M. SOMMER and J. HOORFAR (2010) Collaborative validation of a rapid method for efficient virus concentration in bottled water. *Int. J. Food Microbiol.* http://dx.doi.org/10.1016/j.ijfoodmicro.2010.07.030.

SCHVARTZMAN M. S., X. BELESSI, F. BUTLER, P. SKANDAMIS and K. JORDAN (2010a) Comparison of growth limits of *Listeria monocytogenes* in milk, broth and cheese. *J. Appl. Microbiol.* In Press.

SCHVARTZMAN M. S., A. MAFFRE, F. TENENHAUS-AZIZA, M. SANAA, F. BUTLER and K. JORDAN (2010b) Modelling the fate of *Listeria monocytogenes* during manufacture and ripening of smeared cheese made with pasteurised or raw milk. *Int. J. Food Microbiol.* In Press.

WALCHER G., M. DZICIOL, M. WAGNER, M. LOESSNER and I. HEIN (2010) Evaluation of paramagnetic beads coated with recombinant Listeria phage endolysin-derived cell wall-binding domains for separation of *Listeria monocytogenes* from raw milk in combination with culture- and real-time PCR-based quantification. *Food. Path. Dis.*7: 1019–1024.

WALLIN-CARLQUIST N., R. CAO, D. MÁRTA, A. S. DA SILVA, J. SCHELIN and P. RÅDSTRÖM (2010) Acetic acid increases the phage-encoded enterotoxin A expression in *Staphylococcus aureus*. *BMC Microbiol.* 10: 147.

WIELINGA P. R., R. A. HAMIDJAJA, J. AGREN, R. KNUTSSON, B. SEGERMAN, M. FRICKER, M. EHLING-SCHULZ, A. DE GROOT, J. BURTON, T. BROOKS, I. JANSE and B. VAN ROTTERDAM (2010) A multiplex real-time PCR for identifying and differentiating *B. anthracis* virulent types. *Int. J. Food Microbiol.* doi:10.1016/j.ifoodmicro.2010.09.039.

© Woodhead Publishing Limited, 2011

3

Using stochastic simulation to evaluate the cost-effectiveness of traceability systems: the case of quality control in a fresh produce supply chain

M. Garcia Martinez, A. Silva and J. R. O'Hanley, University of Kent, UK

Abstract: The objective of this chapter is to estimate, at the firm level, the costs and benefits of upgrading existing traceability systems using stochastic simulation. The proposed changes represent potential improvements to the status quo through the use of advanced diagnostic devices, referred to as 'traceability solutions'. The aim is to estimate the impact of each traceability solution on a firm's profitability. To this end, the chapter uses stochastic (Monte Carlo) simulation and firm-level data from a fresh produce distributor in Spain.

Key words: quality control, traceability, cost-benefit analysis, stochastic simulation.

3.1 Introduction

In recent years, the implementation of food traceability systems has become a priority as an attempt to restore consumer confidence after a number of well-published food safety incidents such as bovine spongiform encephalpathy (BSE), *Salmonella* and *E. Coli*, and concerns about production and process attributes such as genetic modification, antibiotics and animal welfare. In the European Union (EU), under the framework of the General Food Law, Regulation (EC) 178/2002 has set new standards to ensure stricter implementation of traceability

© Woodhead Publishing Limited, 2011

in food and feed chains by enforcing mandatory traceability of food and food ingredients as of 1 January 2005. The regulatory requirements necessitate all food and feed business operators to be able to identify the source of all foods and ingredients and to provide the basis for further monitoring throughout supply chains. Traceability systems are thus conceived as a risk managment tool to reduce the impact of product recalls and food safety incidents (Souza-Monteiro and Caswell, 2010) and are expected to increase transparency throughout the food chain and to result in the development and maintenance of consumer trust in food and food producers (Van Rijswijk *et al.*, 2008).

In the United States (US), consumer interest in food traceability has increased after recent food safety incidents such as the discovery of BSE in 2003 and the case of *E. coli* contaminated spinach in 2006. However, progress in the implementation of traceability systems has been slow; the lack of knowledge about the effects of traceability on consumers' welfare and firms' profits has hindered its wider adoption (Pouliot, 2010).

According to Golan *et al.* (2004), the main reasons for adopting food traceability systems are to improve management decisions, to create markets for credence attributes (content and process characteristics), to assure food safety and quality levels and to help minimise the impact of a food crisis at firm and industry levels. However, the benefits of traceability go beyond food safety management; traceability can be used as a tool to monitor and manage information and product flow, which may lead to improved efficiency and ultimately enhanced profitability (Bantham and Oldham, 2003). In addition to food safety control and supply chain management, traceability can also be used in new product development and is increasingly required in securing foreign market access (Hobbs *et al.*, 2005).

From a theoretical standpoint, the perfect scenario would be to trace a product from farm to fork. However, full traceability is normally too costly (Sparling *et al.*, 2006). Hence, a rigorous cost-benefit analysis would help to identify the most cost-effective food traceability level for a given situation. According to Golan *et al.* (2003), the breadth, depth and precision (error tolerance and traceable unit size) of a food traceability system would depend on the attribute and the cost-benefit analysis, whose optimum would in turn vary for each particular case. Even for a single industry, such as food, the level of food traceability required varies across sectors (Wilson *et al.*, 2008).

According to Golan *et al.* (2004), traceability costs include recordkeeping and product differentiation, which to a larger extent are easier to quantify compared to benefits. Some of the factors that would impact traceability costs include the breadth, depth and precision levels and type of industry. On the benefit side, market size, chain coordination level, likelihood of a food crisis and its consequences and expected price premium would determine the level of benefit of a company after a food traceability system is adopted. Evidence supporting a price premium from traceability is uncertain because while consumers often state they would be willing to pay more for safer food, observational data show that economic convenience most affects purchasing decisions

© Woodhead Publishing Limited, 2011

rather than notice of safety or traceability on the product label (Meuwissen *et al.*, 2003; Loureiro and Umberger, 2007).

The objective of this chapter is to estimate, at the firm level, the costs and benefits of upgrading existing traceability systems using stochastic simulation. The proposed changes represent potential improvements to the status quo through the use of advanced diagnostic devices, referred to as 'traceability solutions'. The aim is to estimate the impact of each traceability solution on a firm's profitability. To this end, the paper uses stochastic (Monte Carlo) simulation and firm-level data from a fresh produce distributor in Spain. The simulations were performed as part of the EU-funded project TRACEBACK. Stochastic simulation is a quantitative methodology, well-supported academically, which identifies the likelihood of an event occurring and the magnitude of its consequences. Stochastic simulation offers flexibility and power that combined with empirical data bridges the gap between managerial applications and academic rigor (Shafer and Smunt, 2004). Firm-level operational and business data were collected through semi-structured questionnaires. In addition, device cost and performance data were gathered from the project device developers. Empirical distributions of the variables under study were determined and used to drive our simulation models. Simulation results show the probability of obtaining a certain profit level before and after a traceability solution is implemented.

The remainder of the chapter is organised as follows. Section 3.2 presents an overview of past research on cost-benefit analysis using stochastic simulation. Section 3.3 introduces the case study and simulation results for a fresh produce distributor. In the concluding section, we discuss the business implications of possible changes and propose a number of recommendations.

3.2 Review of cost-benefit analysis of food traceability

Caswell and Jensen (2007) suggest that the economic analysis of intervention measures for food safety controls (such as mandatory traceability) should address important empirical questions including:

1. What are the costs to government, industry, and consumers of implementing a regulatory intervention?
2. How effective is an intervention in reducing the risk of illness in terms of public health outcomes?
3. What is the effect on the regulated supply chain, in terms of behavioural changes upstream or downstream from where the intervention takes place, that will significantly influence the effectiveness of the intervention?
4. What are the net benefits associated with different types of interventions?

The economics of food safety, which is a relatively new but rapidly expanding field (Antle, 1999), attempts by and large to address these questions. This literature suggests that there is a clear rationale for the government regulation of

© Woodhead Publishing Limited, 2011

food safety to protect public health and consumer interest on the grounds of market failure resulting from asymmetric information about food safety attributes (see Ritson and Mai, 1998; Antle, 1999; Unnevehr and Jensen, 1999, 2001; Hennessy *et al.*, 2003). Henson and Traill (1993) identify three primary reasons as to why the market mechanism may fail to guarantee supply safety of foods: (i) the nature of foodborne risks which are not easily detectable; (ii) the public good nature of food safety provision with diverging private and social costs and benefits; and (iii) asymmetries in the supply of food safety information which entails considerable uncertainties.

The efficiency of regulatory interventions is traditionally assessed by comparing costs and benefits of new interventions and/or existing regulations (Stiglitz, 2000). Most methods for measuring the benefits of food safety regulation are based on economic approaches to valuing reductions in health risk in monetary terms, whereas costs mainly consist of estimated (or observed) regulatory and industry compliance costs (Unnevehr, 2000; Unnevehr and Jensen, 1999; Antle, 1999; USDA, 1996).

Cost-benefit studies of food traceability have tried to catalogue the costs and benefits using surveys and interviews. Gellynck *et al.* (2005) focused on the costs of traceability systems and through interviews they distinguished between investment and operation costs, taking into account three company sizes. Mora and Menozzi (2005) used a survey to identify the perceived benefits and costs in the Italian beef sector resulting from the implementation of traceability systems to comply with regulation (EC) 1760/2000. Sparling *et al.* (2006) used in-depth, semi-structured interviews and surveys to catalogue investment and implementation costs and perceived benefits from implementing traceability systems in the Canadian dairy processing sector. Food safety/recall crisis management and fulfilment of trade requirements were the two main reasons for implementing traceability systems. The most important problems found were attitude, motivation and the need to retain production and supervisory staff.

Chryssochoidis *et al.* (2009) developed a questionnaire to determine the perceived costs and benefits of a traceability system, *ex-ante* and *ex-post*, in a mineral water facility. The perceived benefit questionnaire was based on work by DeLone and McLean (2004), which identifies the following dimensions: net benefits, user satisfaction, intention to use, information quality, system quality and service quality. Costs are decomposed into initial investment and ongoing costs. In the questionnaire, the decision maker rates, on a 1 to 7 scale, the attributes of the traceability system, where the *ex-ante* and *ex-post* situation are visually compared.

In the UK meat sector, Hobbs (1996) measured the significance of transaction costs between processor and retailer levels using conjoint analysis. The author identifies important cost attributes in the retailer procurement decision: quality consistency, traceability in the supply chain, and animal welfare. All of the aforementioned attributes are associated to a food traceability system. The author found, in decreasing order, that quality consistency, traceability, farm assurance and price level were the factors for which retailers were willing to pay a premium.

© Woodhead Publishing Limited, 2011

Regarding the application of stochastic simulation to the food safety area, Koutsoumanis *et al.* (2005) assessed the development of microbiological agents in a chilled food chain. The authors simulated pathogen growth from the production point to the selling point, assuming distributions for the transportation time and cabinet temperature. Doing this, it optimised market destination according to the product safety level. The objective was to cross-compare their simulation model with a FIFO (first in, first out) inventory policy with respect microbiological risk. The authors showed the simulation model to be superior at reducing pathogen damage. In other words, the simulation model included pathogen information and time-temperature history that optimises the operation over the food supply chain, reducing the product damages.

Wilson *et al.* (2008) developed a stochastic simulation model to estimate the cost and optimal testing strategy to comply with the EU regulation standards to import non-GM wheat from the US. The authors defined a three-level supply chain (farmer, exporter and importer) and optimised the strategy that minimises costs of wheat traceability systems that met EU requirements. The costs included testing, quality loss, auditing, certifications, traceability and risk compensation. At the end, the authors ran a series of sensitivity analyses on key stochastic variables.

Salmonella contamination has drawn research attention because of its high incidence in human food safety. McNamara *et al.* (2007) modelled a complete pork chain. Characterisation of distributions describing pathogen development allowed the assessment of the expected food safety cost across the chain. Using the same approach, but this time at a plant level, Malcolm *et al.* (2004) used stochastic simulation to measure the economic impact of seven technologies to reduce risk contamination in a slaughterhouse. Pathogen distributions were identified across the plant production chain. The authors were able to compare effectiveness against cost and include plant size in the analysis. Nganje *et al.* (2007) wanted to determine the optimum control level for *Salmonella* in a turkey plant in the US. In addition to estimating the benefits and costs of implementing different levels of safety control at the plant level, the authors developed a stochastic optimisation model that took into account the risk-adverse nature of the decision maker. The results show that costs can increase exponentially depending on the tolerance level.

Van der Gaag *et al.* (2004) developed a state-transition simulation model for a pork supply chain. The model consists of a state vector (i.e., susceptible, infectious and carrier states) and transition probabilities, which depend on the current state. Vosough Ahmadi *et al.* (2007), using the same approach, assessed the impact of food safety measures in the Dutch dairy herd sector. Within an epidemiological focused study, Turner *et al.* (2003) described a disease transmission model for *E. coli* in a dairy farm. The model includes states and identifies direct and indirect transmission vectors.

The application of stochastic simulation to food traceability is a novelty. Stochastic simulation can help to determine the optimal investment decision level of improvements in traceability system at firm level. Managers would be able to see the impacts of devices independently or a combination of them

© Woodhead Publishing Limited, 2011

through the production process in terms of profit/time saving. At the same time, stochastic simulation allows estimation of the impact of the devices before they are actually in place. Finally, early feedback can be used to reallocate resources to develop the devices that are more profitable for each specific company.

3.3 Cost-benefit analysis of traceability solutions in a fresh produce distributor

As part of a TRACEBACK project, a stochastic simulation model was developed in order to assess the impact of traceability solutions on the profitability of a large Spanish fresh produce distributor. Figure 3.1 shows the products flowing from the farmer to the end consumer. Farmers produce fresh fruit and vegetables and deliver them in boxes to the distributor. Boxes are packed into larger containers referred to as 'pallets'. The number of units per box and the number of boxes per pallet depend on the particular product and variety. Pallets are delivered to the distibutor by trucks throughout the day.

On arrival at the distributor's warehouse, products are sampled to certify the quality of the consignment according to a standardised sampling procedure. Normally, three fruits per pallet are sampled. The sampled fruits are analysed in the lab to measure internal and external attributes related to the quality and shelf-life, including colouration, external aspect, decomposition, size, internal aspect, sugar content (brix) and firmness. If testing results are below accepted standards, boxes are rejected and sent back to the farmer. Rejections can range from 6 to 33% of the total consignment. The remaining boxes enter the warehouse as stock. Following the initial testing programme, fruit and vegetables are also sampled while in stock. Up to 24 hours after receipt, fruit and vegetables can still be rejected and sent back to the farmer at his own expense. Rejections at this stage typically represent 1 to 5% of the remaining consignment. Finally, the distributor sends the produce to retail outlets around the region. Stores need to open every box to display the products. Each store has

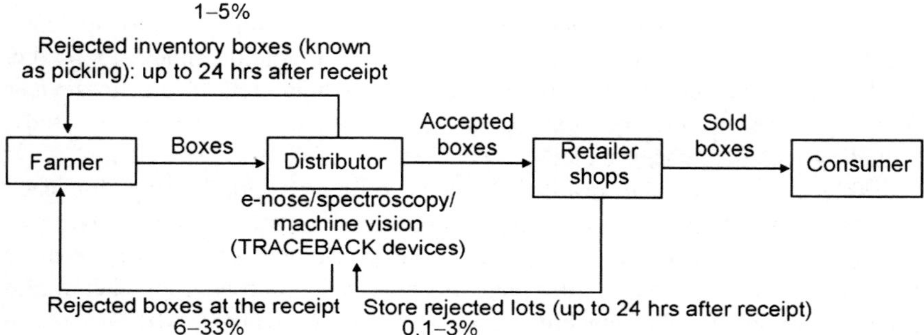

Fig. 3.1 Product flow from farmers to the consumers.

© Woodhead Publishing Limited, 2011

up to 24 hours to return the produce to the distributor. In this case, the cost of produce rejection is borne by the distributor.

The distributor would like to have more certainty regarding the quality of products received. This could be done by increasing the sample size or increasing the accuracy of the current testing devices. Increasing the sample size would involve sampling more fruit and vegetable units during each inspection. Using current quality assessment techniques, which involve low-cost instruments, sampled items of produce are destroyed during testing and then disposed of. Alternatively, improved sampling devices can offer a non-destructive means of testing. This represents a potentially signficant cost saving for the distributor, not only due to the fact that tested fruit can subsequently be sold, but also due to the avoided disposal costs, which can be substantial during the summer period and for bulky units like water melon.

Model development began by collecting data and specifying distributions for key input variables as described in Table 3.1. Modelling was performed using Simetar simulation software. Simetar is a low-cost Microsoft Excel add-in package with numerous user-friendly functionalities for distribution fitting and result generation.

Having built our simulation model, we subsequently assessed each diagnostic device's impact on the distributor's operations and profitiability both individually and in combination. In performing our simulations,the following assumptions were made:

- Accuracy for each device was assessed based on live tests using tomatoes only. Acuracy for all other products and varieties is assumed to be the same as for tomatoes.
- The testing plan remains constant through time. In practice, testing plans can change according to the season, past incidents, etc.
- There are no interaction effects among devices. In other words, assessments about produce quality and the accuracy of assessments for each device are entirely independent.
- When two traceability devices measure the same attritibute, the higher accuracy value is used.

To help support improvements to the distributor's quality control management programme, we developed a Monte Carlo simulation model to predict produce flow impacts and profitiability of various high-tech diagnostic devices used to assess product quality. A list of devices considered in this study, including technical specifications, is provided in Table 3.2.

Table 3.3 shows the results of mean estimated profitability for each traceability solution. Machine vision would cause a monthly cost saving of €87.9, NIR €93.3 and e-nose −€16.1. If the distributor decides to have more than one traceability solution in place, machine vision and NIR together would produce savings of €9.8, machine vision and e-nose −€114.0, NIR and e-nose −€112.6 and the three devices together −€192.0. Negative numbers mean that the company would increase its testing cost.

© Woodhead Publishing Limited, 2011

Table 3.1 Simulation product variables

Product	Data	Level
Tomato (2 varieties)	Price (euros per box) and quantity (boxes)	Rejected boxes at the reception, picking and stores levels
Melon	Price (euros per box) and quantity (boxes)	Rejected boxes at the reception, picking and stores levels
Mango	Price (euros per box) and quantity (boxes)	Rejected boxes at the reception, picking and stores levels
Papaya	Price (euros per box) and quantity (boxes)	Rejected boxes at the reception, picking and stores levels

Table 3.2 Selected devices

Device	Description
Machine vision	It takes a picture of the product under particular light conditions. The acquired image is processed in terms of colour, shape and size. Later on, this information can be used to determine the ripening stage.
Near Infra-red Spectrometer (NIR)	It uses a sensor to measure the absorption of near infra-red light. The results are processed and compared against benchmark patterns in order to characterise the measured sample.
Electronic nose (e-nose)	It is an instrument sensitive to volatile organic compounds. The results can be used to estimate ripening stage, ageing and product deterioration.

In this case, a traceability solution alone generates bigger savings than when they are combined. This is due to two factors: firstly, traceability solutions require a high up-front investment, in particular, the e-nose has an investment cost of €15 000 (almost three times more expensive than the other two devices), which is fairly expensive in comparison to the company's current, low-tech testing procedures. Secondly, even when traceability devices overlap in their functionality (e.g., machine vision and NIR both test for firmness), because of

Table 3.3 Average monthly cost savings

Machine vision	€87.9
NIR	€93.3
e-nose	−€16.1
M-vision and NIR	€9.8
M vision and e-nose	−€114.0
NIR and e-nose	−€112.6
Total	−€192.0

© Woodhead Publishing Limited, 2011

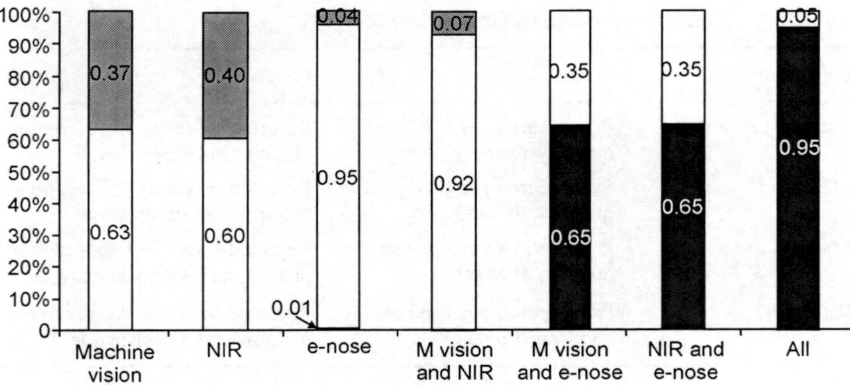

Fig. 3.2 Comparison of traceability solutions. Note: distribution of cost savings for each traceability solution in euros per month. Black area means that the value is less than the lower cut-off value (−€100). Mid-grey area means that the value is between cut-off values (−€100 and €100). Light grey area means that the value is higher than the upper cut-off value (€100).

our independent assumptions, there is no added value from testing with two devices. In all cases, with the exception of the e-nose, the traceability solutions on average save costs for the distributor.

Figure 3.2 shows the distribution of mean cost savings presented in Table 3.3. The cut-off values are −€100 and €100. The black area shows when the distributor platform would increase cost more than €100. For instance, for the machine vision, the distributor would have a 63% chance of saving between −€100 and €100, and a 37% chance of saving more than €100. For the NIR and

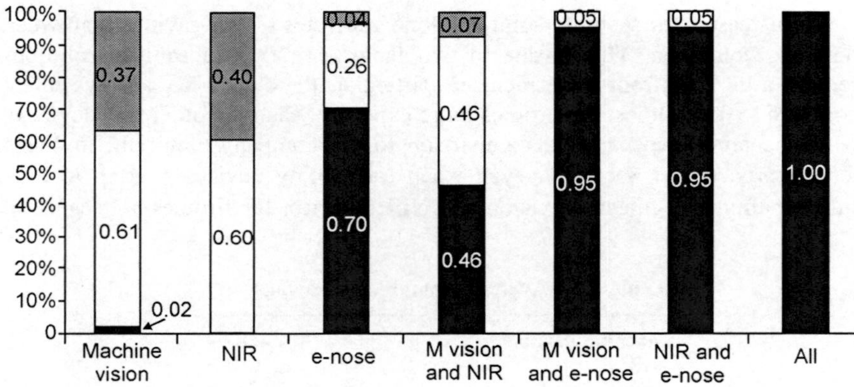

Fig. 3.3 Comparison of traceability solutions. Note: distribution of cost savings for each traceability solution in euros per month. The cut-off values are €0 and €100. Black area means that the value is less than the lower cut-off value (€0, cost increases). Mid-grey area means that the value is between cut-off values (€0 and €100). Light grey area means that the value is higher than the upper cut-off value (€100).

© Woodhead Publishing Limited, 2011

e-nose together, the distributor platform has a 65% chance of losing more than €100, a 35% chance of saving between −€100 and €100 per month.

In Fig. 3.3, the cut off values are €0 and €100. The black area shows when the distributor would increase cost (no cost savings). The e-nose or more than one traceability solution in place would increase the chances of increasing cost. When the three traceability solutions are in place, there is a 100% chance that the cost will increase with respect to the current situation (before implementation of a new traceability system).

As Figs 3.2 and 3.3 show, the last three combinations are most likely to result in a cost increase. The two most profitable alternatives in Fig. 3.3 (machine vision and NIR separately) are selected for sensitivity analysis.

Table 3.4 and Fig. 3.4 show the sensitivity analysis with respect to their price penalty (a reduction in price when a box is rejected). After a box is rejected, the producer or distributor needs to find an alternative outlet to sell their rejected

Table 3.4 Price penalty hypothetical scenarios

	Rejected box location		
Price penalty scenarios	At the reception	Picking	At the store
A	20%	30%	40%
B	30%	40%	50%
C	40%	50%	60%

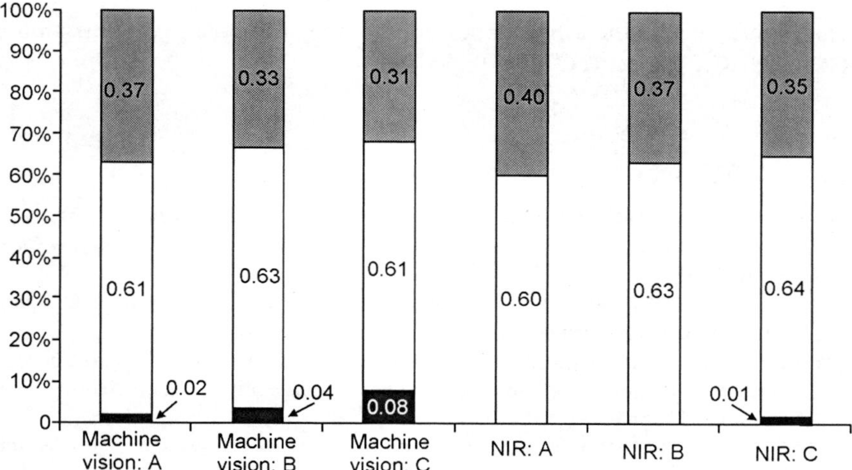

Fig. 3.4 Comparison of traceability solution price penalty scenarios. Note: distribution of cost savings for each traceability solution in euros per month. The cut-off values are €0 and €100. Black area means that the value is less than the lower cut-off value (€0, cost increases). Mid-grey area means that the value is between cut-off values (€0 and €100). Light grey area means that the value is higher than the upper cut-off value (€100).

© Woodhead Publishing Limited, 2011

product. However, the rejected product would have less value, since it has deteriorated (perishability) and it is not in optimal condition (it does not fulfil at least one acceptance criterion). We identify three possible scenarios (A, B and C), which differ in the level of penalty price (Table 3.4). For instance, in scenario A, we assume a 20% price penalty at the reception level, 30% at the picking level and 40% at the store level.

3.4 Conclusions and recommendations

Device profitability is highly dependent on the conditions of the specific application, such as the current technological level of the company, type of operation (farmer, distributor or retailer), product damage responsibility and consumer willingness to pay a price premium for a higher level of food safety. This chapter presents a stochastic simulation as an approach to conducting cost-benefit analysis and assist decision makers to evaluate alternative traceability solutions. The exercise shows how stochastic simulation can be used to include some measure of risk associated with a set of potential new traceability devices. With respect to future research, this analysis can be expanded to include other players in the food chain, include some market structure determinants (market power), different geographical areas and possible price premiums.

3.5 Acknowledgement

This work is financially supported by the European Commission's TRACEBACK Project (FOOD-CT-036300).

3.6 References

ANTLE, J. M. (1999) Benefits and costs of food safety regulation. *Food Policy*, 24, 605–623.

BANTHAM, A. & OLDHAM, C. (2003) *Creating Value Through Traceability Solutions: A Case Study*. IAMA.

CASWELL, J. A. & JENSEN, H. H. (2007) Introduction: economic measures of food safety interventions. *Agribusiness*, 23, 153–156.

CHRYSSOCHOIDIS, G., KARAGIANNAKI, A., PRAMATARI, K. & KEHAGIA, O. (2009) A cost-benefit evaluation framework of an electronic-based traceability system. *British Food Journal*, 111, 565–582.

DELONE, W. H. & McLEAN, E. R. (2004) Measuring e-commerce success: applying the DeLone & McLean information systems success model. *International Journal of Electronic Commerce*, 9, 31–47.

GELLYNCK, X., JANUSZEWSKA, R., VERBEKE, W. & VIAENE, J. (2005) Firm's costs of traceability confronted with consumer requirements. *92th EAAE-Seminar on Quality Management and Quality Assurance in Food Chains*. Göttingen, Germany, European Association of Agricultural Economists (EAAE).

© Woodhead Publishing Limited, 2011

GOLAN, E., KRISSOFF, B., KUCHLER, F., NELSON, K., PRICE, G. & CALVIN, L. (2003) Traceability in the US food supply: dead end or superhighway? *Choices*, 17–20.

GOLAN, E., KRISSOFF, B., KUCHLER, F., NELSON, K. & PRICE, G. (2004) Traceability in the US food supply: economic theory and industry studies. *Agricultural Economic Report* Washington, DC, US Department of Agriculture.

HENNESSY, D. A., ROOSEN, J. & JENSEN, H. H. (2003) Systemic failure in the provision of safe food. *Food Policy*, 28, 77–96.

HENSON, S. & TRAILL, B. (1993) The demand for food safety: market imperfections and the role of government. *Food Policy*, 18, 152–162.

HOBBS, J. E. (1996) A transaction cost analysis of quality, traceability and animal welfare issues in UK beef retailing. *British Food Journal*, 98, 16–26.

HOBBS, J. E., BAILEY, D. V., DICKINSON, D. L. & HAGHIRI, M. (2005) Traceability in the Canadian red meat sector: do consumers care? *Canadian Journal of Agricultural Economics*, 53, 47–65.

KOUTSOUMANIS, K., TAOUKIS, P. S. & NYCHAS, G. J. E. (2005) Development of a safety monitoring and assurance system for chilled food products. *International Journal of Food Microbiology*, 100, 253–260.

LOUREIRO, M. L. & UMBERGER, W. J. (2007) A choice experiment model for beef: what US consumer responses tell us about relative preferences for food safety, country-of-origin labeling and traceability. *Food Policy*, 32, 496–514.

MALCOLM, S. A., NARROD, C. A., ROBERTS, T. & OLLINGER, M. (2004) Evaluating the economic effectiveness of pathogen reduction technologies in cattle slaughter plants. *Agribusiness*, 20, 109–123.

McNAMARA, P. E., MILLER, G. Y., LIU, X. & BARBER, D. A. (2007) A farm-to-fork stochastic simulation model of pork-borne salmonellosis in humans: lessons for risk ranking. *Agribusiness*, 23, 157–172.

MEUWISSEN, P. M. M., VELTHUIS, A. G. J., HOGEVEEN, H. & HUIRNE, R. B. M. (2003) Traceability and certification in meat supply chains. *Journal of Agribusiness*, 21, 167–181.

MORA, C. & MENOZZI, D. (2005) Vertical contractual relations in the Italian beef supply chain. *Agribusiness*, 21, 213–235.

NGANJE, W. E., KAITIBIE, S. & SORIN, A. (2007) HACCP implementation and economic optimality in turkey processing. *Agribusiness*, 23, 211–228.

POULIOT, S. (2010) Welfare effects of mandatory traceability when firms are heterogenous. Paper presented at the Agricultural & Applied Economics Association 2010 AAEA, CAES & WAEA Annual Meeting. Denver, Colorado.

RITSON, C. & MAI, L. W. (1998) The economics of food safety. *Nutrition & Food Science*, 98, 253–259.

SHAFER, S. M. & SMUNT, T. L. (2004) Empirical simulation studies in operations management: context, trends, and research opportunities *Journal of Operations Management*, 22, 345–354.

SOUZA-MONTEIRO, D. M. & CASWELL, J. A. (2010) The economics of voluntary traceability in multi-ingredient food chains. *Agribusiness*, 26, 122–142.

SPARLING, D., HENSON, S., DESSUREAULT, S. & HERATH, D. (2006) Costs and benefits of traceability in the Canadian dairy-processing sector. *Journal of Food Distribution Research*, 37, 154–160.

STIGLITZ, J. E. (2000) *Economics of the public sector*, New York, W. W. Norton and Company.

TURNER, J., BEGON, M., BOWERS, R. G. & FRENCH, N. P. (2003) A model appropriate to the transmission of a human food-borne pathogen in a multigroup managed herd.

© Woodhead Publishing Limited, 2011

Preventive Veterinary Medicine, 57, 175–198.

UNNEVEHR, L. J. (2000) *The economics of HACCP*, St. Paul, MN, Eagan Press.

UNNEVEHR, L. J. & JENSEN, H. H. (1999) The economic implications of using HACCP as a food safety regulatory standard. *Food Policy*, 24, 625–635.

UNNEVEHR, L. J. & JENSEN, H. H. (2001) Industry compliance costs: what would they look like in a risk-based integrated food system? Paper presented at the Resources for the Future conference, 'Risk-Based Priority Setting in an Integrated Food Safety System: Current Knowledge and Research Needs'. Washington, DC.

USDA (1996) (US Department of Agriculture), Food Safety and Inspection Service, Pathogen Reduction Hazard Analysis and Critical Control (HACCP) Systems: Final Rule, Docket no. 93-016F. *Federal Register* 61 (144), 38805–38989.

VAN DER GAAG, M. A., VOS, F., SAATKAMP, H. W., VAN BOVEN, M., VAN BEEK, P. & HUIRNE, R. B. M. (2004) A state-transition simulation model for the spread of Salmonella in the pork supply chain. *European Journal of Operational Research*, 156, 782–798.

VAN RIJSWIJK, W., FREWER, L. J., MENOZZI, D. & FAIOLI, G. (2008) Consumer perceptions of traceability: a cross-national comparison of the associated benefits. *Food Quality and Preference*, 19, 452–464.

VOSOUGH AHMADI, B., FRANKENA, K., TURNER, J., VELTHUIS, A. G. J., HOGEVEEN, H. & HUIRNE, R. B. M. (2007) Effectiveness of simulated interventions in reducing the estimated prevalence of *E. coli* O157:H7 in lactating cows in dairy herds. *Veterinary Research*, 38, 775–771.

WILSON, W. W., HENRY, X. & DAHL, B. L. (2008) Costs and risks of conforming to EU traceability requirements: the case of hard red spring wheat. *Agribusiness*, 24, 85–101.

© Woodhead Publishing Limited, 2011

4

Preventing and mitigating food bioterrorism

N. Marmiroli, E. Maestri and M. Marmiroli, University of Parma, Italy, R. Onori, Italian National Institute of Health, Italy, R. Setola, Campus Bio-medico University of Rome, Italy and V. Krivilev, Academy for Geopolitical Problems, Russia

Abstract: Preservation of food integrity in the face of intentional and accidental threats is a major issue in modern society and an essential aspect of international trade. Food production chains are vulnerable to intentional acts aimed at undermining consumer confidence and causing health problems. Terrorists may interfere with food production and distribution chains, with the aim of causing diseases to plants, animals and humans through the use of biotic and abiotic agents. Traceability measures coupled with analytical detection and fingerprinting of food components can be applied effectively to prevent and mitigate the effects of these threats. Innovative research in the field of devices and microdevices for food chain analysis offers fresh hope that the phenomenon can be more effectively addressed.

Key words: food safety and security, food defence, traceability, devices and microdevices.

4.1 Introduction

Recent international security reports assume that environments in both rural and anthropogenic settings will be increasingly threatened by terrorism, or more generally by criminal organisations (Nadasi *et al.*, 2007). Structures and critical infrastructures including water supply and distribution, agricultural soils, agricultural chemical depots, farms and crops, urban infrastructures and agri-food

© Woodhead Publishing Limited, 2011

factories are all vulnerable to attack (Chalecki, 2002). The strategic importance of critical infrastructures and of natural resources has been recognised and their protection considered from different points of view (Marmiroli *et al.*, 2010). Recent evidence of accidental contamination of vital resources has shown their vulnerability and the impact that this may have on citizens and consumers. This was absolutely devastating when food products were involved (Chalk, 2004). Imagine what the public reaction would have been had events such as the Chernobyl disaster of 1986 or the Gulf of Mexico oil spill of 2010 been the result of terrorist attacks, or even had there been a suspicion of terrorist involvement. Moreover, epidemics such as BSE or the spread of avian influenza could be triggered by terrorists once the appropriate biological agents were available to them.

Terrorism's principal aim is to attack resources which are not only critical and vulnerable, but which also have the highest emotional impact: food supply chains certainly belong to all of these categories. The term 'food bioterrorism' is generally defined as 'an act or threat of deliberate contamination of food for human consumption with chemical, biological or radionuclear agents for the purpose of causing injury or death to civilian populations and/or disrupting social, economic or political stability' (WHO, 2002). Food safety and integrity concerns the population at large, and specifically the weakest, the young and the old. Food is a fundamental human requirement, and any alteration or adulteration of it has the highest psychological impact (Just *et al.*, 2009). The protection of the food supply chain could be considered as the mission of a nation, and different countries have started developing 'food defence' strategies to prevent and mitigate against any malicious manipulation or alteration of food carried out by terrorists, criminals and fantasists, and to counteract the adverse effects. Food defence, or biosecurity, includes the use of advance measures designed to reduce the risk of deliberate attack on the food chain. Any disruption of food chain integrity can impact adversely on consumer trust, and can lead to negative economic impacts, even if the incident has no bearing on public health.

4.1.1 Economic and political impact

In the age of the global economy, any attack on the agri-food system could have such a severe psychological and emotional impact that it could trigger a change in household spending patterns and business investment decisions. Deliberate food contamination has enormous economic implications, regardless of whether or not it affects public health. Disruption of the economic network has been the motive behind some deliberate acts, targeted at products, manufacturers, or industries: one such example was the poisoning of citrus fruits from Israel in 1978, to decrease the exports (Pate and Cameron, 2001). Real or perceived terrorist threats also have a major impact on tourism and the movement of people in general. Costs resulting from the consequences of these threats and from measures taken against them are listed in Table 4.1.

© Woodhead Publishing Limited, 2011

Table 4.1 Economic costs incurred in cases of terrorism to food supply chains

Category of costs	Notes
Diagnosis and surveillance	Related to the specific event
Health care	For those consumers that have been affected by the contaminated food
Production losses, quarantine and eradication	Due to the introduction of diseases
Decontamination	At farm, processing, or distribution levels
Destruction of contaminated crops and animals	To contain the spread of disease or toxication
Compensation to producers	For the destroyed goods
International prohibitions on exports or increased regulation on the domestic markets	Can be amongst the reaction measures
Income losses of farmers, food processors, distributors and exporters	Due to drop in consumer confidence in the safety of food products
Income losses of consumers	Affected by higher prices of food products

4.1.2 Disruption of the food supply chain

Disrupting or damaging the integrity of the food supply chain is possible at several levels, directly or indirectly. In addition to primary production, production lines, processing structures, the logistics system, stocks and retailers are all major targets in the food supply chain. But food safety can also be compromised through water resources, including inland waters, freshwater reserves, municipal facilities and utilities, or through agricultural soils, targeting crop cultivation and animal farming (Stamm, 2002). In addition, air contamination can lead to fallout of hazardous agents on crops and food products.

Data regarding about 450 malicious incidents that occurred worldwide from 1950 to 2008 have been investigated (SecuFood, http://secufood.unicampus.it). The analysis showed that in the data under investigation, the only evidence of any terrorist attack against the food supply chain was the grapefruit contamination that occurred in 1989 in Israel (Pate and Cameron, 2001 and references therein). However, in recent years there has been a steady increase in malicious incidents caused by chemical, biological, radiological or physical contaminants. Concerning contamination along the food supply chains, the highest percentage happens at consumer level (58%), whereas 38% takes place within retail and food service structures while the remaining 4% occur at harvesting level (SecuFood, 2010). No incidents at production level were reported. According to the results of interviews and questionnaires at industries that have suffered sabotage incidents, the reported cases of attacks were due to: (i) fantasists, who want attention, (ii) initiatives which pose no risk to human health, carried out by

© Woodhead Publishing Limited, 2011

non-governmental organisations to prove that sabotage is possible, (iii) claims by internal company staff, and, less frequently (iv) terrorist attacks.

Coordinated intervention at different levels – international, national, regional, and local – along with the role played by public and private stakeholders, is an essential factor in counteracting food terrorism. A detailed analysis of the legislation and organisations concerned with food defence showed that all countries have specific legislation and agencies devoted to food safety that are involved in early warning alert networks for food contamination. However, with the exception of the UK, no country has specific legislation for food defence. Though there are several mandatory constraints on food operators to improve and guarantee food safety, specific requirements concerning food defence do not exist. All the activities performed by food operators are on a voluntary basis but their efforts are not coordinated. This means that the level of protection varies significantly along the food supply chain. All sectors in the major food industries have a high level of awareness of the damage that can be caused by any deliberate attacks by criminals or terrorists. The more effective counter-measures in preventing food attacks are considered to be area control, video-surveillance and, mainly, traceability. However, the development of operational procedures to promote and to facilitate effective coordination is essential, as is the exchange of information between different types of stakeholders involved in food defence, including within the government and between public and private enterprise.

Table 4.2 Main biological threats for food and environmental terrorism

Category	Examples	Possible origin	Dispersion means
Bacteria	*Bacillus anthracis, Francisella tularensis, Salmonella, Vibrio cholerae, Yersinia pestis*	Contaminated soil or water, infected animals, foecal contamination	Aerial dispersion, water, sprinkling dried bacteria, spores
Viruses	Alphaviruses, noroviruses, arenaviruses, hantaviruses, Variola, avian influenza, calicivirus	Infected animals, insect bites, direct contact	Aerial dispersion, infected arthropods, direct contact, infection of livestock and crops
Fungi	*Coccidioides, Phytophthora, Puccinia, Xanthomonas*	Soil	Inhalation of spores, infection of crops
Protozoa	*Giardia lamblia*	Water	Food contamination
Toxins	Botulinum toxin, aflatoxins, ricin, saxitoxin	Inadequate processing, fungal contamination	Aerial dispersion, food contamination

© Woodhead Publishing Limited, 2011

Frequent examples of intentional contamination of food products derive essentially from economic fraud and adulteration. But there are other examples of accidental contamination which show that the extent of possible terrorist threats may include: environmental contamination of agricultural land, chemical contamination of raw materials and commodities, sabotage of production chains, contamination during storage, logistics, and delivery of goods, contamination at the level of retail distribution, on-the-shelf single product tampering and adulteration (Manning *et al.*, 2005). The effects of accidental contamination are reminiscent of the effects caused by intentional contamination. It follows therefore that all the existing controls and protection that can be applied to the former can also be applied to the latter. Table 4.2 lists some examples of biological agents that could be used for the purposes of terrorism, classified into main categories, bacteria, viruses, fungi, protozoa and toxins of biological origin. These can be responsible for accidental or intentional contamination of food products and of food-producing animals or plants (Marmiroli *et al.*, 2010). Depending on the nature of the potential hazards, the detection methods deployed can be directed at specific contaminating organisms, at classes of contaminants, or at the products of contaminating organisms.

4.2 Contamination of biological origin

Food supply chains are particularly sensitive and vulnerable as their constituents come from a number of different natural sources with human intervention involved. When foods are the target of intentional contamination, the most vulnerable are ready-to-eat products in retail outlets or in public locations such as cafeterias, bars and restaurants. In the USA in 1984 a radical group contaminated salad bars in several restaurants with the pathogenic bacteria *Salmonella*. The bacteria for the contamination had been isolated from standard hospital laboratory material, and were purchased through a medical supply company. The intentional nature of the outbreak was discovered only because the perpetrators confessed, otherwise it would have simply been considered a case of accidental food poisoning (Carus, 1998).

Agents for potential biological contamination of food supply chains (Table 4.2) include organisms responsible for diseases of humans, animals and plants, from all biological groups: viruses, bacteria, fungi, protozoa and small invertebrates. Additionally, toxins produced by some of these organisms can be used as contaminants. Biological agents can be used to contaminate food products, or in an attack on crops and farms, or can be dispersed in air and water. When considering agents that could affect the integrity of a food supply chain, we can also take into account examples which are not contaminants *per se*: undeclared, unknown or undesirable ingredients (e.g. allergens).

All biological entities involved in production or processing, as well as many contaminants, can be recognised by analysing nucleic acids or proteins present in the product. Methods for identifying contaminants require accepted com-

© Woodhead Publishing Limited, 2011

ponents or ingredients to be distinguished from unwanted or noxious components. Recent research activities have focused on the development and testing of innovative and reliable analytical methods for identifying biological components present in food (Bufflier *et al.*, 2007; Marmiroli and Maestri, 2007; Malorny *et al.*, 2008; Poms *et al.*, 2009; Reynisson *et al.*, 2008; Carcea *et al.*, 2009). The application of detection methods can be useful in the control of food supply chains, in assessing conformity to specifications and labelling, and in assisting traceability. At the same time, these methods can be of help in preventing and discovering forms of intentional contamination.

4.3 Detection methods for specific organisms

The identification of anomalous or unwanted biological components in food products can be performed through the identification and classification of the genetic material belonging to animals, plants or microorganisms (fingerprint), utilised in producing or processing the food product. The search for pathogen contaminants in food conventionally involves the isolation of microorganisms from food, their growth or enrichment in selective media and their identification with biochemical and immunological assays. The process can be long and biased by false negatives. A different approach is the analysis of the nucleic acids, either DNA or RNA, of the microorganisms that are believed to be contaminants. The use of quantitative real-time PCR (polymerase chain reaction) (qtRT-PCR) can allow quantification of bacteria, and the unambiguous identification of the bacterial species (example in Reynisson *et al.*, 2008). Methods for identification of *Salmonella enterica*, *Listeria monocytogenes*, *Campylobacter jejuni* and *Staphylococcus aureus* could be applied to raw broiler meat (Fig. 4.1), poultry carcasses, raw milk and other food products. Interest in qtRT-PCR for food security purposes is dependent on their deployment in field conditions, as with some portable RT-PCR systems. One example of these is the thermocycler R.A.P.I.D. (Ruggedised Advanced Pathogen Identification Device) which has been used successfully in field detection of pathogens such as *Francisella tularensis* (McAvin *et al.*, 2004). An essential step towards the implementation of these applications is the development of DNA-extraction-free RT-PCR methods, or of solid-state DNA extraction methods, to avoid lengthy procedures before a diagnostic PCR analysis. An assay which does not require DNA extraction has been proposed (Fyske *et al.*, 2008): bacteria are collected from the air into liquid substrate, lysed with silica beads and used for RT-PCR with primer sets and probes specific for 16S rRNA or with species-specific primers. Organisms that produce spores, such as *Bacillus anthracis*, the agent for anthrax, present a particular problem; the extraction of DNA from spores requires harsh treatments, such as sonication, freeze-thaw cycles, or hot detergents (Saikaly *et al.*, 2007).

Several innovative devices based on sampling and analysis of nucleic acids or proteins have been recently developed for monitoring high-risk locations,

© Woodhead Publishing Limited, 2011

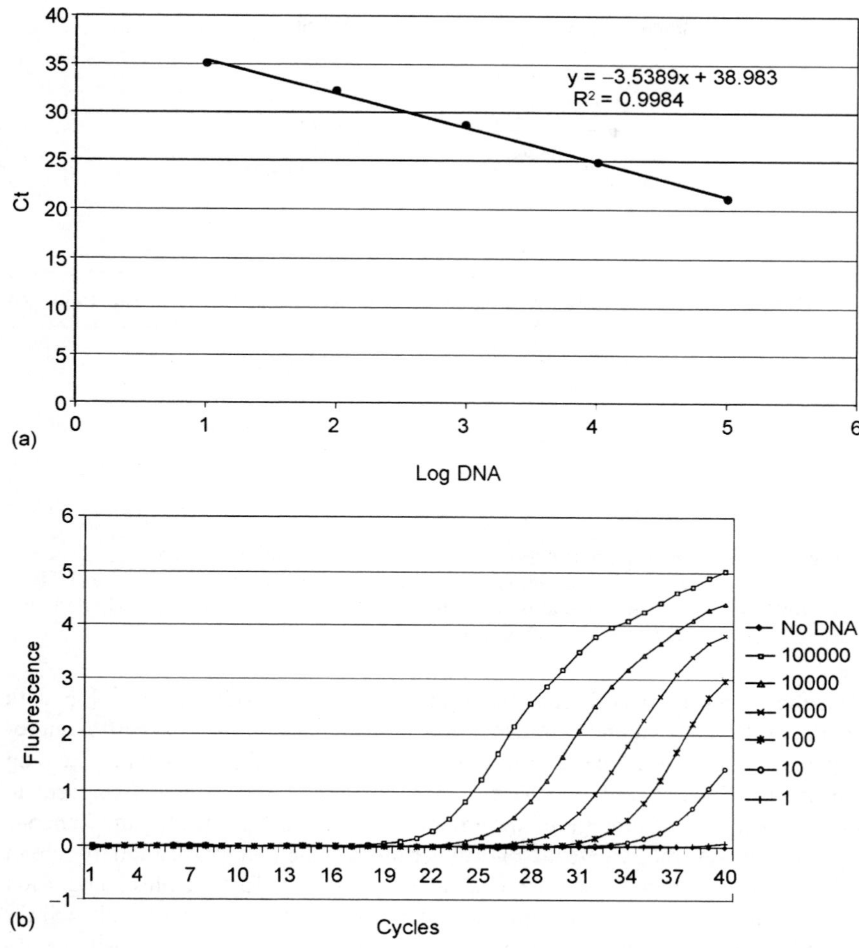

Fig. 4.1 Real-time PCR analysis of *Salmonella enterica* with primers SA4, based on the invA gene. (a) standard curve obtained with serial dilutions of *Salmonella* genomic DNA. The equation of the regression line is reported. (b) Amplification plots of samples containing from 100 000 to 1 copies of *Salmonella* genome.

especially for defence against bioterrorism (Regan *et al.*, 2008). The Autonomous Pathogen Detection System (APDS) periodically concentrates airborne particles (1–10 μm) into a liquid. It then, performs multiloci detection of several biothreat agents with PCR followed by hybridisation to pathogen-specific microspheres recognised with fluorescence (Fig. 4.2). The system operates autonomously for a week performing seven analyses each day.

Several different methods are available for rapid detection of multiple pathogenic components at the same time. One of the most advanced is the use of DNA microarrays for detection of pathogens (Sergeev *et al.*, 2004). Target

© Woodhead Publishing Limited, 2011

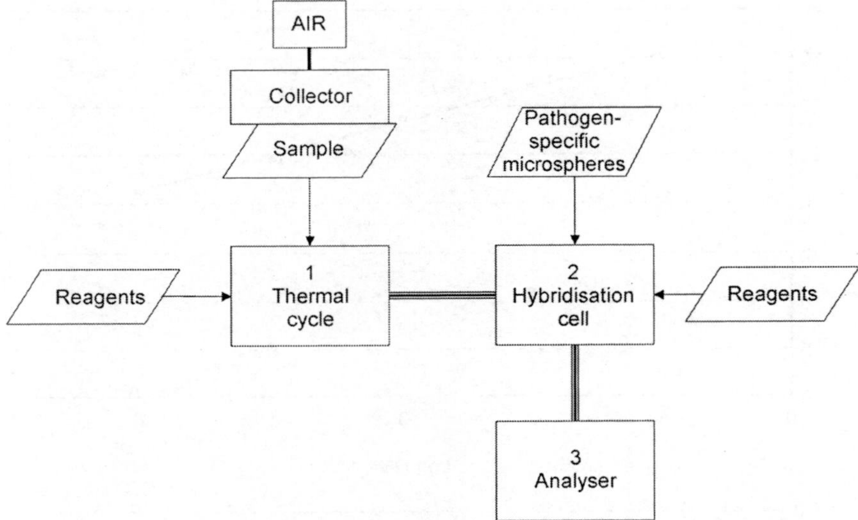

Fig. 4.2 Schematic representation of the Autonomous Pathogen Detection System (Regan *et al.*, 2008). The air sample prepared in the collector is subjected to PCR amplification (1) and the amplicons are then hybridised to microspheres (2) labelled with fluorescent dyes, which are then detected in the analyser (3).

sequences amplified with specific primer pairs are hybridised on a chip with several different oligonucleotides, each one diagnostic for a specific micro-organism. The specific pattern of hybridisation in the different spots of the chip surface provides information for the identification of the pathogens present in the sample. The advantage of the microarray approach is in the simultaneous analysis of several targets in one experiment, where hybridisation of target DNAs to specific probes confirms the identity and therefore enables recognition of the contaminant. A similar approach, but modified with a microfluidic cell for hybridisation to reduce reaction volumes and times, has been developed to identify *L. monocytogenes*, and *S. aureus* (Cretich *et al.*, 2010). The target gene for recognition was the 16S rDNA gene amplified with consensus primers designed on conserved regions. Specific oligonucleotides targeted to variable regions of the gene were instead used on the chip, printed at a concentration of 20 μM in 5 × 5 sub-arrays and inserted in a polydimethylsiloxane microchamber connected to a fluidic system (Fig. 4.3). Based on the same approach, a microarray in which results are shown with colorimetric reactions has been developed, eliminating the need for an expensive fluorescence reader (Cretich *et al.*, 2010): amplified products carry biotin linked to one of the primers, and this is recognised by streptavidin linked to alkaline phosphatase, which reacts with the chromogenic substrate. Coupling microfluidics for DNA hybridisation and colorimetric reading can provide innovative solutions for deployment of devices near-line in critical points of the food supply chain and fast analytical devices for screening in case of terrorist threats.

© Woodhead Publishing Limited, 2011

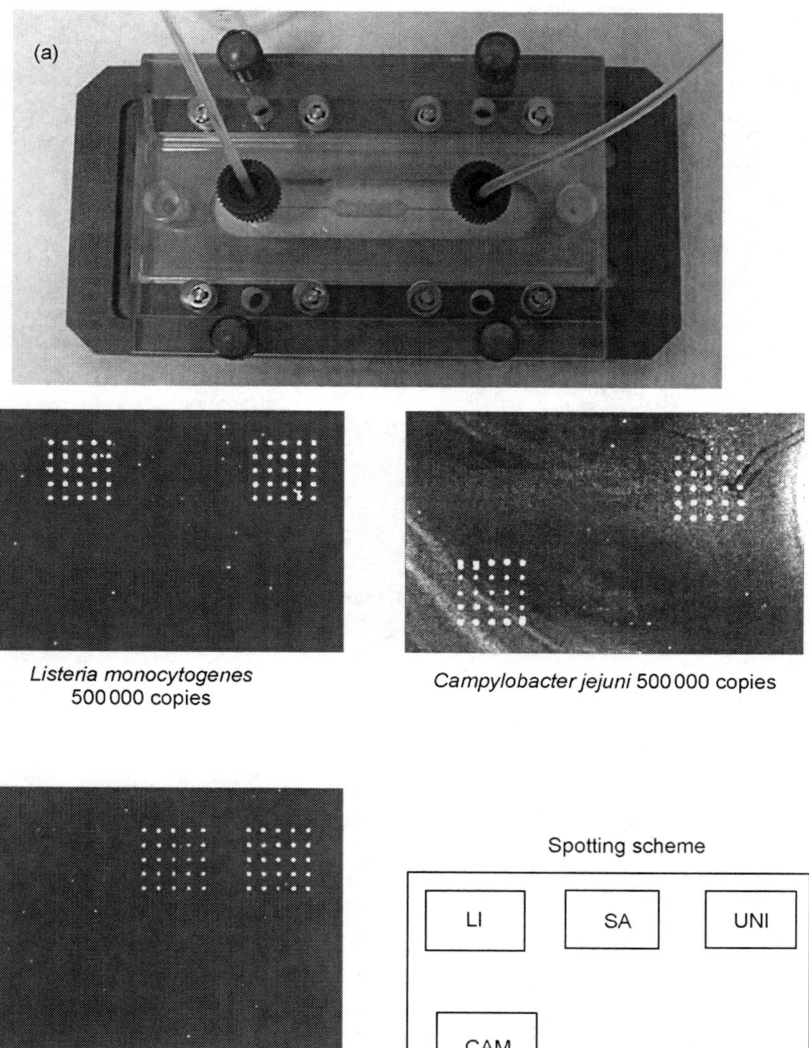

Fig. 4.3 Microfluidic microarray system for detection of pathogen bacteria (Cretich *et al.*, 2010). (a) Microfluidic cell for hybridisation of DNA microarray or immunoarray. (b) Results of hybridisation of genomic DNA from bacteria on the microarray chip. The spotting scheme shows the position of probes specific for *Listeria monocytogenes* (LI), *Staphylococcus aureus* (SA), *Campylobacter jejuni* (CAM) and of universal probes complementary to 16S rDNA (UNI).

© Woodhead Publishing Limited, 2011

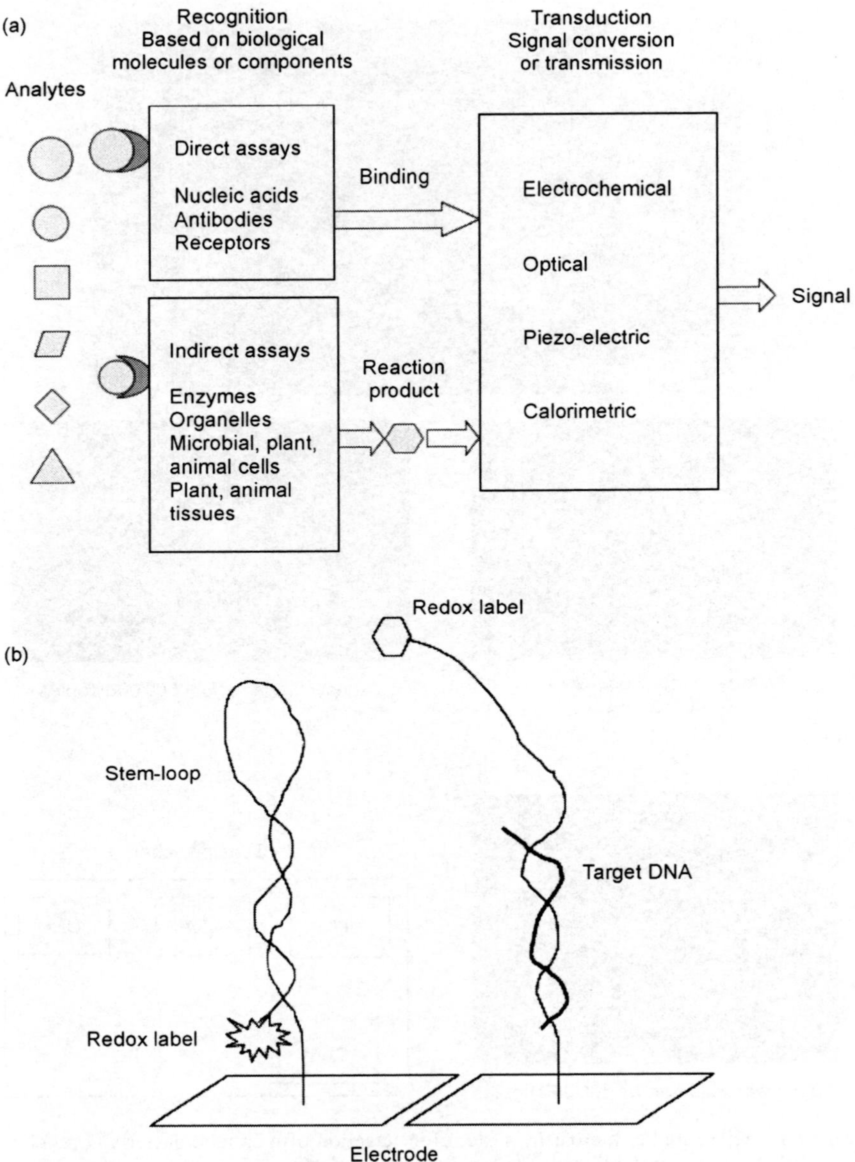

Fig. 4.4 Sensors for detection of contaminants. (a) General principle of biosensors, showing the recognition element based on biomolecular components and the transducing element converting the signal for data analysis. (b) Schematic depiction of the functioning of the E-DNA sensor (Lai *et al.*, 2006).

© Woodhead Publishing Limited, 2011

Recently, DNA applied to biosensors (Fig. 4.4a) has been exploited as an active component in the generation of the electric signal (Lai *et al.*, 2006). The E-DNA is an electrochemical biosensor, in which a DNA stem-loop labelled with a molecule such as ferrocene is covalently attached to an electrode (Fig. 4.4b). The DNA strands are stabilised by thermostable GC pairs, and thus both ends are near to the gold surface of the electrode. When the probe recognises fragments amplified from the DNA extracted from the food products, the stem-loop is disrupted producing a semi-quantitative electrochemical signal. Hepatitis B virus has been detected with such systems, involving a 21 nt probe, methylene blue as the hybridisation detector and voltammetry for signal detection (Erdem *et al.*, 2000). The interaction of methylene blue with immobilised ssDNA gives enhanced voltammetric signals, whereas a decrease signals hybrid formation. Detection of the pathogenic *Mycobacterium tubercolosis* with a carbon paste electrode and 27–36 nt DNA has been reported (Wang *et al.*, 1997).

Lab-on-chips are the most advanced applications of portable biosensors, reproducing complex analytical techniques on small slides of a few square centimetres (Whitesides, 2006). Their development was started by the Defence Advanced Research Projects Agency (DARPA) of the US Department of Defence in the 1990s, with the aim of developing field-deployable microfluidic systems designed to serve as detectors for chemical and biological threats (Cheng *et al.*, 2009). Nowadays, many analytical techniques have been integrated or associated with Lab-on-chip sensors including PCR, RT-PCR and microarray. An extensive review of the application of lab-on-chips for detection of pathogens has recently been published (Mairhofer *et al.*, 2009). It is recognised that the main advantages are found in the low limits of detection, in the capacity for analysing complex matrixes, and in the portability of the devices.

4.4 Detection methods for classes of contaminants

Electronic noses and tongues are sensors designed to mimic the sensorial abilities of humans in detection of complex mixtures of chemical substances, also of biological origin. They operate in a non-destructive procedure, assessing properties of the sample related to volatile molecules and to dissolved compounds, respectively. These devices work with a matrix of broadly selective chemical sensors supported by software modules for intelligent data analysis. Their structure mimics that of the mammalian sensory system, and can be divided into three different parts: the identification system, the pattern recognition system, and the sampling system with the matrix of sensors (Ciosek and Wroblewski, 2007). In the electronic nose, each aroma is pumped across the array of non-selective sensors, which are coated with different sensing materials. The aroma induces a unique pattern of responses that is fed into a computer and identified through algorithms, comparing the pattern with *ad-hoc* memory banks. New e-nose models adopt nanogravimetric sensors, exploiting the

© Woodhead Publishing Limited, 2011

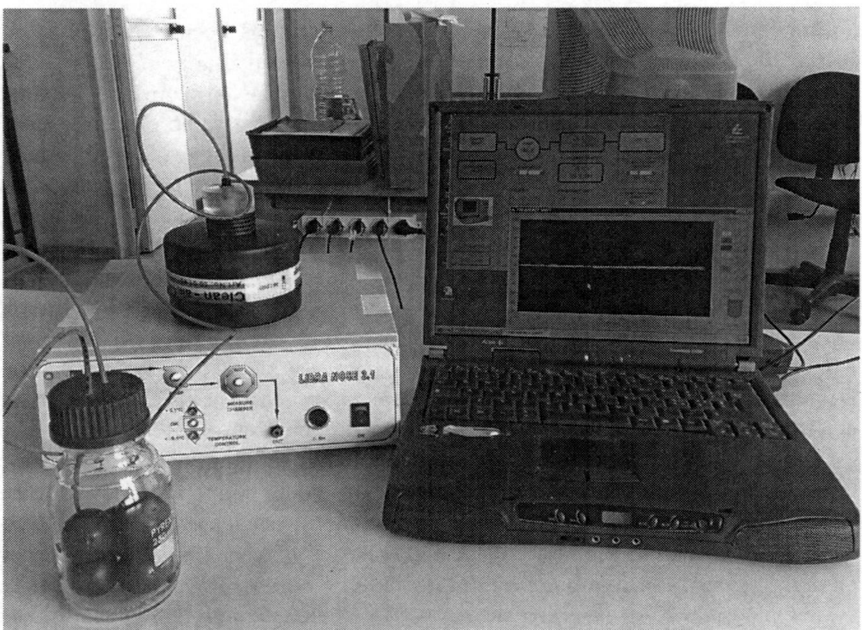

Fig. 4.5 Assembly of the electronic nose (Libra Nose) produced by Technobiochip during tomato measurements in the TRACEBACK project.

properties of piezo-electric quartz crystals in order to measure very small mass variation (nanogram order) on the surface of the quartz. The quartz-based transducer, which can be used in both gas and liquid environments, consists of an oscillator scale made of a piezo-electric system, which is in turn made of a quartz crystal covered by a chemically sensitive material (Deisingh *et al.*, 2004).

The Libra Nose detector has been applied to the tomato food supply chain for evaluation of the maturity stage (Fig. 4.5). This instrument consists of a thermostatic measuring chamber with an array of eight sensors, completed by a pump with an adjustable flow rate which is controlled by software. The sensors are resonant quartzes coated with patented poly-pyrrole-derivatives detecting volatile organic compounds. With respect to applications of such technology in food defence, electronic noses could be trained to classify volatile substances associated with the presence of specific pathogens, as an aid in diagnosing contamination (Wilson and Baietto, 2009).

4.5 Detection of products of biological contamination

In some cases it would be faster to directly detect the product of a contamination, rather than identifying or detecting its source, be it biotic or abiotic. If a toxin is a biologically active compound it is appropriate to develop methods

© Woodhead Publishing Limited, 2011

for the recognition of that compound; recognition of the organism is still a viable option when detection of the toxic product is not feasible.

For faster analyses, which can also be carried out in field conditions for some organisms, a lateral flow device can be of some use. This device is a combination of ELISA (enzyme-linked immunosorbent analysis) immunochromatographic methodology and a detection signal based on changes in colour. It employs colloidal, Au-labelled antibodies for pathogen (antigen) capture on a pad: the antigen-antibody complex migrates on a membrane to the reagent zone, which contains an anti-antigen- antibody. The antigen-antibody complex forms a distinct red line. A second binding reagent is a control zone and forms a second red line. The device is normally inserted in a plastic housing. This device allows manual rapid screening for the presence or absence of some food-borne pathogens, based on recognition of their proteins or other surface molecules. A recent review by Ngom *et al.* (2010) reports on the applicability of such devices to many organisms and toxins of interest to bioterrorism, with limits of detection calculated in different exprimental conditions.

An immunoarray sensor for direct detection of bacterial toxins has been developed by exploiting a microfluidic approach reaching limit of detection (LOD) values of 14–147 pg/mL (Cretich *et al.*, 2009). The targets were *S. aureus* enterotoxins A and B recognised through interaction with polyclonal antibodies linked to a glass chip. After binding, secondary monoclonal antibodies labelled with a fluorochrome allowed detection in a fluorescence reader. Similar applications have been reported for detection of bacterial toxins in sandwich immunoassays: cholera toxin, ricin, shiga-like toxin 1 and staphylococcal enterotoxin B, all of great interest in antiterrorism operations, which were detected in mixtures containing 30 ng/mL of each toxin. The method utilised luminescent semiconductor nanocrystals (quantum dots) conjugated to antibodies in fluoroimmunoassays (Goldman *et al.*, 2004).

Detection of biological contaminants in liquid samples can also be performed without the use of labelling, with sensors based on surface plasmon resonance (SPR) utilising specific capture by biomolecular recognition elements, i.e. antibodies. SPR characterises innovative label-free optical immunosensors with high selectivity and sensitivity; Taylor *et al.* (2008) report an increase of sensitivity up to 50-fold in the recognition of analytes by immobilised antibodies as compared to other immunosensing techniques, because of decrease in non-specific binding by other components in the matrix. It is an optical sensor exploiting electromagnetic waves (plasmons) on a normal range or on a long range, to probe interactions between an analyte and the corresponding recognition element immobilised on the sensor surface. The binding occurs on a metallic sensor surface, inducing an increase in the refractive index of the underlying polymer. Portable instruments are currently being developed for different purposes, including environmental monitoring (Taylor *et al.*, 2008). SPR sensors can be functionalised with antibodies, oligonucleotides, peptide nucleic acids, proteins and whole cells. Several applications of bacteria detection have been reported (Taylor *et al.*, 2008), including *Escherichia coli*

© Woodhead Publishing Limited, 2011

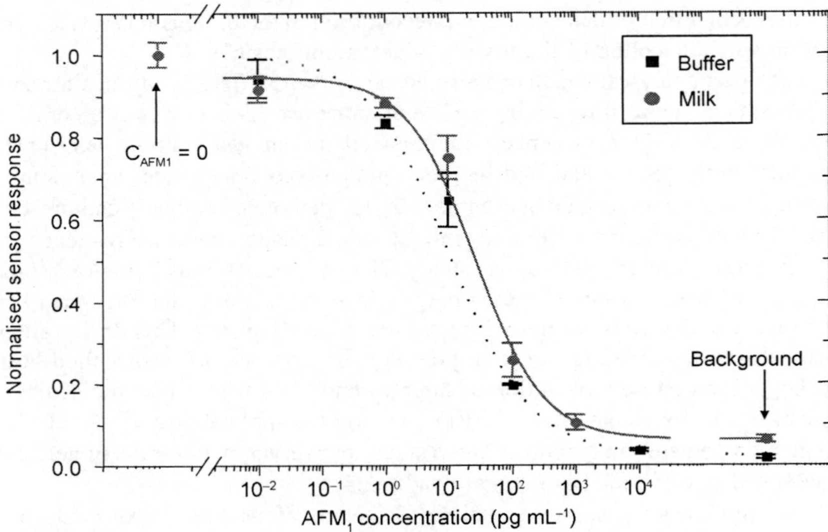

Fig. 4.6 Immunoassay for aflatoxin M_1 (AFM_1) implemented in Long Range Surface Plasmon Resonance (Wang *et al.*, 2009). The graph shows the response of the sensor to increasing concentrations of AFM_1 dissolved in buffer (square symbol) or in raw milk (circle symbol).

O157:H7 (LOD 10^2 CFU/mL), *Salmonella* (LOD 10^2 CFU/mL), *Listeria* (LOD 3.5×10^3 CFU/mL), *Yersinia* (LOD 10^2 CFU/mL), *Vibrio cholerae* (LOD 3.7×10^5 CFU/mL) and others. Extension to other organisms related to bioterror can be foreseen. Enhancement of the signal for detection of low concentration analytes can be obtained with fluorescent labelling and Long Range (LR) surface plasmons (Wang *et al.*, 2009). Recently, an LR-SRP sensor has been applied to detection of the aflatoxin M_1 in milk using antibodies as recognition elements (Fig. 4.6, Wang *et al.*, 2009). Aflatoxin M_1 is the metabolite found in secretions of animals which have ingested feed contaminated by aflatoxin B_1, produced by some *Aspergillus* strains. The limit of detection in milk was lower by two orders of magnitude with respect to the level of aflatoxin prescribed by legislation. Therefore this sensor is particularly well adapted for detection of potential terrorist activity.

4.6 Conclusions and recommendations

Recently, reports of terrorist attacks have been steadily increasing, but the analyses show that the targets of these attacks are shifting from high-profile locations such as airports or stations, to so-called 'soft' targets, such as those represented by private properties, or individual citizens (Mohtadi and Murshid, 2009). Food supply chains are certainly amongst these soft targets, and therefore forecasts point to an increased risk. Modelling of plausible scenarios has shown

© Woodhead Publishing Limited, 2011

that in cases of contamination of products with a short shelf-life, the effect may extend to a higher number of people. This is a result of the fact that the time required for forensic analyses is quite long and by the time an alert is issued most of the contaminated products would already have been consumed (Hartnett *et al.*, 2009).

The time required for identification of pathogens or chemicals is a critical issue, because it determines the speed and effectiveness of product withdrawal and effective recall from the market or from availability to the public. In particular, it is the identification of the affected foods which causes the longest delay. A particularly clear example of this was seen in a 2008 event in the USA – from April of that year several hundred people over 43 states were affected by symptoms consistent with contamination by a *Salmonella* Saintpaul strain, which was initially linked to consumption of raw tomatoes that according to reports by patients were imported from Mexico. After an initial recall of tomatoes from several supermarkets, authorities failed to match the bacterial fingerprinting to any tomato samples. The genetic match was finally found in jalapeno and serrano peppers, which were recalled starting from July 2008. But the real cause of the outbreak was not discovered, and may in fact have been related to several products. The episode highlighted some of the major problems in ascertaining the origin of outbreaks:

- the length of time between contamination and the earliest reports of poisoning,
- the false alarm over the wrong food product, and
- the difficulty in finding the real contaminated products and therefore in issuing effective measures to deal with the problem.

It is recognised that the two major strategies against food terrorism are prevention and response (WHO, 2002). The EC has issued a Green Paper on Bio-preparedness (EC, 2007) stating that the existing framework for food safety must be complemented by a new framework aimed at increasing security, preventing terrorist acts and responding to both intentional and accidental outbreaks. According to this document, the actions required are 'early detection, sound traceability systems, rapid control and eradication measures, contingency plans and overall coordination' (EC, 2007).

One way to address the prevention issue has been identified in the form of the HACCP procedure, where technology is used to detect pathogens or other hazards. It remains to be seen whether this and other systems, developed to prevent accidental contamination, can be equally effective against intentional contamination.

As compared to other terrorist threats, such as dissemination of aerosols or liquids, food contamination may seem to be more difficult to achieve, because several links in the food chain should be subject to control and monitoring. However, food is an ideal means of delivering contaminants to humans, since it can reach a plethora of targets in many different ways, and previous cases have shown that the control measures can be easily bypassed, either deliberately or

© Woodhead Publishing Limited, 2011

accidentally. Most countries have some form of prevention and response systems in place in relation to food safety (e.g. general food law and Rapid Alert System for Food and Feed – RASFF in Europe), public health (e.g. European Communicable Diseases Network) or terrorism (e.g. European Programme for Critical Infrastructure Protection – EPCIP and Critical Infrastructures Warning Information Network – CIWIN). However, these systems rarely include any consideration of food as a vehicle for delivering harmful agents for terrorism, and even more rarely include any consideration of agro-terrorism.

It is certain that traceability along the entire food supply chain is essential to complement any prevention measures. Speed in locating the source of contamination and the identification of the product flow along the entire supply chain is of utmost importance in performing an effective recall of contaminated products. A complete integrated system which links on one side any information about the history and movement of the products along all the steps of the chain, and on the other side objective parameters of quality and safety measured at critical points, would certainly increase preparedness towards intentional and accidental threats (Opara, 2003). Monitoring of critical environmental resources and food production areas in rural environments is also possible with remote sensing coupled with artificial intelligence, and can provide additional information for the security of the food supply chains (Wang *et al.*, 2006).

A weakness in the system is the identification of a terrorist act, when this is not clearly indicated as such. Time can be lost in collecting data from the public health system, connecting instances of disease symptoms and tracing them to a common source. It is very difficult to recognise symptoms of foodborne pathogens or toxins, and an improved network of information collection is required, involving hospitals, family doctors and pharmacists. This is outside the scope of the present review, but it can be argued that improving the analytical methods for detection (as described earlier) and awareness, can provide impetus for the construction of these networks. Situational centres can provide the environment for collecting data, sharing information and implementing systems for support in decision making (Marmiroli *et al.*, 2010).

4.7 Acknowledgements

The authors thank their colleagues in the TRACEBACK project from Technobiochip, AINIA, Max Planck Institute, and National Research Council. The authors wish to acknowledge the support of the European Commission in the research projects CHILLON FP6-016333 (NM, EM), TRACEBACK FP6-036300 (NM, EM, MM), and SecuFood JLS/2008/CIPS/022 (RO, RS). They also acknowledge the support of the NATO Science for Peace project SITCEN SFP982498 (NM, EM, MM, VK).

© Woodhead Publishing Limited, 2011

4.8 References

BUFFLIER, E., A. SUSCA, M. BAUD, G. MULÈ, K. BRENGEL and A. LOGRIECO (2007) Detection of *Aspergillus carbonarius* and other black aspergilli from grapes by DNA OLISA™ microarray. *Food Addit. Contam. A* 24: 1138–1147.

CARCEA, M., P. BRERETON, R. HSU, S. KELLY, N. MARMIROLI, F. MELINI, C. SOUKOULIS and D. WENPING (2009) Food authenticity assessment: ensuring compliance with food legislation and traceability requirements. *Qual. Assur. Safety Crops Foods* 1: 93–100.

CARUS, W. S. (1998) *Bioterrorism and biocrimes. The illicit use of biological agents since 1900.* Center for Counterproliferation Research, Washington, DC.

CHALECKI, E. L (2002) A new vigilance: identifying and reducing the risks of environmental terrorism. *Global Environ. Politics* 2: 46–64.

CHALK, P. (2004) *Hitting America's soft underbelly. The potential threat of deliberate biological attacks against the US agricultural and food industry.* RAND Corporation, Santa Monica CA [www.rand.org].

CHENG, X., G. CHEN and W.R. RODRIGUEZ (2009) Micro- and nanotechnology for viral detection. *Anal. Bioanal. Chem.* 393: 487–501.

CIOSEK, P. and W. WROBLEWSKI (2007) Sensor arrays for liquid sensing – electronic tongue systems. *Analyst* 132: 963–978.

CRETICH, M., G. DI CARLO, R. LONGHI, C. GOTTI, N. SPINELLA, S. COFFA, C. GALATI, L. RENNA and M. CHIARI (2009) High sensitivity protein assays on microarray silicon slides. *Anal. Chem.* 81: 5197–5203.

CRETICH, M., V. SEDINI, F. DAMIN, M. PELLICCIA, L. SOLA and M. CHIARI (2010) Coating of nitrocellulose for colorimetric DNA microarrays. *Anal. Biochem.* 397: 84–88.

DEISINGH, A.K., D.C. STONE and M. THOMPSON (2004) Applications of electronic noses and tongues in food analysis. *Int. J. Food Sci. Technol.* 39: 587–604.

ERDEM, A., K. KERMAN, B. MERIC, U.S. AKARCA and M. OZSOZ (2000) Novel hybridization indicator methylene blue for the electrochemical detection of short DNA sequences related to the hepatitis B virus. *Anal. Chim. Acta* 422: 139–149.

EUROPEAN COMMISSION (2007) Green paper on bio-preparedness. COM(2007) 399 final.

FYSKE, E.M., B. LANGSETH, J.S. OLSEN, G. SKOGAN and J.M. BLATNY (2008) Detection of bioterror agents in air samples using real-time PCR. *J. Appl. Microbiol.* 105: 351–358.

GOLDMAN, E.R., A.R. CLAPP, G.P. ANDERSON, H.T. UYEDA, J.M. MAURO, I.L. MEDINTZ and H. MATTOUSSI (2004) Multiplexed toxin analysis using four colors of quantum dots fluoroagents. *Anal. Chem.* 76: 684–688.

HARTNETT, E., G.M. PAOLI and D.W. SCHAFFNER (2009) Modeling the public health system response to a terrorist event in the food supply. *Risk Anal.* 29: 1506–1520.

JUST, D.R., B. WANSINK and C.G. TURVEY (2009) Biosecurity, terrorism, and food consumption behavior: using experimental psychology to analyze choices involving fear. *J. Agric. Res. Econ.* 34: 91–108.

LAI, R.Y., E.T. LAGALLY, S.-H. LEE, H.T. SOH, K.W. PLAXCO and A.J. HEEGER (2006) Rapid, sequence-specific detection of unpurified PCR amplicons via a reusable, electrochemical sensor. *Proc. Natl. Acad. Sci. U.S.A.* 103: 4017–4021.

MAIRHOFER, J., K. ROPPERT and P. ERTL (2009) Microfluidic systems for pathogen sensing: A review. *Sensors* 9: 4804–4823.

MALORNY, B., C. LÖFSTRÖM, M. WAGNER and J. HOORFAR (2008) Enumeration of Salmonella bacteria in food and feed samples by real-time PCR for quantitative microbial risk

© Woodhead Publishing Limited, 2011

assessment. *Appl. Environ. Microb.* 74: 1299–1304.

MANNING, L., R.N. BAINES and S.A. CHADD (2005) Deliberate contamination of the food supply chain. *Br. Food J.* 107: 225–245.

MARMIROLI, N. and E. MAESTRI (2007) Polymerase chain reaction (PCR), pp.147–187. In Y. Picò (ed.), *Food Toxicants Analysis. Techniques, Strategies and Development.* Elsevier, Amsterdam.

MARMIROLI, N., M. MARMIROLI and E. MAESTRI (2010) Monitoring of environmental resources against intentional threats, pp. 51–73. In V. Koukouliou, M. Langer and O. Premstaller (eds), *Threats to Food and Water Chain Infrastructure.* NATO Science for Peace and Security Series – C: Environmental Security. Springer, Dordrecht.

McAVIN, J.C., M.M. MORTON, R.M. ROUDABUSH, D.H. ATCHLEY and J.R. HICKMAN (2004) Identification of *Francisella tularensis* using real-time fluorescence polymerase chain reaction. *Milit. Med.* 169: 330–333.

MOHTADI, H. and A. P. MURSHID (2009) Risk analysis of chemical, biological, or radionuclear threats: implications for food security. *Risk Anal.* 29: 1317–1335.

NADASI, E., T. VARJAS, I. PRANTNER, V. VIRAG and I. EMBER (2007) Bioterrorism: warfare of the 21st century. *Gene Ther. Mol. Biol.* 11: 315–320.

NGOM, B., Y. GUO, X. WANG and D. BI (2010) Development and application of lateral flow test strip technology for detection of infectious agents and chemical contaminants: a review. *Anal. Bioanal. Chem.* 397: 1618–2642.

OPARA, L.U. (2003) Traceability in agriculture and food supply chain: a review of basic concepts, technological implications, and future prospects. *J. Food Agric. Environ.* 1: 101–106.

PATE, J. and G. CAMERON (2001) Covert biological weapons attacks against agricultural targets: assessing the impact against US agriculture. BCSIA Discussion Paper 2001-9, ESDP Discussion paper ESDP-2001-05. John F. Kennedy School of Government, Harvard University.

POMS, R., M. THOMAS, P. FINGLAS, S. ASTLEY, D. SPICHTINGER, M. ROSE, B. POPPING, A. ALLDRICK, H. VAN EGMOND, M. SOLFRIZZO, E.N.C. MILLS, W. KNEIFEL, S. PAULIN, V. OREOPOULOU, K.A. TO, M. CARCEA, H. TURESKJA, M. SAARELA, J.-E. HAUGEN and M. GROSS (2009) MoniQA (Monitoring and Quality Assurance): an EU-funded network of excellence working towards the harmonization of worldwide food quality and safety monitoring and control strategies status report 2008. *Qual. Assur. Safety Crops Foods*, 1: 9–22.

REGAN, J.F., A.J. MAKAREWICZ, B.J. HINDSON, T.R. METZ, D.M. GUTIERREZ, T.H. CORZETT, D.R. HADLEY, R.C. MAHNKE, B.D. HENDERER, J.W. BRENEMAN IV, T.H. WEISGRABER and J.M. DZENITIS (2008) Environmental monitoring for biological threat agents using the autonomous pathogen detection system with multiplexed polymerase chain reaction. *Anal. Chem.* 80: 7422–7429.

REYNISSON, E., H.L. LAUZON, H. MAGNÚSSON, G.Ó. HREGGVIÐSSON and V.Þ. MARTEINSSON (2008) Rapid quantitative monitoring method for the fish spoilage bacteria Pseudomonas. *J. Environ. Monit.* 10: 1357–1362.

SAIKALY, P.E., M.A. BARLAZA and F.L. DE LOS REYES III (2007) Development of quantitative Real-Time PCR assays for detection and quantification of surrogate biological warfare agents in building debris and leachate. *Appl. Environ. Microbiol.* 73: 6557–6565.

SECUFOOD (2010) Security of European Food Supply Chain – Project summary – 27 April 2010 [available at http://secufood.unicampus.it, accessed 6 August 2010].

© Woodhead Publishing Limited, 2011

SERGEEV, N., M. DISTLER, S. COURTNEY, S. F. AL-KHALDI, D. VOLOKHOV, V. CHIZHIKOV and A. RASOOLY (2004) Multipathogen oligonucleotide microarray for environmental and biodefense applications. *Biosens. Bioelectron.* 20: 684–698.

STAMM, B.H. (2002) Terrorism risks in rural and frontier America. *IEEE Eng. Med. Biol. Mag.* 21: 100–111.

TAYLOR, A.D., J. LADD, J. HOMOLA and S. JIANG (2008) Surface plasmon resonance (SPR) sensors for the detection of bacterial pathogens, pp. 83–108. In M. Zourob, S. Elwary and A.P.F. Turner (eds), *Principles of Bacterial Detection: Biosensors, Recognition Receptors and Microsystems*. Springer, Berlin.

WANG, J., G. RIVAS, X. CAI, N. DONTHA, H. SHIRAISHI, D. LUO and F. S. VALERA (1997) Sequence-specific electrochemical biosensing of *M. tubercolosis* DNA. *Anal. Chim. Acta* 337: 41–48.

WANG, N., N. ZHANG and M. WANG (2006) Wireless sensors in agriculture and food industry – recent development and future perspective. *Comput. Electron. Agr.* 50: 1–14.

WANG, Y., J. DOSTÁLEK and W. KNOLL (2009) Long range surface plasmon-enhanced fluorescence spectroscopy for the detection of aflatoxin M_1 in milk. *Biosens. Bioelectron.* 24: 2264–2267.

WHITESIDES, G. M. (2006) The origins and the future of microfluidics. *Nature* 442: 368–373.

WILSON, A.D. and M. BAIETTO (2009) Applications and advances in electronic-nose technologies. *Sensors* 9: 5099–5148.

WORLD HEALTH ORGANIZATION (2002) *Terrorist threats to food: guidance for establishing and strengthening prevention and response systems*. WHO Library, Geneva.

© Woodhead Publishing Limited, 2011

Part II

Food safety and quality

© Woodhead Publishing Limited, 2011

5

Understanding and monitoring pathogen behaviour in the food chain

M. Jakobsen, University of Copenhagen, Denmark, J. Verran, Manchester Metropolitan University, UK, L. Jacxsens, University of Ghent, Belgium, B. Biavati, University of Bologna, Italy, J. Rovira, University of Burgos, Spain and L. Cocolin, University of Turin, Italy

Abstract: Understanding and monitoring of microbial behaviour of pathogens in the entire food chain is seen as one of the most important prerequisites for a lower prevalence of foodborne pathogens. For this purpose, novel methods have been developed to analyse the interactions of pathogens and food and feed matrices at cellular and molecular levels as well as contact surfaces in the food chain, including the intestinal tract of farm animals. The methods are mainly based upon advanced techniques of microscopy allowing analysis of single cells in time lapse studies and the rapidly growing information available in the field of functional genomics. Further emphasis has been given to development of functional mammalian cell models to study host pathogen interactions and to select novel probiotic cultures. Molecular culture-independent methods and microarrays are useful in monitoring microbial behaviour in the food chain and are often superior to the classical methods for detection of pathogens. Effective interventions to break the transmission of pathogens along the food chain are addressed, with special attention given to the present gap between food technology, hygienic design and food safety. Finally, the fact that effective food safety management is necessary for safe food is emphasised.

Key words: food safety, microbial behaviour, hygienic design, food safety management.

© Woodhead Publishing Limited, 2011

5.1 Introduction

Food production in Europe is generally very high-tech and has never been more stringently controlled. However, the incidence of foodborne diseases is still increasing. The significant investments to secure food safety by governments and the food industry have not been fully successful according to the statistics supplied by the European Food Safety Agency (EFSA, 2010). The provision of new and essential information and methods on reduction in the prevalence of pathogens in food has been a focus of PathogenCombat (www.pathogencombat.com). Following a study of the establishment and behaviour of pathogens in the food chain it has been concluded that risk assessment should not only be based upon numbers of pathogens but also their behaviour. New and more rapid methods are required to monitor and understand microbial behaviour, providing valuable information. Understanding of the presence and behaviour of pathogens should be linked with interventions to break the transmission of pathogens along the food chain. In this regard, attention must be given to the present significant gap between technology and hygienic design in food processing and food safety. Interventions such as novel processing techniques (Rajkovic *et al.*, 2010) and use of protective and probiotic cultures (Gaggia *et al.*, 2010a,b) are not described in detail here but examples of successful applications in the food industry are mentioned in the two references quoted. It should be emphasised that regardless of new scientific information, novel detection methods and interventions in the food chain, safe food and appropriate consumer protection can only be assured by preventive measures applied in the entire food chain and the maintainence of effective food safety management systems.

In this chapter three different approaches to improving food safety are taken:

- pathogens and pathogen detection methods
- hygienic design factors
- effective food management systems.

5.2 Novel platforms to understand the presence and behaviour of pathogens in the food chain

The life of a pathogen in the food chain is predominantly a life within or on structured matrices and seldom a planctonic life in free water or other liquids. Furthermore, it is the life of the individuals where single cells have to adapt, survive and in some cases initiate growth and form a micro colony within a defined space of the structured matrix. Such a life is in strong contrast to studies of pathogens in liquid laboratory media and analyses of pathogens relying upon transfer from the food matrix to a laboratory substrate with subsequent counting of colony-forming units. It is also in contrast to studies carried out with populations of pathogens in laboratory media and laboratory incubation conditions. Nevertheless, research on foodborne pathogens is mainly carried out under such artificial conditions, and results from such studies may not reflect the true

© Woodhead Publishing Limited, 2011

behaviour of the organisms in the food chain. This is still a significant gap in European research in microbial food safety, but it has been addressed by development of new methods to analyse the interactions at cellular and molecular levels between pathogens and food and feed matrices, as well as contact surfaces in the food chain, including the intestinal tract of farm animals. Single-cell microbiology, as reviewed by Brehm-Stecher and Johnson (2004), is seen as an important discipline behind these studies, resulting in a better understanding of when pathogens attach and de-attach, adapt, persist and express virulence and resistance in the food chain. Methods using advanced microscopy, (like optical tweezers, fluorescence ratio imaging microscopy (FRIM) and atomic force microscopy (AFM)), and functional genomics can now be used. Optical tweezers allows single microbial cell studies of attachment and de-attachment to surfaces and has been used in food microbiology (Rasmussen *et al.*, 2008). FRIM, which allows *in-situ* time lapse studies of single cells, has been extensively used to monitor viability and activity of cells of pathogens by determination of intracellular pH (Budde and Jakobsen, 2000; Nielsen *et al.*, 2010; Gaggia *et al.*, 2010b; Smigic *et al.*, 2010; Kastbjerg *et al.*, 2009). A particular use of FRIM has been developed for rapid monitoring of the viability of *Mycobacterium avium* ssp. *paratuberculosis* exposed to antimicrobial compounds (Gaggia *et al.*, 2010b). It avoids the extensive incubation times needed for cultivation of this bacterium. AFM has been applied to studies of interactions between pathogens and food contact surfaces to obtain information on the effect of defined topographic features on surface hygiene and cleanability (Verran *et al.*, 2010c). The results obtained have challenged the traditional use of colony-forming units to determine survival of pathogens.

Mammalian functional cell models based upon true epithelial cells have been used to study host–pathogen interactions and for selection of protective and probiotic cultures to inactivate pathogens, including viruses as reviewed by Cenciè and Langerholc (2010). An example of actual use of the cell models is described by Maragkoudakis *et al.* (2010a,b) and Pogacar *et al.* (2010).

The functional genomic platform has a strong impact on the molecular techniques for determining the behaviour of pathogens and the various detection methods, as described in the following.

5.3 Importance of number and behaviour of pathogens for risk assessment

The new challenge for food microbiologists that emerges in the meta-genomic era is understanding of the molecular mechanisms of virulence of foodborne pathogens *in vitro* and *in vivo*. Previous knowledge, coupled with data from the recently completed and ongoing sequencing projects (http://www.ncbi.nlm.nih.gov/sites/genome), have identified several potential virulence factors. As virulence factors, we consider both pathogenicity characteristics that are directly related to disease symptoms (e.g. the Shiga-like toxin produced by pathogenic *Escherichia coli*) as

© Woodhead Publishing Limited, 2011

well as attributes of foodborne pathogens that allow them to survive and respond to the environmental stresses encountered during food processing, distribution and most importantly, after ingestion and within the human body. These factors will determine if a bacterium will eventually be able to cause a disease. Based on DNA sequences and genome mining, several genes, encoding potential virulence factors, can be identified. The goal now is to determine if these genes are expressed in any of the stages 'from farm to fork', and if so, whether the synthesised proteins are acting in the predicted way and contribute to the disease.

Global approaches are now available that permit scientists to look into the total RNA synthesised in the cells (transcriptomics, Chatterjee *et al.*, 2006) and the total proteins produced by the cells (proteomics), at any given moment or condition. These are powerful techniques that allow target genes that may require further investigation to be pinpointed.

Some of the approaches that scientists have available to study aspects of the behaviour of pathogenic microorganisms are reverse transcription – quantitative polymerase chain reaction (RT-qPCR) and microarrays. RT-qPCR offers the possibility to quantify the expression of target genes in an accurate manner (Kubista *et al.*, 2006; Sharkey *et al.*, 2004). It is a qPCR method that is preceded by an enzymatic reaction, carried out by a reverse transcriptase, an enzyme that synthesises complementary DNA (cDNA) from an RNA template. The most common approach, employed for the study of the expression of a potential virulence target gene, is the definition of the comparative (or relative) gene expression (Sue *et al.*, 2004, Werbrouck *et al.*, 2006). The expression of the target gene(s) is measured, by qPCR, under controlled conditions that resemble the food environment or the human intestinal environment. The expression of the target gene may be increased, decreased or remain stable. Up-regulation of a target gene under conditions that resemble the food environment can be a strong indication that this gene is important for the adaptation of the microorganism in this particular environment. On the contrary, a gene that is down-regulated is considered not essential, under these conditions. This strategy has been used to study virulence and stress responses of foodborne pathogens in synthetic media (Olesen *et al.*, 2009; Critzer *et al.*, 2008), as well as in real food matrices. Although expression studies in foods are scarce, the first evidence gathered strongly suggests the need to perform gene expression experiments in real food samples instead of standard broth systems (Rieu *et al.*, 2010; Olesen *et al.*, 2010). The general trend emerging from these studies is that relative transcription of certain virulence genes is higher in laboratory broths compared to real food matrices. Furthermore, previous studies suggest that a food matrix, in particular a meat-based food matrix, may influence virulence potential of *Listeria monocytogenes*, possibly through down-regulation of virulence genes (Mahoney and Henriksson, 2003).

Powerful tools to study expression behaviour of foodborne pathogens are also represented by microarrays that allow global monitoring and description of the adaptation strategies and the virulence potential in one specific environment. So far, microarrays have been used extensively *in vitro* (Wang *et al.*, 2010; Zhang *et*

© Woodhead Publishing Limited, 2011

al., 2010), but *in situ* applications are still very rare. One example includes a sub-array containing 54 genes of *L. monocytogenes*, encoding for virulence, adhesion and stress response genes, to study the effect of different environmental conditions and food matrices on the expression of its virulence genes (Rantsiou *et al.*, 2010a). To achieve this, different strains of *L. monocytogenes* were grown *in vitro*, at different temperatures, pH and salt concentrations. Following this, they were artificially inoculated in different food matrices, to simulate the conditions of 'virulence' of this microorganism at time of consumption. The results obtained in this study underlined extreme heterogeneity within the *L. monocytogenes* species and differences in terms of number and expression of genes when strains were inoculated in fresh meat or in fermented sausages.

The possibility to study the behaviour of foodborne pathogens *in situ* will allow a better comprehension of the risks associated with a specific category of food, in terms of composition (pH, A_w, salt), storage temperature and atmosphere. New information can be generated that in the future will substitute the classical risk assessment based on the numbers of pathogenic micro-organisms. Application of these approaches has highlighted differences in virulence expression between strains of the same species, thereby introducing new parameters that must be evaluated rather than determining the colony-forming units per gram or millilitre.

Availability of such data will elucidate the molecular mechanism by which foodborne bacteria elicit disease and furthermore, will allow: (i) design of tools in order to predict the risk related with the presence of foodborne pathogens in food, and (ii) to develop intervention strategies to reduce such risk.

5.4 Rapid detection methods

In the last 10 years, new approaches to study microorganisms in food systems have taken advantage of technological advancements in the field of molecular biology. In the late 1980s, the polymerase chain reaction (PCR) was invented, thereby opening a new area in the microbiological analysis, especially when pathogens have to be detected. The philosophy of the molecular methods is not cultivation, but the direct analysis of the nucleic acids. In this way, total DNA and/or RNA are extracted from the sample and subsequently subjected to analysis in which a specific target, often solely present in the microorganism under investigation, is searched for and amplified. Methods that are based on PCR can be culture-independent since the nucleic acids can be isolated without any cultivation. However, there is a risk of detecting dead cells that must be mitigated against. If prior to DNA or RNA extraction, an enrichment step is performed, the method can be considered partially culture-independent. Enriching the sample is a common practice in microbiological analysis, including food examinations, when PCR methods are used as detection systems. It facilitates an increase in the number of target species and, in the case of food systems, it helps by diluting compounds present in the samples that are inhibitory towards the

© Woodhead Publishing Limited, 2011

activity of the DNA polymerase, the enzyme responsible for the synthesis of the DNA molecules during PCR.

The application of PCR-based methods in food microbiology allows the development of protocols that are faster, more sensitive and more specific than traditional microbiological methods for the detection and identification of foodborne pathogens. Normally, the result on the presence or absence of a specific bacterium is obtained after 3–4 hours or at most 24–36 hours if an enrichment step is carried out.

'Conventional' PCR protocols, where the amplification products are detected by gel electrophoresis, have a big disadvantage, being exclusively qualitative. Following specific technological advancements that allow detection of a PCR product while it is being produced and amplified, PCR became a method that allows quantification. This technique, called quantitative PCR (qPCR) or real time PCR (Rt-PCR), revolutionised molecular approaches in food microbiology. While 'conventional' PCR can be used only to determine the presence or the absence of the specific pathogen in food, with qPCR it is also possible to perform a quantification (in terms of cells per g or ml). Furthermore, in the quantification assays, both DNA and RNA can be quantified. If DNA is extracted and analysed, qPCR can be used for quantification of the total number of the cells of the pathogen without differentiation between dead and live cells. On the other hand, if the RNA is extracted, qPCR gives the possibility of quantifying exclusively cells that are alive. Of course in this last case, the extracted RNA will have to be reverse transcribed and then amplified. This last possibility is important in the case of foodborne pathogens. The approach of food industries towards foodstuffs should be different depending on whether pathogens are live or dead.

As a basis of quantification there is the construction of a calibration curve. In the case of food samples, it should be pointed out that, because of the extreme heterogeneity of the food matrices, it is not possible to construct a generic calibration curve to use for quantification. Specific curves, one for each food matrix to be considered in the analysis, should be prepared. One major problem related with the construction of calibration curves in food samples is the availability of nucleic acid extraction methods that will result in PCR-grade DNA and RNA. The necessity of removing all inhibitory compounds from the extract becomes extremely important when considering qPCR. Because of its great sensitivity, the presence of small quantities of inhibitors may result in lower efficiencies, thereby underestimating the real quantity of nucleic acids present in the sample.

In the last five years a number of papers were published on the application of qPCR to quantify foodborne pathogens in foods (Rantsiou *et al.*, 2008, 2010b; He and Chen, 2010; Kawasaki *et al.*, 2010; Josefsen *et al.*, 2010; Wang and Mustapha, 2010). It is interesting to note that apart from the use of qPCR in food analysis, it is only very recently that the method has been proposed as a tool to detect pathogenic microorganisms in food processing plants. Alessandria *et al.* (2010) and Osés *et al.* (2010) exploited qPCR for quantification of *L.*

© Woodhead Publishing Limited, 2011

monocytogenes in a dairy plant and Shiga-like Toxin producer *Escherichia coli* (STEC) in the lamb food chain, respectively. In both studies it was shown that traditional microbiological methods underestimated the numbers of pathogens contaminating environments. When surfaces of cheese packaging machines were analysed, no isolates could be detected by traditional microbiological methods. However, the application of qPCR highlighted populations of *L. monocytogenes* higher than 10^5 colony-forming units/cm^2 (Alessandria *et al.*, 2010). Processing conditions are often stressful for microorganisms, thereby they can provoke the loss of capability to grow in synthetic media. This state, described as viable but not culturable (VBNC), is of particular concern in the food industry because it is impossible to control with traditional, culture-dependent methods (Rowan, 2004). The application of second-generation PCR methods can help food business operators by contributing to a better understanding of persistence of foodborne pathogens in processing plants, and of possible routes of transmission.

5.5 Hygienic design factors affecting fouling and cleanability of food contact surfaces

Many of the basic concepts of food hygiene, hygienic design and hygienic practice in food production are well known and easily accessed (for example, European Hygienic Engineering and Design Group (EHEDG) guidelines: www.ehedg.org). However, during extensive visits to food-processing organisations around Europe, it was noted that many new CE-marked food process equipment did not comply with the hygiene provisions of the Machinery Directive which has been in force for more than a decade (Cocker, 2010). This supports indications that about 30% of outbreaks of foodborne disease are linked to cross-contamination in the food chain. Efforts therefore should be aimed at reinforcing the hygiene provisions of the Directive and improving the environment so that the risks associated are minimised. Faulty hygienic design and poor hygienic practices will lead to fouling or formation of biofilms and two aspects affecting surface hygiene will be explored in more detail here: surface topography, and the presence of organic food soil.

5.5.1 Surface topography and cleanability

Novel information obtained in this field is based upon newly developed methods as described in the following. The wear of a hygienic surface can alter its topography and reduce cleanability. Earlier work has shown that typical wear of stainless steel did not affect retention of microorganisms, but increased retention of food soil (Verran *et al.*, 2001). Focusing on specific features ('pits') it was subsequently demonstrated that the relationship between feature diameter, size of microorganisms and contact area was key in determining the strength of retention of cells on the surface (Whitehead *et al.*, 2005). Further, a novel

© Woodhead Publishing Limited, 2011

impression technique has been developed, whereby acetate sheets softened with acetone were pressed onto a range of new and used stainless steel surfaces, allowed to harden, and then removed and examined using scanning electron microscopy and AFM, thus enabling characterisation of surface features *in situ* (Verran *et al.*, 2010a).

The retention of *L. monocytogenes*, *Escherichia coli*, and *Staphylococcus sciuri*, a species related to *Staphylococcus aureus*, on test surfaces presenting regular linear features of 1.0 and 0.5 micron diameter (dimensions that related to features identified on in-use surfaces) was assessed. Briefly, the method involved incubating the surfaces with standardised cell suspensions for one hour, after which the substrata were removed from the cell suspension and rinsed. Retained cells were observed either using scanning electron microscopy (SEM), or were stained with acridine orange and observed via epifluorescence microscopy, and counted, to give an indication of the amount of retention (Verran *et al.*, 2010b), thus enabling relationships between cell size and surface feature size to be identified, and facilitating more detailed specification of the properties of hygienic surfaces.

The strength of attachment of cells to surface was assessed using AFM, whereby an increasing lateral force is applied via the cantilever until cells are removed from the surface (Verran *et al.*, 2010c). The method enables properties of cells, surfaces and cleaning regimes to be compared regarding surface cleanability and hygiene.

It was shown that the retention of cells on surfaces was affected by the area of contact between cells and features. Thus the parameters used to describe topography were not sufficient to explain observations – data on the cell size and shape, and their relationship with features on the surfaces were required. Interestingly, *E. coli* was very easily detached from the titanium surfaces. To explore this further, stainless steel surfaces were coated with titanium, enabling comparison of surface chemistry with the same underlying topography, in contrast to previous work where the topography differed but the chemistry was the same. Again, *E. coli* was not retained on the titanium coated surface (Verran *et al.*, 2010a).

5.5.2 Presence of organic food soil

In situ, it is highly unlikely that microbial cells will be found on a food contact surface in isolation from food material. Therefore, assessment of the removal of cells from surfaces via cleaning and disinfection regimes should include some indication of the interaction of organic material with the cells, and with the cleaning and disinfection agents themselves. As reported by Verran *et al.* (2006), methods for detection of microorganisms do not include assessment of organic material, and methods for detection of organic material do not include microorganisms. Hygiene monitors, for example via bioluminescence, give an indication of the status of the surface, but do not discriminate between cells and soil. Thus a method using fluorescent stains that would differentiate between

© Woodhead Publishing Limited, 2011

cells and soil on surfaces and enable, via image analysis, separate assessment of the presence and removal of these two components from the surface was developed and evaluated. The method proved successful in differentiation. Not surprisingly, soil covered the surface more extensively, and was harder to remove (Whitehead *et al.*, 2008, 2010).

The differences observed between titanium and stainless steel in terms of cleanability also indicate avenues for future work in the development of novel surfaces. The impression technique as an indirect method for assessing topography, and the importance of feature/cell size helps clarify characteristics of cleanable surfaces. The evaluation of novel methods for visualisation of soil has been of direct benefit to food companies involved in method development. A range of methods are described and compared by Whitehead *et al.* (2008).

5.6 Effective food management systems for ensuring safe food

Nowadays, companies have implemented various Good Manufacturing Practices (GMP) and Hazard Analysis Critical Control Point (HACCP) guidelines into their own food safety management system (FSMS). Important guidelines and quality assurance standards are General Principles of food hygiene (Codex Alimentarius 2003), ISO 9001:2008, ISO 22000:2005 (http://www.iso.org/iso), BRC (http://www.brc.org.uk), and IFS (http://food-care.info) (Jacxsens *et al.*, 2009; Luning and Marcelis, 2009a).

The performance of such systems in practice is, however, still variable. Moreover, the continuous pressure on FSMS performance and the dynamic environment wherein the systems operate (such as emerging pathogens, changing consumer demands, developments in preservation techniques) require that they can be systematically analysed to determine opportunities for improvement (Luning *et al.*, 2009). There are various tools that support food companies in systematically analysing their FSMS and its microbiological performance as a basis for strategic choices on interventions to improve the FSMS performance. The major tools that have been developed are described in the following.

5.6.1 Food safety management system diagnostic instrument (FSMS-DI)

A diagnostic tool (FSMS-DI) has been developed using a techno-managerial research approach to consider both technological factors and people behaviour in the performance of food safety management systems (Luning and Marcelis, 2009a,b; Luning *et al.*, 2008, 2009; Jacxsens *et al.*, 2010). The FSMS-DI is a tool that enables a systematic analysis and assessment of a company's unique FSMS independent of the Quality Assurance (QA) standards and/or guidelines that have been implemented. The instrument consists of comprehensive lists with sets of indicators to analyse, respectively, which core control and core

© Woodhead Publishing Limited, 2011

assurance activities are addressed in the company-specific FSMS, which major contextual factors could affect FSMS performance, and to analyse the microbiological safety performance of the system.

The principle behind the FSMS-DI is that companies that operate in a more vulnerable, uncertain, and ambiguous situation (due to, e.g., highly risky products and processes, less supporting organisational conditions) require a more advanced (higher level) FSMS (i.e. based on precise information, scientifically underpinned) to be able to realise and ensure stakeholders that safety requirements on their products are met (Luning *et al.*, 2009). Different scenarios can take place, i.e. a company in a simple, undemanding environment (i.e. low risk context) has no need for a complex and sophisticated FSMS (i.e. basic FSMS) in order to obtain a high food safety output (i.e. high food safety level). While a company operating in a complex demanding environment needs to have a robust, tailored and sophisticated FSMS in order to obtain a high food safety level. The food safety level can be analysed by using the food safety performance indicators (part of FSMS-DI) (Jacxsens *et al.*, 2010) and can be really measured by using a microbiological assessment scheme (MAS) as described below.

5.6.2 Microbiological assessment scheme (MAS)

The MAS tool is a procedure that defines the identification of critical sampling locations (CSL), the selection of microbiological parameters, the assessment of sampling frequency, the selection of sampling method and method of analysis, and finally data processing and interpretation. The principle behind the MAS tool is that low numbers of microorganisms and small variations in microbial counts indicate an effective FSMS (Jacxsens *et al.*, 2009). Also, a MAS analysis method selection tool has been developed, which can assist the process of decision making regarding selection of microbial analysing methods in specific situations. A comprehensive review of literature regarding different enumeration and detection methods was performed. Based on this review, specific method characteristics were determined that have been used as parameters in the selection tool (Jasson *et al.*, 2010).

The two outputs of MAS (i.e. evaluation of each microbiological parameter in a specific CSL and the microbial safety level profiles) give insight into the actual microbiological status/performance of the current FSMS of a company. First, the MAS results give information concerning the performance of a specific control activity in a FSMS. Secondly, the microbial safety profiles provide additional information concerning the nature of the microbiological problem. The microbial safety profiles can also be used to compare the current microbiological performance of different companies with the same type of production processes and food products as benchmarking tools. As such, microbiological problems in a sector can be identified, independent from company level. For more details the reader is referred to Jacxsens *et al.* (2009, 2010), and Sampers *et al.* (2010).

© Woodhead Publishing Limited, 2011

5.6.3 Food safety management system support web tool

To support companies in improving their FSMS, they need to have access to information, knowledge, and experience about these tools. In this perspective, a FSMS support web tool was developed to provide information in a systematic way. Information related to control and assurance principles is available, supporting tools to improve control and assurance activities (like new enumeration, detection and monitoring techniques for pathogens, new intervention techniques and methods, protocols and procedures on sanitation, validation, verification, microbial assessment, etc.), principles and structure of acknowledged quality assurance standards and guidelines, and legislative requirements. All tools are available online for SMEs to use (www.pathogencombat.com).

5.7 Conclusions

These scientific achievements and novel methods and technologies are expected to have a substantial impact on food safety if they are implemented in the entire food chain. This applies to the control systems used in the public and private sectors. At present they rely heavily, and sometimes exclusively, on conventional microbiological testing and sampling, which in many cases make the testing inefficient and for some samples result in false negative results. With the availability of rapid molecular methods for detection of pathogens and their behaviour, at the critical points of the entire food chain, the control systems can be made more effective and hence less costly. In particular, a change is introduced from food safety initiatives being largely retrospective, to becoming preventive and predictive with less emphasis on end product testing and more emphasis on critical points in the food chain. Information on the behaviour of pathogens and their numbers is important to improved food safety, but it is also imperative that interventions are implemented to prevent the transmission of viable and virulent pathogens along the food chain. In this regard, priority should be given to hygienic design and processing systems including cleaning and disinfection, novel processing technologies including the potential use of probiotic and protective cultures and not the least effective food safety management systems. 'Human behaviour' is as important as 'microbial behaviour' and needs to be addressed in food safety management to be effective. It should also be mentioned that food safety management can be applied by a company with the purpose to provide documentation for a competitive advantage in claiming that a food will be produced with low or 'zero' level of a defined pathogen. This has been done successfully by Scandinavian companies offering *Samonella* and *Campylobacter* 'free' fresh chicken, which also demonstrates that it is possible to ensure food safety at farm level including supplies of feed. It is a good example of the necessity of the involvement of the entire food chain.

Unfortunately innovation and technology transfer to public and private organisations is often hindered by a variety of issues. Nowadays, removal of

© Woodhead Publishing Limited, 2011

these barriers and the improvement of science-based knowledge transfer are seen as the main driver of economy growth (Baardseth *et al.*, 1999). It is estimated that a major fraction of the food sector does not carry out any research activity, because a large number of businesses are small and medium sized enterprises with no appropriate facilities, and the employees are entirely focussed on the production chain. In addition, such companies have to deal with the demand for safer products and higher quality, as well as an increased competition. To overcome these challenges, new technologies have to be introduced and the dialogues between research and industry need to be improved (Anonymous, 2009; Braun and Hadwiger, 2010).

5.8 References

ALESSANDRIA, V., K. RANTSIOU, P. DOLCI and L. COCOLIN (2010) Molecular methods to assess *Listeria monocytogenes* route of contamination in a dairy processing plant. *Int. J. Food Microbiol.*, 141: S156–S162.

ANONYMOUS (2009) TrueFood – Traditional United Europe Food. Guidelines on effective knowledge and technology transfer activities to SMEs in the food sector with particular focus on traditional food manufactures. Prepared by Campden BRI Magyarorszag Nonprofit Kft.

BAARDSETH, P., G.A. DALE and A. TANDBERG (1999) Innovation/technology transfer to food SMEs. *Trends Food Sci. Tech.* 10: 234–238.

BRAUN, S. and K. HADWIGER (2010) Obstacles and solutions for the knowledge transfer between science and industry (p. 87–91). In: *How to Improve the Food Safety of my Products.* Instituto Tomás Pascual, Madrid.

BREHM-STECHER, B.F. and E.A. JOHNSON (2004) Single cell microbiology: tools, technologies and applications. *Microbiol. Mol. Biol. Rev.* 68: 538–559.

BUDDE, B.B. and M. JAKOBSEN (2000) Real-time measurements of the interaction between single cells of *Listeria monocytogenes* and nisin on a solid surface. *Appl. Environ. Microbiol.* 66: 3586–3591.

CENCIČ, A. and T. LANGERHOLC (2010) Functional cell models of the gut and their applications in food microbiology – a review. *Int. J. Food Microbiol.*, 141: S4–S14.

CHATTERJEE, S.S., H. HOSSAIN, S. OTTEN, C. KUENNE, K. KUCHMINA, S. MACHATA, E. DOMANN, T. CHAKRABORTY and T. HAIN (2006) Intracellular gene expression profile of *Listeria monocytogenes*. *Infect. Immun.*, 74: 1323–1338.

COCKER, R. (2010) Prevention of biofilm formation and foodborne infections by control of moisture management. In: *How to Improve the Food Safety of my Products.* Instituto Tomás Pascual, Madrid.

CRITZER, F.J., D.H. DSOUZA and D.A. GOLDEN (2008) Transcription analysis of *stx1*, *marA*, and *eaeA* genes in *Escherichia coli* O157:H7 treated with sodium benzoate. *J. Food Prot.*, 71: 1469–1474.

EFSA (2010) The Community Summary Report on Trends and Sources of Zoonoses, Zoonotic Agents and food-borne outbreaks in the European Union in 2008. *The EFSA Journal*. 1496.

GAGGÌA, F., P. MATTARELLI and B. BIAVATI (2010a) Probiotics and prebiotics in animal feeding for safe food production. *Int. J. Food Microbiol.*, 141: S15–S28.

© Woodhead Publishing Limited, 2011

GAGGÌA, F., D.S. NIELSEN, B. BIAVATI and H. SIEGUMFELDT (2010b) Intracellular pH of *Mycobacterium avium* subsp. *paratuberculosis* following exposure to antimicrobial compounds monitored at the single cell level. *Int. J. Food Microbiol.*, 141: S188–S192.

HE, Y. and C.Y. CHEN (2010) Quantitative analysis of viable, stressed and dead cells of *Campylobacter jejuni* strain 81-176. *Food Microbiol.*, 27: 439–446.

JACXSENS, L., F. DEVLIEGHERE and M. UYTTENDAELE (2009) *Quality Management Systems in the Food Industry*. Book in the framework of Erasmus, ISBN 978-90-5989-275.

JACXSENS, L., M. UYTTENDAELE, F. DEVLIEGHERE, J. ROVIRA, S. OSES GOMEZ and P.A. LUNING (2010) Food safety performance indicators to benchmark food safety output of food safety management systems. *Int. J. Food Microbiol.*, 141: S180–S187.

JASSON, V. L. JACXSENS, P.A. LUNING, A. RAJKOVIC and M. UYTTENDAELE (2010) Alternative microbial methods: an overview and selection criteria. *Food Microbiol.*, 27: 710–730.

JOSEFSEN, M.H., C. LÖFSTRÖM, T.B. HANSEN, L.S. CHRISTENSEN, J.E. OLSEN and J. HOORFAR (2010) Rapid quantification of viable *Campylobacter* on chicken carcasses by real-time PCR and propidium monoazide as a tool for quantitative risk assessment. *Appl. Environ. Microbiol.*, 76: 5097–5104.

KASTBJERG, V. G., D.S. NIELSEN, N. ARNEBORG and L. GRAM (2009) Response of *Listeria monocytogenes* to disinfection stress at the single-cell and population levels as monitored by intracellular pH measurements and viable-cell counts. *Appl. Environ. Microbiol.*, 75: 4550–4556.

KAWASAKI, S., P.M. FRATAMICO, N. HORIKOSHI, Y. OKADA, K. TAKESHITA, T. SAMESHIMA and S. KAWAMOTO (2010) Multiplex real-time polymerase chain reaction assay for simultaneous detection and quantification of *Salmonella* species, *Listeria monocytogenes*, and *Escherichia coli* O157:H7 in ground pork samples. *Foodborne Pathog. Dis.*, 7: 549–554.

KUBISTA, M., J.M. ANDRADE, M. BENGTSSON, A. FOROOTAN, J. JONÁK, K. LIND, R. SINDELKA, R. SJÖBACK, B. SJÖGREN, L. STRÖMBOM, A. STÅHLBERG and N. ZORIC (2006) The real-time polymerase chain reaction. *Mol. Aspects Med.*, 27: 95–125.

LUNING, P.A. and W.J. MARCELIS (2009a) *Food Quality Management: technological and managerial principles and practices*. Wageningen Academic Publishers, Wageningen, The Netherlands.

LUNING, P.A. and W.J. MARCELIS (2009b) A food quality management research methodology. Integrating technological and managerial theories. *Trends Food Sci. Technol.*, 20: 35–44.

LUNING, P. A., L. BANGO, J. KUSSAGA, J. ROVIRA and W.J. MARCELIS (2008) Comprehensive analysis and differentiated assessment of food safety control systems: a diagnostic instrument. *Trends Food Sci. Technol.*, 19: 522–534.

LUNING, P. A., W.J. MARCELIS, J. ROVIRA, M. VAN DER SPIEGEL, M. UYTTENDAELE and L. JACXSENS (2009) Systematic assessment of core assurance activities in a company specific food safety management system. *Trends Food Sci. Technol.*, 20: 300–312.

MAHONEY, M. and A. HENRIKSSON (2003) The effect of processed meat and meat starter cultures on gastrointestinal colonization and virulence of *Listeria monocytogenes* in mice. *Int. J. Food Microbiol.*, 84: 255–261.

MARAGKOUDAKIS, P., W. CHINGWARU, L. GRADISNIK, E. TSAKALIDOU and A. CENCIČ (2010a) Lactic acid bacteria efficiently protect human and animal intestinal epithelial and immune cells from enteric virus infection. *Int. J. Food Microbiol.*, 141: S91–S97.

© Woodhead Publishing Limited, 2011

MARAGKOUDAKIS, P., K. MOUNTZOURIS, C. ROSU, G. ZOUMPOPOULOU, K. PAPADIMITRIOU, E. DALAKA, A. HADJIPETROU, G. THEOFANOUS, G.P. STROZZI, N. CARLINI, G. ZERVAS and E. TSAKALIDOU (2010b) Feed supplementation of *Lactobacillus plantarum* PCA 236 modulates gut microbiota and milk fatty acid composition in dairy goats – a preliminary study. *Int. J. Food Microbiol.*, 141: S109–S116.

NIELSEN, D.S., G.-S. CHO, A. HANAK, M. HUCH, C.M.A.P. FRANZ and N. ARNEBORG (2010) The effect of bacteriocin-producing *Lactobacillus plantarum* strains on the intracellular pH of sessile and planktonic *Listeria monocytogenes* single cells. *Int. J. Food Microbiol.*, 141: S53–S59.

OLESEN, I., F.K. VOGENSEN and L. JESPERSEN (2009) Gene transcription and virulence potential of *Listeria monocytogenes* strains after exposure to acidic and NaCl stress. *Foodborne Pathog. Dis.*, 6: 669–680.

OLESEN, I., L. THORSEN and L. JESPERSEN (2010) Relative transcription of *Listeria monocytogenes* virulence genes in liver pâtés with varying NaCl content. *Int. J. Food Microbiol.*, 141: S60–S68.

OSÉS, S.M., K. RANTSIOU, L. COCOLIN, I. JAIME and J. ROVIRA (2010) Prevalence and quantification of Shiga-toxin producing *Escherichia coli* along the lamb food chain by quantitative PCR. *Int. J. Food Microbiol.*, 141: S163–S169.

POGACAR, M. S., A. KLANCNIK, S.S. MOZINA and A. CENCIČ (2010) Attachment, invasion, and translocation of *Campylobacter jejuni* in pig small-intestinal epithelial cells. *Foodborne Pathog. Dis.*, 7: 589–595.

RAJKOVIC, A., N. SMIGIC and F. DEVLIEGHERE (2010) Contemporary strategies in combating microbial contamination in food chain. *Int. J. Food Microbiol.*, 141: S29–S42.

RANTSIOU, K., V. ALESSANDRIA, R. URSO, P. DOLCI and L. COCOLIN (2008) Detection, quantification and vitality of *Listeria monocytogenes* in food as determined by quantitative PCR. *Int. J. Food Microbiol.*, 121: 99–105.

RANTSIOU, K., V. ALESSANDRIA and L. COCOLIN (2010a) Virulence transcriptome analysis of *Listeria monocytogenes* by application of microarrays *in vitro* and *in situ*. XVII International Symposium on Problems of Listeriosis (ISOPOL), Porto.

RANTSIOU, K., C. LAMBERTI and L. COCOLIN (2010b) Survey of *Campylobacter jejuni* in retail chicken meat products by application of a quantitative PCR protocol. *Int. J. Food Microbiol.*, 141: S75–S79.

RASMUSSEN, M. B., L. B. ODDERSHEDE and H. SIEGUMFELDT (2008) Optical tweezers cause physiological damage to *Escherichia coli* and *Listeria* bacteria. *Appl. Environ. Microbiol.*, 74: 2441–2446.

RIEU, A., J. GUZZO and P. PIVETEAU (2010) Sensitivity to acetic acid, ability to colonize abiotic surfaces and virulence potential of *Listeria monocytogenes* EGD-e after incubation on parsley leaves. *J. Appl. Microbiol.*, 108: 560–570.

ROWAN, N.J. (2004) Viable but non culturable forms of food and waterborne bacteria: *Quo Vadis? Trends Food Sci. Technol.*, 15: 462–467.

SAMPERS, I., L. JACXSENS, P.A. LUNING, W.J. MARCELIS, A. DUMOULIN and M. UYTTENDAELE (2010) Relation between *Campylobacter* contamination and performance of Food Safety Management Systems in poultry meat industries. *J. Food Prot.*, 73: 1447–1457.

SHARKEY, F. H., I. M. BANAT and R. MARCHANT (2004) Detection and quantification of gene expression in environmental bacteriology. *Appl. Environ. Microbiol.*, 70: 3795–3806.

SMIGIC, N., A. RAJKOVIC, D.S. NIELSEN, N. ARNEBORG, H. SIEGUMFELDT and F. DEVLIEGHERE (2010) Survival of lactic acid and chlorine dioxide treated *Campylobacter jejuni*

© Woodhead Publishing Limited, 2011

under suboptimal conditions of pH, temperature and modified atmosphere. *Int. J. Food Microbiol.*, 141: S140–S146.

SUE, D., D. FINK, M. WIEDMANN and K.J. BOOR (2004) σ^B-dependent gene induction and expression in *Listeria monocytogenes* during osmotic and acid stress conditions simulating the intestinal environment. *Microbiology*, 150: 3843–3855.

VERRAN, J. and K.A. WHITEHEAD (2006) Assessment of organic materials and microbial components on hygienic surfaces. *Trans. Inst. Chem. Eng. C: Food Bioprod. Process.*, 84: 260–265.

VERRAN, J., D.L. ROWE and R.D. BOYD (2001) The effect of nanometer dimension surface topography on the cleanability of stainless steel. *J. Food Prot.*, 64: 1183–1187.

VERRAN, J., A. PACKER, P. KELLY and K.A. WHITEHEAD (2010a) Titanium-coating of stainless steel as an aid to improved cleanability. *Int. J. Food Microbiol.*, 141: S134–S139.

VERRAN, J., A. PACKER, P. KELLY and K.A. WHITEHEAD (2010b) The retention of bacteria on hygienic surfaces presenting scratches of microbial dimensions. *Lett. Appl. Microbiol.,* 50: 258–263.

VERRAN, J., A. PACKER, P. KELLY and K.A. WHITEHEAD (2010c) Use of the atomic force microscope to determine the strength of bacterial attachment on grooved surface features. *J. Adhes. Sci. Technol.*, 24: 2271–2285.

WANG, L. and A. MUSTAPHA (2010) EMA-real-time PCR as a reliable method for detection of viable *Salmonella* in chicken and eggs. *J. Food Sci.* 75: M134–139.

WANG, S., A.M. PHILLIPPY, K. DENG, X. RUI, Z. LI, M.L. TORTORELLO and W. ZHANG (2010) Transcriptomic response of *Salmonella enterica* Enteritidis and Typhimurium to chlorine based oxidative stress. *Appl. Environ. Microbiol.*, 76: 5013–5024.

WERBROUCK, H., K. GRIJSPEERDT, N. BOTTELDOORN, E. VAN PAMEL, N. RIJPENS, J. VAN DAMME, M. UYTTENDAELE, L. HERMAN and E. VAN COILLIE (2006) Differential *inlA* and *inlB* expression and interaction with human intestinal and liver cells by *Listeria monocytogenes* strains of different origins. *Appl. Environ. Microbiol.*, 72: 3862–3871.

WHITEHEAD, K. A., J. COLLIGON and J. VERRAN (2005) Retention of microbial cells in substratum surface features of micrometer and sub-micrometer dimensions. *Colloid Surfaces B*, 41: 129–138.

WHITEHEAD, K. A., L.A. SMITH and J. VERRAN (2008) The detection of food soils and cells on stainless steel using industrial methods: UV illumination and ATP bioluminescence. *Int. J. Food Microbiol.*, 127: 121–128.

WHITEHEAD, K.A., L.A. SMITH and J. VERRAN (2010) The detection and influence of food soils on microorganisms on stainless steel using scanning electron microscopy and epifluorescence microscopy. *Int. J. Food Microbiol.*, 141: S125–S133.

ZHANG, D.L., T. ROSS and J.P. BOWMAN (2010) Physiological aspects of *Listeria monocytogenes* during inactivation accelerated by mild temperatures and otherwise non-growth permissive acidic and hyperosmotic conditions. *Int. J. Food Microbiol.*, 141: S177–S185.

© Woodhead Publishing Limited, 2011

6

Foodborne viruses: understanding the risks and developing rapid surveillance and control measures

J. Hoorfar and A. C. Schultz, Technical University of Denmark, Denmark, D. N. Lees, Centre for Environment, Fisheries and Aquaculture Science, UK and A. Bosch, University of Barcelona, Spain

Abstract: Viruses are increasingly recognized as pathogens involved in foodborne infections. The viruses most adapted and likely to be carried in this way are those transmitted by the faecal-oral route. These include viral agents causing gastro-intestinal disease in humans such as noroviruses (NoV), but also agents such as hepatitis A virus (HAV), which, although being transmitted by the faecal-oral route, exhibit their classical clinical symptoms elsewhere in the body. Data from a Europe-wide study of NoV outbreaks illustrate that outbreaks are commonly detected in all countries that have sufficient diagnostic capacity and undertake systematic evaluation of outbreaks of gastro-enteritis other than for bacterial infections. Experience with other pathogen/food matrix combinations has demonstrated the necessity to establish sampling strategies in order to undertake effective monitoring of foodstuffs for risk management and control purposes. The ultimate goal in source tracing is the possibility to rapidly type the isolates. This is particularly relevant for the highly variable NoV, which together with HAV, is the main target for detection in food.

Key words: norovirus, hepatitis A, shellfish, fresh fruits, berries, irrigation water, food processing, surveillance, methods, standardization.

© Woodhead Publishing Limited, 2011

6.1 Introduction

Viruses are obligate intracellular parasites depending on living cells for replication. Viruses cannot multiply in dead or processed food and therefore are not responsible for food spoilage. Viral problems are thus limited to the role of food in passive transfer of viruses to humans. The viruses most adapted and likely to be carried in this way are those transmitted by the faecal-oral route. These include viral agents causing gastro-intestinal disease in humans such as norovirues (NoV), but also agents such as hepatitis A virus (HAV), which although being transmitted by the faecal-oral route exhibit their classical clinical symptoms elsewhere in the body.

Infected individuals shed large amounts of gastroenteric viruses (such as NoV), which are infective at low doses hence the danger from infected food handlers, from faecal contamination of irrigation water, or shellfish harvesting area by sewage (Formiga-Cruz *et al.*, 2002; Lodder and de Roda Husman, 2005). Viruses can contaminate food at source through faecal pollution of the environment, through poor hygiene during food handling or processing.

Since viruses are inactivated by thorough cooking, the foods often associated with viral illness are ready-to-eat or fresh produce contaminated during food handling, or uncooked or lightly cooked foods exposed to faecal contamination during primary production. Examples of the latter are salad crops or berries exposed to, for example, contaminated irrigation water (Bosch, 1998; Vasickova *et al.*, 2005) and filter-feeding shellfish (such as oysters) exposed to sewage contamination in their growing areas (Lees, 2000; Pinto *et al.*, 2009; Sanchez *et al.*, 2002). Zoonotic viruses of food animals, such as HEV, are also a potential risk either through direct consumption of inadequately cooked infected meat products or following environmental contamination of food products with animal faeces during primary production (Feagins *et al.*, 2007; Li *et al.*, 2005; Tei *et al.*, 2003).

Viruses are increasingly recognized as pathogens involved in foodborne infections. In 1999, a report from the Centers for Disease Control and Prevention (CDC) estimated that 67% of all foodborne illness in the US could be attributed to viruses, particularly NoV and HAV (Mead *et al.*, 1999). For HAV, the proportion of infections that is foodborne was estimated to be 5%, but the potential for larger outbreaks is increasing as a result of decreasing population immunity to these viruses. For NoV, the proportion estimated to be foodborne was 40%. These viruses have a significant impact due to the high incidence of infection in the community, their propensity to cause illness in people of all age groups (Green, 2007), and the potential for causing outbreaks that can be highly disruptive when they occur in institutions such as nursing homes or hospitals.

Large international foodborne outbreaks of NoV and HAV have occurred. Such outbreaks may be difficult to detect, in the absence of sustained surveillance and international sharing of data (Koopmans *et al.*, 2003). Data from surveillance of outbreaks of viral gastro-enteritis have suggested that NoV illness is increasing in incidence, possibly due to emergence of viruses that have

© Woodhead Publishing Limited, 2011

acquired mutations providing an advantage over resident virus populations (Kroneman *et al.*, 2006; Widdowson *et al.*, 2005).

While there is consensus about the importance of NoV as a cause of food-related illness, estimating its true extent is difficult due to its transmissibility. As a result, outbreaks resulting from a common source event will rapidly present as secondary person-to-person infections when the initial seeding of infection has not been detected. This is where the use of molecular tracing will provide valuable information for estimating the true size of foodborne outbreaks. A recent study on NoV genotype profiles seen in all outbreaks suggests that a foodborne attribution may be significantly under reported (Verhoef *et al.*, 2010). In this approach, a key property that is used is that viruses with an RNA genome change during subsequent rounds of replication by accumulation of mutations. Therefore, based on the genetic diversity observed in viruses detected in cases and in food items, their level of relatedness can be estimated, provided this approach has been validated in known transmission chains.

The other faecal-oral transmitted virus of major significance in human outbreaks is HAV. HAV is a common endemic infection in developing countries with most children being seropositive by 6 years of age. However, improving sanitary conditions in developed countries have lead to declining prevalence and resulted in large sectors of the population being susceptible to infection. HAV can be readily demonstrated in stools by molecular techniques (Yotsuyanagi *et al.*, 1996) and has also been demonstrated in sewage effluents and polluted receiving waters (Tsai *et al.*, 1994). Thus food products susceptible to contamination, such as bivalve molluscs, berries and lettuce, have frequently been implicated as food vehicles in outbreaks (Bosch *et al.*, 2001; Conaty *et al.*, 2000; Desenclos *et al.*, 1991; Ethelberg *et al.*, 2010; Falkenhorst *et al.*, 2005, Le Guyader *et al.*, 2004; Pinto *et al.*, 2009; Sanchez *et al.*, 2002). The main question the present review is addressing is what do we miss in order to understand the risk and develop rapid methods to improve the control and surveillance strategies for foodborne viruses. Part of this question, which was related to norovirus in bottled water, was addressed in the EU funded project www.biotracer.org.

6.2 Occurrence and epidemiology

The most commonly identified virus health risk in foods is from human enteric viruses spread by the faecal-oral route. The viruses most commonly identified in illness reports world-wide are NoV causing gastro-enteritis and HAV. Threats from new or emerging zoonotic viruses, such as hepatitis E virus (HEV), are also a cause for concern. These viruses present a new significantly underestimated challenge for food safety regulators, scientists and producers world-wide. They present a hazard in very small numbers, they cannot be detected or measured by conventional food microbiology techniques, and no (or very little) information is available on their behaviour, occurrence or control in food commodities. Even

© Woodhead Publishing Limited, 2011

knowledge about the extent of human illness is patchy and subject to large variation in estimates. Because of the technical limitations for virus detection and measurement, food controls cannot currently directly address the virus risk, even where this is explicitly identified for a food commodity. Instead the risk is tackled through non-specific faecal indicator controls, such as setting food quality standards for *E. coli* (as a faecal contamination indicator). However, it is now well recognized that this approach is inadequate to contain the risk from human enteric viruses. Many documented outbreaks have occurred, where the foods are fully compliant with the legislative faecal indicator sanitary standards (Lees, 2000; Pinto *et al.*, 2009; Sanchez *et al.*, 2002).

However, recent technical breakthroughs, particularly in molecular methods for viruses, offer new possibilities for improving scientific understanding and thus building, for the first time, specific defences against viral contamination of food supplies. PCR has offered the requisite sensitivity for detection of the low levels of virus commonly present in at risk food commodities such as bivalve shellfish or raspberries contaminated during primary production or following contamination by a food handler (Costafreda *et al.*, 2006; Jothikumar *et al.*, 2005b; Le Guyader *et al.*, 2009; Pinto *et al.*, 2009). Molecular sequencing has enabled virus strain characterisation and facilitated epidemiological insights into routes and causes of foodborne disease outbreaks, human disease burden and impact of the foodborne transmission route.

6.3 Clinical manifestations

Noroviruses are infectious to humans by the oral route and as low as 18–1000 particles can cause infection (Teunis *et al.*, 2008), replicate in an as yet unknown location (probably in the intestinal tract) and are shed in amounts of up to 10^{11} viral particles per gram of stool. Peak shedding is seen in the first days following infection, but can continue for over 3 weeks, especially in young children. Asymptomatic infection is common, and matters are complicated further by the fact that people without symptoms may shed similar amounts of virus as ill persons (Graham *et al.*, 1994). NoV cause gastro-enteritis, often including nausea, diarrhoea, vomiting, fever and abdominal pain (Kaplan *et al.*, 1982). The incubation period is 1 to 4 days with a duration of about 2 days and generally followed by complete recovery. Overall, the illness associated with NoV infections may be mild, but more severe illness can be observed when outbreaks occur in risk groups such as the elderly or persons with cardiovascular or renal disease (Mattner *et al.*, 2006). Protracted illness has been observed in transplant recipients persons with impaired immunity (Huffman *et al.*, 2003; Nilsson *et al.*, 2003). Once the virus exits the host, via stool or vomit, it remains infectious for periods up to several months depending on the circumstances. Finally, many distinct lineages of NoV exist, that differ in the make-up of their genes and proteins, their capacity to bind to the host cells, and the immunity that they induce (if any) (Green *et al.*, 1998).

© Woodhead Publishing Limited, 2011

Based on these basic parameters, it is easy to understand why NoV infections are so difficult to control. Important missing data are the levels of shedding of NoV in people at different intervals following infection, and what proportion of those infected develop a protracted form of infection. Cases that shed NoV for months have anecdotally been reported, and understanding how common this is may be crucial to understanding the epidemiology of NoV (Nilsson *et al.*, 2003). Also, since there is no established cell culture system for routine NoV assay, studies providing estimates for the viability of viruses that are shed are crucial (Duizer *et al.*, 2004).

HAV infection mostly develops asymptomatically or subclinically among young children (under 5), while in older children and in adulthood the infection usually proceeds with symptoms (Previsani *et al.*, 2004). In this latter case, the clinical course of HAV is indistinguishable from that of other types of acute viral hepatitis. The clinical case definition for HAV is an acute illness with moderate onset of symptoms (fever, malaise, anorexia, nausea, abdominal discomfort, dark urine) and jaundice, and elevated serum bilirubin and aminotransferases levels later on. The HAV infectious dose is unknown but according to the US Food and Drug Administration presumably is around 10–100 virus particles (http://www.fda.gov/Food/FoodSafety/FoodborneIllness/FoodborneIllnessFoodbornePathogensNaturalToxins/BadBugBook/ucm071294.htm). The incubation period of HAV ranges from 15 to 50 days and clinical illness usually does not last longer than 2 months, although 10–15% of patients have prolonged or relapsing signs and symptoms for up to 6 months (Glikson *et al.*, 1992; Sjogren *et al.*, 1987). In fact, with the advent of new highly sensitive techniques, even in normal clinical courses a high and long-lasting viremia has been detected (Costafreda *et al.*, 2006), with the peak (up to 107 genome; copies/ml of sera) occurring at two weeks after the onset of symptoms and lasting up to an average of six weeks after the start of symptoms (Bower *et al.*, 2000; Costafreda *et al.*, 2006; Sjogren *et al.*, 1987). There is no evidence of chronicity of the infection; however, occasionally the infection may proceed to a fulminant hepatitis, mainly among patients with underlying chronic liver diseases (Akriviadis and Redeker, 1989; Previsani *et al.*, 2004).

6.4 Risk assessment

Noroviruses that affect people are uniquely human viruses with no known animal reservoir. Consequently, all sources of human infection ultimately derive directly or indirectly from previous human infections, particularly, for example, through contaminated foods. Practically, any foodstuff appears capable of providing a vehicle of infection for this virus, but particularly fresh fruit and vegetables, and filter-feeding shellfish (predominantly raw oysters and clams). Food stuffs can become contaminated with virus at many and different stages along the food chain, from primary production through to preparation immediately prior to consumption (e.g. infected food handlers). The probability

© Woodhead Publishing Limited, 2011

of this occurring and the likely consequences to human health, therefore, vary greatly for each type of food, depending on how it is grown, handled and processed. Indeed, food production processes are key determinants, since they affect viral viability, survival and persistence. A previous quantitative microbial risk assessment (QMRA) feasibility study by Food Standards Agency, UK illustrated that full parameterization of all (or any) of the different food chains from primary production to consumption, would currently be unfeasible, even in the cases of those food chains most relevant to human disease risk (Lees, 2010). A more feasible approach might be to consider a more specified type of food, such as oysters. However, even here there are likely to be considerable differences in probabilities and contamination levels at the point of consumption depending, for example, on production area, spatial and temporal variation in marine water quality, processing and industry practice between countries.

Although risk flow schema were developed that might provide the basis for future QMRAs for fresh fruit and vegetables and for bivalve molluscs, it was identified that there were currently large data gaps that, even if addressed, were likely to be subject to considerable variation and uncertainty because of the factors alluded to above. Other key issues impeding parameterization were the current inability to grow the virus, the lack of animal models, and potential uncertainties in virus quantification methods, particularly concerning infectivity. The QMRA study (Lees, 2010) recommended a more feasible approach which would be to focus on point-of-sale studies on particular food items. An estimation of the transmissibility of the virus from these routes could then be compared with that of person-to-person transmission in the general population and to data on the overall burden of infection in the population arising from all potential infection routes.

6.5 Fast and efficient source tracing

Data from a Europe-wide study of NoV outbreaks illustrate that outbreaks are commonly detected in all countries that have sufficient diagnostic capacity and undertake systematic evaluation of outbreaks of gastro-enteritis other than for bacterial infections (Lopman *et al.*, 2003). This includes foodborne outbreaks. Here, food has been contaminated by one of three routes, namely an infected food handler, contact with sewage, or contact with (sewage-)contaminated water. Contamination of food may happen at the end of the chain with the typical examples being a food handler with poor hand-washing hygiene or persons who were sick inside a room where food was prepared or stored. In these instances, the eventual size of the outbreak depends on the number of persons served by this particular food handler, and such outbreaks are relatively easily unravelled by a proper outbreak investigation (Tauxe, 2002).

Much more difficult to detect are the outbreaks in which the contamination event occurs early in the food chain, for instance with berries that are harvested under poor hygienic conditions, or that are irrigated or washed with contaminated

© Woodhead Publishing Limited, 2011

water (Koopmans and Duizer, 2004). In these examples, diffuse outbreaks may occur over a vast geographic region, depending on the final destination of the food products and the amount of processing that is done prior to consumption. It is these types of outbreaks that the Foodborne Viruses in Europe Network (FBVE), NoroNet (www.nornet.nl) and CaliciNet (www.cdc.gov) in the US are trying to detect (Koopmans *et al.*, 2003). In Europe, there were 26 alerts through the EU Rapid Alert System for Food and Feed (RASFF) between 2000 and 2007, in which international foodborne transmission of NoV or HAV was suspected.

Thus, the ultimate goal in source tracing is the possibility to rapidly type the isolates. This is particularly relevant for the highly variable NoV, which together with HAV is the main target for detection in food (Mead *et al.*, 1999). Despite the preponderant role of these two viruses as food contaminants, methods aimed at other viral agents, notably emergent or re-emergent viruses should be developed. One important issue in food virology is the significance of genome copies in terms of infectivity. The lack of cell culture systems applicable to NoV and wildtype (wt) HAV propagation has long prevented the development of infectivity assays for these agents. However, the *in vitro* propagation of NoV has recently been reported in a 3-D organ culture system (Straub *et al.*, 2007). Additionally, a cell line with enhanced sensitivity to wt HAV has been described (Konduru and Kaplan, 2006). The validity of these new methodologies for the detection of infectious NoV and HAV in food is yet to be ascertained. The availability of infectious assays for these viral pathogens may fill the gap regarding the meaning of genome copy number in foodstuffs. Additionally it is important to utilize epidemiological data as has recently been attempted for consumption of oysters in a restaurant setting (Lowther *et al.*, 2010).

Challenges here are the lack of internationally standardized molecular typing of viruses, and the need for defining cut-offs that can be used to group strains as linked (Vinje *et al.*, 2003). These cut-offs differ for each of the NoV genotypes, and validation of their use should come from investigating known transmission chains.

6.6 Detection bottlenecks

The norovirus genus belongs to the Caliciviridae family and comprises a genetically diverse group of viruses, which can be separated beneath this level into genogroups, clusters or genotypes, and individual strains (Hansman *et al.*, 2006; Zheng *et al.*, 2006). NoV strains that infect humans are found in genogroups I, II and IV. NoV genogroups III and V contain strains that infect cows and mice, respectively. Porcine strains are found in genogroup II (II.11, II.18, and II.19) and strains that infect feline and canine species are found in genogroup IV (IV.2). Genetically related animal NoV strains have also been described (Oliver *et al.*, 2006). However, currently there is no evidence that they are capable of directly infecting man. The genetic diversity of NoV strains

© Woodhead Publishing Limited, 2011

presents a difficult challenge for the design of molecular diagnostics capable of detecting all strains of health significance and for strain typing.

Hepatitis A virus is a positive-strand RNA virus classified in its own genus of *Hepatovirus* within the Picornaviridae. There is only a single major serotype of HAV with three human antigenic variants and a number of genotypes identified by sequence analysis (Costa-Mattioli *et al.*, 2003). Specific virus detection is accomplished by amplification of targeted nucleic acids and real-time reverse transcription-polymerase chain reaction (RT-PCR) assays are becoming the methods of choice since they enable not only detection but also virus quantifications (Costafreda *et al.*, 2006; Jothikumar *et al.*, 2005a, 2005b; Kageyama *et al.*, 2003; Loisy *et al.*, 2005). These procedures are fast, do not require confirmation, may be accommodated to handle a large number of samples and enable the adoption of QA/QC practices. For this latter purpose, reference materials must be developed. The delivery systems employed for the reference materials will depend on their nature, i.e. viruses, plasmids, nucleic acids, etc. Reference materials can support the wide-scale adoption of these methods in the more routine food control laboratory context.

At present, virus analysis in food consists of the molecular detection of the virus targets extracted from the assayed food matrix. A most critical step in these procedures is the virus and nucleic acid extraction, since its efficiency greatly influences the final outcome of the entire virus detection assay (Costafreda *et al.*, 2006). The type of food being analysed, and even its state, i.e. fresh, frozen, etc., also influences the difficulty in producing reliable data. Although efforts are aimed to develop horizontal methods applicable to a wide variety of food matrices, in particular, bivalve shellfish, salad crops and soft fruits, some specific method adjustments may be required to enhance virus detection in a given foodstuff.

Recent methodology developments and international cooperation through the European Committee for Standardization (CEN) have facilitated the establishment of a standardized and well-controlled quantitative and qualitative method for detection of NoV and HAV in food stuffs (salad crops, soft fruit, bottled water, bivalve shellfish and hard surfaces). A draft of the method has already been submitted and following that an international method validation exercise is planned (Lees and CEN WG6 TAG4, 2010). EU legislative text foreshadows the adoption of virus controls when the method is sufficiently developed, standardized, and available for use. In preparation for the availability of the CEN standardized method, with its potential to be adopted into legislation, it becomes essential to gain more information about the practical application of the method to product safety assessment.

6.7 Zoonotic and emerging viruses via food-producing animals

In the past decade, the SARS, avian influenza, Nipah virus, tick-borne encephalitis virus and filovirus outbreaks have, among others, raised issues about their

© Woodhead Publishing Limited, 2011

potential for foodborne transmission. Expert opinion agrees that it is likely that new viruses will continue to emerge in coming years, especially RNA viruses. Operational contingency procedures and principles of early warning, in particular at the early steps of the farm-to-fork chain, are essential to guard against emerging viruses, their potential as pathogens being unpredictable. In addition, a systematic approach is needed to define which generic properties of currently known foodborne viruses contribute to their success. From that, the potential for foodborne transmission could be assessed with limited pathogen-specific data when new pathogens arise. Most zoonotic viruses such as West Nile virus, tick-borne encephalitis virus and cowpox virus originate in the environment and domestic animals such as horses, cattle and cats, respectively, are exposed to such viruses (Acha and Szyfres, 2003). Often, infection and clinical affection are seen in domestic animals before they are seen in humans. Thus, domestic animals might be regarded not only as source and vector of adventitious virus but may also serve as sentinels for timely diagnosis. Domestic animals therefore serve not only as a guard against exposure from food but also as an early warning against viruses emerging in the environment.

Hepatitis E virus (HEV) belongs to a newly created viral family, *Hepeviridae*, in the prototypic genus *Hepevirus*. HEV is a spherical, non-enveloped viral particle of around 32–34 nm in diameter (Emerson and Purcell, 2007). The genome is a positive ssRNA molecule of around 7.2 Kb containing three overlapping open reading frames (ORF) and a 3′poly (A) tail (Emerson and Purcell, 2003; Worm *et al.*, 2002).

Irrigation of vegetables as well as washing and processing of any food that is consumed raw with HAV or HEV-contaminated water could lead to enteric hepatitis outbreaks. Food manipulation by an HAV or HEV-infected person may also transmit the disease. However, in the particular case of HEV the infection has also been transmitted through the consumption of uncooked pig liver, wild boar and deer meat in Japan (Li *et al.*, 2005; Tei *et al.*, 2003; Yazaki *et al.*, 2003).

In fact, anti-HEV antibodies as well as HEV have been detected in pigs (Balayan *et al.*, 1990; Meng *et al.*, 1997), rodents (Clayson *et al.*, 1995; Kabrane-Lazizi *et al.*, 1999), wild boar (Matsuda *et al.*, 2003), donkeys (Guthmann *et al.*, 2006), chickens, cattle and dogs (Goens and Perdue, 2004). Additionally, infectious HEV has been detected from commercial pig livers sold in local grocery stores in the US (Feagins *et al.*, 2007). All these observations have strengthened the zoonotic potential of HEV.

6.8 Modelling development and behaviour of viruses

The initial contact of the virus with these foods may occur at any time during production, including before harvest, during processing and at the time of preparation. It is important to know how events leading to food contamination may influence the persistence and the recovery of viral particles for a successful

© Woodhead Publishing Limited, 2011

identification. For example, the fact that in shellfish most of viruses are concentrated in digestive tissues explains not only the inefficacy of depuration but also the persistence of infectious virus even after cooking, as has been demonstrated in some outbreaks (Bosch *et al.*, 2001; McDonnell *et al.*, 1997; Sasaki *et al.*, 2006). It has been demonstrated that Norwalk virus (prototype of genogroup I) binds specifically to a HBGA type A-like carbohydrate ligand present in shellfish tissues in two oyster species (Le Guyader *et al.*, 2006; Tian *et al.*, 2006). These observations are similar to those made in virus-like particle (VLP) binding studies with human intestinal tissues (Tan and Jiang, 2005). *In vitro* studies with additional NoV genotypes are needed to determine whether binding to shellfish tissues is genotype-specific. With such information, epidemiologic studies of shellfish-associated outbreaks can be designed to address the importance of specific binding of NoVs to shellfish tissues.

However, very little is known about the impact of food on viral behaviour. Different types of binding on food surfaces may explain why some foods are more often implicated in outbreaks than others (i.e. raspberries/strawberries, oysters/mussels). Factors such as complex surfaces, pH, sugar content, and presence of polysaccharides may influence virus binding, as might the temperature and moisture required for growing conditions. Up to now, very few experiments have been carried out to explain whether or not foods have a major impact on viral behaviour. The recent advances in methods and quantification by molecular biology should facilitate such type of studies. This issue is important as it will help to solve the problem of transmission within the food chain, either by preventing contamination or by removing the food before marketing.

6.9 Production and processing control strategies

Eating soft fruit and lettuce is generally considered healthy and provides the consumer with important minerals and vitamins. However, many fruits and vegetables are produced in countries with a warm climate, and often with low hygiene conditions, reduced sanitary installations, poor water quality and a high prevalence of transmissible gastrointestinal diseases. All these factors, including the use of contaminated fertilizers or waste-water (Steele and Odumeru, 2004; Tierney *et al.*, 1977), can contribute to the risk of contaminating the produce with pathogenic microorganisms. Therefore these products are considered particularly hazardous to consumers when they are eaten raw (Pebody *et al.*, 1998; Rosenblum *et al.*, 1990).

Bivalve molluscan shellfish feed by filtering large volumes of seawater and accumulating food particles from their surrounding environment. When that environment is contaminated by sewage, shellfish will also accumulate human pathogenic bacteria and viruses during filter-feeding and present a health risk when consumed raw or only lightly cooked. In order to render such shellfish fit for consumption, a commonly used commercial treatment process is immersion in tanks of clean seawater to allow sewage contaminants to be purged. This

© Woodhead Publishing Limited, 2011

process is called depuration and is widely used for species such as oysters, clams and mussels which are traditionally marketed live. In EU legislation all shellfish that are intended for direct human consumption must meet the end-product standard of <230 *E. coli* per 100 g shellfish flesh. Thus this is the standard required of shellfish depurated in an approved shellfish purification system. Depuration plants are also subject to control and approval through the HACCP approach.

Unfortunately the evidence shows that depuration plants, functioning satisfactorily by faecal indicator criteria, often fail to effectively remove enteric viruses. Most significant is the epidemiological evidence demonstrating that infection can occur following consumption of depurated shellfish. This was documented in Australia during volunteer trials to assess the safety of depurated shellfish (Grohmann *et al.*, 1981) and has since been documented in outbreaks in the UK the US, and elsewhere (Ang, 1998; Chalmers and McMillan, 1995; Gill *et al.*, 1983; Heller *et al.*, 1986; Murphy *et al.*, 1979; Perrett and Kudesia, 1995; Richards, 1985). The epidemiological evidence is supported by numerous laboratory studies which have examined elimination rates of human enteric viruses (such as poliovirus) or possible models for the behaviour of human enteric viruses (such as various bacteriophage species), from shellfish during the depuration process (Dore and Lees, 1995; Power and Collins, 1989, 1990; Richards, 1988; Sobsey and Jaykus, 1991). Although elimination rates in individual studies vary significantly, the overwhelming finding from these studies is that viruses are eliminated from bivalve shellfish at a slower rate than are faecal indicator bacteria (faecal coliforms or *E. coli*). Similar findings are reported for cell-adapted HAV which, unlike human NoV, can be cultured in the laboratory (Abad *et al.*, 1997; Chironna *et al.*, 2002; De Medici *et al.*, 2001) and for NoV using PCR. NoV was found to efficiently accumulate in shellfish (oysters and clams); however, it was only poorly removed by depuration compared to *E. coli* (Cheng *et al.*, 2005; Costantini *et al.*, 2006; Schwab *et al.*, 1998).

6.10 Sampling

Experience with other pathogen/food matrix combinations has demonstrated the necessity to establish sampling strategies in order to undertake effective monitoring of foodstuffs for risk management and control purposes. This is particularly important when methods are applied within a legislative context for public health protection. It is well established that the distribution and titre of pathogens may vary markedly within a ‘sample’ and also between samples taken from, for example, a large batch or consignment. Sampling during primary production (for example, in bivalve shellfish/salad crop/soft fruit production areas) introduces further geographic and environmental variables. Thus, the establishment of effective pathogen monitoring strategies depends on an understanding of the nature and distribution of pathogens within foods and,

© Woodhead Publishing Limited, 2011

consequently, the derivation of appropriate sampling protocols. There is a lack of data available on sampling strategies to enable the effective application of virological methods for monitoring virus contamination in food stuffs.

6.11 Acknowledgements

The work was supported in part by the European Union-funded Integrated Project BIOTRACER (Contract 036272) under the 6th RTD Framework and in part by grant Consolider-Ingenio 2010 CSD2007-00016 from the Spanish Ministry of Science and 2009SGR0024 from the Generalitat de Catalunya.

6.12 References

ABAD, F. X., R. M. PINTO, R. GARJADO and A. BOSCH (1997) Viruses in mussels – public health implications and depuration. *J. Food Prot.* **60**: 681.

ACHA, P. N. and B. SZYFRES (2003) *Zoonoses and communicable diseases common to men and animals. Third Edition, Volume II: Chlamydioses, Rickettsioses and Viroses.* Pan American Health Organization, World Health Organization, Washington, DC.

AKRIVIADIS, E. A. and A. G. REDEKER (1989) Fulminant hepatitis A in intravenous drug users with chronic liver disease. *Ann. Intern. Med.* **110**: 838–839.

ANG, L. H. (1998) An outbreak of viral gastroenteritis associated with eating raw oysters. *Commun. Dis. Public Health.* **1**: 38–40.

BALAYAN, M. S., R. K. USMANOV, N. A. ZAMYATINA, D. I. DJUMALIEVA and F. R. KARAS (1990) Brief report: experimental hepatitis E infection in domestic pigs. *J. Med. Virol.* **32**: 58–59.

BOSCH, A. (1998) Human enteric viruses in the water environment: a minireview. *Int. Microbiol.* **1**: 191–196.

BOSCH, A., G. SANCHEZ, G. F. LE, H. VANACLOCHA, L. HAUGARREAU and R. M. PINTO (2001) Human enteric viruses in Coquina clams associated with a large hepatitis A outbreak. *Water Sci. Technol.* **43**: 61–65.

BOWER, W. A., O. V. NAINAN, X. HAN and H. S. MARGOLIS (2000) Duration of viremia in hepatitis A virus infection. *J. Infect. Dis.* **182**: 12–17.

CHALMERS, J. W. and J. H. McMILLAN (1995) An outbreak of viral gastroenteritis associated with adequately prepared oysters. *Epidemiol. Infect.* **115**: 163–167.

CHENG, P. K., D. K. WONG, T. W. CHUNG and W. W. LIM (2005) Norovirus contamination found in oysters worldwide. *J. Med. Virol.* **76**: 593–597.

CHIRONNA, M., C. GERMINARIO, M. D. DE, A. FIORE, P. S. DI, M. QUARTO and S. BARBUTI (2002) Detection of hepatitis A virus in mussels from different sources marketed in Puglia region (South Italy). *Int. J. Food Microbiol.* **75**: 11–18.

CLAYSON, E. T., B. L. INNIS, K. S. MYINT, S. NARUPITI, D. W. VAUGHN, S. GIRI, P. RANABHAT and M. P. SHRESTHA (1995) Detection of hepatitis E virus infections among domestic swine in the Kathmandu Valley of Nepal. *Am. J Trop. Med. Hyg.* **53**: 228–232.

CONATY, S., P. BIRD, G. BELL, E. KRAA, G. GROHMANN and J. M. McANULTY (2000) Hepatitis A in New South Wales, Australia from consumption of oysters: the first reported outbreak. *Epidemiol. Infect.* **124**: 121–130.

© Woodhead Publishing Limited, 2011

COSTA-MATTIOLI, M., N. A. DI, V. FERRE, S. BILLAUDEL, R. PEREZ-BERCOFF and J. CRISTINA (2003) Genetic variability of hepatitis A virus. *J. Gen. Virol.* **84**: 3191–3201.

COSTAFREDA, M. I., A. BOSCH and R. M. PINTO (2006) Development, evaluation, and standardization of a real-time TaqMan reverse transcription-PCR assay for quantification of hepatitis A virus in clinical and shellfish samples. *Appl. Environ. Microbiol.* **72**: 3846–3855.

COSTANTINI, V., F. LOISY, L. JOENS, F. S. LE GUYADER and L. J. SAIF (2006) Human and animal Enteric caliciviruses in oysters from different coastal regions of the United States. *Appl. Environ. Microbiol.* **72**: 1800–1809.

DE MEDICI, D., M. CICCOZZI, A. FIORE, S. DI PASQUALE, A. PARLATO, P. RICCI-BITTI and L. CROCI (2001) Closed-circuit system for the depuration of mussels experimentally contaminated with hepatitis A virus. *J. Food Prot.* **64**: 877–880.

DESENCLOS, J. C., K. C. KLONTZ, M. H. WILDER, O. V. NAINAN, H. S. MARGOLIS and R. A. GUNN (1991) A multistate outbreak of hepatitis A caused by the consumption of raw oysters. *Am. J. Public Health* **81**: 1268–1272.

DORE, W. J. and D. N. LEES (1995) Behavior of *Escherichia coli* and male-specific bacteriophage in environmentally contaminated bivalve molluscs before and after depuration. *Appl. Environ. Microbiol.* **61**: 2830–2834.

DUIZER, E., K. J. SCHWAB, F. H. NEILL, R. L. ATMAR, M. P. KOOPMANS and M. K. ESTES (2004) Laboratory efforts to cultivate noroviruses. *J. Gen. Virol.* **85**: 79–87.

EMERSON, S. U. and R. H. PURCELL (2003) Hepatitis E virus. *Rev. Med. Virol* **13**: 145–154.

EMERSON, S. U. and R. H. PURCELL (2007) Hepatitis E virus, p. 3047–3058. *In* P. M. H. D.M.Knipe (ed.), *Fields Virology*. Lippincott, Williams and Wilkins, Philadephia.

ETHELBERG, S., M. LISBY, B. BOTTIGER, A. C. SCHULTZ, A. VILLIF, T. JENSEN, K. E. OLSEN, F. SCHEUTZ, C. KJELSO and L. MULLER (2010) Outbreaks of gastroenteritis linked to lettuce, Denmark, January 2010. *Euro Surveill.* **15**: 6.

FALKENHORST, G., L. KRUSELL, M. LISBY, S.B. MADSEN, B. BÖTTIGER and K. MØLBAK (2005) Imported frozen raspberries cause a series of norovirus outbreaks in Denmark, 2005. *Euro Surveill. Wkly.* **10**: 9.

FEAGINS, A. R., T. OPRIESSNIG, D. K. GUENETTE, P. G. HALBUR and X. J. MENG (2007) Detection and characterization of infectious Hepatitis E virus from commercial pig livers sold in local grocery stores in the USA. *J. Gen. Virol.* **88**: 912–917.

FORMIGA-CRUZ, M., G. TOFINO-QUESADA, S. BOFILL-MAS, D. N. LEES, K. HENSHILWOOD, A. K. ALLARD, A. C. CONDEN-HANSSON, B. E. HERNROTH, A. VANTARAKIS, A. TSIBOUXI, M. PAPAPETROPOULOU, M. D. FURONES and R. GIRONES (2002) Distribution of human virus contamination in shellfish from different growing areas in Greece, Spain, Sweden, and the United Kingdom. *Appl. Environ. Microbiol.* **68**: 5990–5998.

GILL, O. N., W. D. CUBITT, D. A. McSWIGGAN, B. M. WATNEY and C. L. BARTLETT (1983) Epidemic of gastroenteritis caused by oysters contaminated with small round structured viruses. *Br. Med. J (Clin. Res. Ed)* **287**: 1532–1534.

GLIKSON, M., E. GALUN, R. OREN, R. TUR-KASPA and D. SHOUVAL (1992) Relapsing hepatitis A. Review of 14 cases and literature survey. *Medicine (Baltimore)* **71**: 14–23.

GOENS, S. D. and M. L. PERDUE (2004) Hepatitis E viruses in humans and animals. *Anim. Health Res. Rev.* **5**: 145–156.

GRAHAM, D. Y., X. JIANG, T. TANAKA, A. R. OPEKUN, H. P. MADORE and M. K. ESTES (1994) Norwalk virus infection of volunteers: new insights based on improved assays. *J. Infect. Dis.* **170**: 34–43.

GREEN, J., K. HENSHILWOOD, C. I. GALLIMORE, D. W. BROWN and D. N. LEES (1998) A nested reverse transcriptase PCR assay for detection of small round-structured viruses in

© Woodhead Publishing Limited, 2011

environmentally contaminated molluscan shellfish. *Appl. Environ. Microbiol.* **64**: 858–863.

GREEN, K. Y. (2007) *Caliciviridae:* The Noroviruses, p. 949–979. In D. M. Knipe, P. M. Howley, D. E. Griffin, R. A. Lamb, M. A. Martin, B. Roizman and S. E. Straus (eds), *Fields Virology*. Lippincott Williams and Wilkins. Phildelphia, PA.

GROHMANN, G. S., A. M. MURPHY, P. J. CHRISTOPHER, E. AUTY and H. B. GREENBERG (1981) Norwalk virus gastroenteritis in volunteers consuming depurated oysters. *Aust. J. Exp. Biol. Med. Sci.* **59**: 219–228.

GUTHMANN, J. P., H. KLOVSTAD, D. BOCCIA, N. HAMID, L. PINOGES, J. Y. NIZOU, M. TATAY, F. DIAZ, A. MOREN, R. F. GRAIS, I. CIGLENECKI, E. NICAND and P. J. GUERIN (2006) A large outbreak of hepatitis E among a displaced population in Darfur, Sudan, 2004: the role of water treatment methods. *Clin. Infect. Dis.* **42**: 1685–1691.

HANSMAN, G. S., K. NATORI, H. SHIRATO-HORIKOSHI, S. OGAWA, T. OKA, K. KATAYAMA, T. TANAKA, T. MIYOSHI, K. SAKAE, S. KOBAYASHI, M. SHINOHARA, K. UCHIDA, N. SAKURAI, K. SHINOZAKI, M. OKADA, Y. SETO, K. KAMATA, N. NAGATA, K. TANAKA, T. MIYAMURA and N. TAKEDA (2006) Genetic and antigenic diversity among noroviruses. *J. Gen. Virol.* **87**: 909–919.

HELLER, D., O. N. GILL, E. RAYNHAM, T. KIRKLAND, P. M. ZADICK and R. STANWELL-SMITH (1986) An outbreak of gastrointestinal illness associated with consumption of raw depurated oysters. *Br. Med. J. (Clin. Res. Ed)* **292**: 1726–1727.

HUFFMAN, D. E., K. L. NELSON and J. B. ROSE (2003) Calicivirus – an emerging contaminant in water: state of the art. *Environ. Eng. Sci.* **20**: 503–515.

JOTHIKUMAR, N., T. L. CROMEANS, M. D. SOBSEY and B. H. ROBERTSON (2005a) Development and evaluation of a broadly reactive TaqMan assay for rapid detection of hepatitis A virus. *Appl. Environ. Microbiol.* **71**: 3359–3363.

JOTHIKUMAR, N., J. A. LOWTHER, K. HENSHILWOOD, D. N. LEES, V. R. HILL and J. VINJE (2005b) Rapid and sensitive detection of noroviruses by using TaqMan-based one-step reverse transcription-PCR assays and application to naturally contaminated shellfish samples. *Appl. Environ. Microbiol.* **71**: 1870–1875.

KABRANE-LAZIZI, Y., J. B. FINE, J. ELM, G. E. GLASS, H. HIGA, A. DIWAN, C. J. GIBBS, JR., X. J. MENG, S. U. EMERSON and R. H. PURCELL (1999) Evidence for widespread infection of wild rats with hepatitis E virus in the United States. *Am. J Trop. Med. Hyg.* **61**: 331–335.

KAGEYAMA, T., S. KOJIMA, M. SHINOHARA, K. UCHIDA, S. FUKUSHI, F. B. HOSHINO, N. TAKEDA and K. KATAYAMA (2003) Broadly reactive and highly sensitive assay for Norwalk-like viruses based on real-time quantitative reverse transcription-PCR. *J. Clin. Microbiol.* **41**: 1548–1557.

KAPLAN, J. E., R. A. GOODMAN, L. B. SCHONBERGER, E. C. LIPPY and G. W. GARY (1982) Gastroenteritis due to Norwalk virus: an outbreak associated with a municipal water system. *J. Infect. Dis.* **146**: 190–197.

KONDURU, K. and G. G. KAPLAN (2006) Stable growth of wild-type hepatitis A virus in cell culture. *J. Virol.* **80**: 1352–1360.

KOOPMANS, M. and E. DUIZER (2004) Foodborne viruses: an emerging problem. *Int. J. Food Microbiol.* **90**: 23–41.

KOOPMANS, M., H. VENNEMA, H. HEERSMA, S. E. VAN, D. Y. VAN, D. BROWN, M. REACHER and B. LOPMAN (2003) Early identification of common-source foodborne virus outbreaks in Europe. *Emerg. Infect. Dis.* **9**: 1136–1142.

KRONEMAN, A., H. VENNEMA, J. HARRIS, G. REUTER, C. H. VON BONSDORFF, K. O. HEDLUND, K. VAINIO, V. JACKSON, P. POTHIER, J. KOCH, E. SCHREIER, B. E. BOTTIGER and M.

© Woodhead Publishing Limited, 2011

KOOPMANS (2006) Increase in norovirus activity reported in Europe. *Euro Surveill.* **11**: E061214.

LE GUYADER, F.S., C. MITTELHOLZER, L. HAUGARREAU, K.O. HEDLUND, R. ALSTERLUND, M. POMMEPUY and L. SVENSSON (2004) Detection of noroviruses in raspberries associated with a gastroenteritis outbreak. *Int. J. Food Microbiol.* **97**: 179–186.

LE GUYADER F., F. LOISY, R. L. ATMAR, A. M. HUTSON, M. K. ESTES, N. RUVOEN-CLOUET, M. POMMEPUY and P. J. LE (2006) Norwalk virus-specific binding to oyster digestive tissues. *Emerg. Infect. Dis.* **12**: 931–936.

LE GUYADER, F. S., S. PARNAUDEAU, J. SCHAEFFER, A. BOSCH, F. LOISY, M. POMMEPUY and R. L. ATMAR (2009) Detection and quantification of noroviruses in shellfish. *Appl. Environ. Microbiol.* **75**: 618–624.

LEES, D. (2000) Viruses and bivalve shellfish. *Int. J. Food Microbiol.* **59**: 81–116.

LEES, D. and CEN WG6 TAG4 (2010) International standardisation of a method for detection of human pathogenic viruses in molluscan shellfish. *Food Environ. Virol.* **2**: 146–155.

LI, T. C., K. CHIJIWA, N. SERA, T. ISHIBASHI, Y. ETOH, Y. SHINOHARA, Y. KURATA, M. ISHIDA, S. SAKAMOTO, N. TAKEDA and T. MIYAMURA (2005) Hepatitis E virus transmission from wild boar meat. *Emerg. Infect. Dis.* **11**: 1958–1960.

LODDER, W. J. and A. M. DE RODA HUSMAN (2005) Presence of noroviruses and other enteric viruses in sewage and surface waters in The Netherlands. *Appl. Environ. Microbiol.* **71**: 1453–1461.

LOISY, F., R. L. ATMAR, P. GUILLON, P. LE CANN, M. POMMEPUY and F. S. LE GUYADER (2005) Real-time RT-PCR for norovirus screening in shellfish. *J. Virol. Methods* **123**: 1–7.

LOPMAN, B. A., M. H. REACHER, D. Y. VAN, F. X. HANON, D. BROWN and M. KOOPMANS (2003) Viral gastroenteritis outbreaks in Europe, 1995–2000. *Emerg. Infect. Dis.* **9**: 90–96.

LOWTHER, J. A., J. M. AVANT, K. GIZYNSKI, R. E. RANGDALE and D. N. LEES (2010) Comparison between quantitative real-time reverse transcription PCR results for norovirus in oysters and self-reported gastroenteric illness in restaurant customers. *J.Food Prot.* **73**: 305–311.

MATSUDA, H., K. OKADA, K. TAKAHASHI and S. MISHIRO (2003) Severe hepatitis E virus infection after ingestion of uncooked liver from a wild boar. *J. Infect. Dis.* **188**: 944.

MATTNER, F., D. SOHR, A. HEIM, P. GASTMEIER, H. VENNEMA and M. KOOPMANS (2006) Risk groups for clinical complications of norovirus infections: an outbreak investigation. *Clin. Microbiol. Infect. Dis.* **12**: 69–74.

McDONNELL, S., K. B. KIRKLAND, W. G. HLADY, C. ARISTEGUIETA, R. S. HOPKINS, S. S. MONROE and R. I. GLASS (1997) Failure of cooking to prevent shellfish-associated viral gastroenteritis. *Arch. Intern. Med.* **157**: 111–116.

MEAD, P. S., L. SLUTSKER, V. DIETZ, L. F. MCCAIG, J. S. BRESEE, C. SHAPIRO, P. M. GRIFFIN and R.V. TAUXE (1999) Food-related illness and death in the United States. *Emerg. Infect. Dis.* **5**: 607–625.

MENG, X. J., R. H. PURCELL, P. G. HALBUR, J. R. LEHMAN, D. M. WEBB, T. S. TSAREVA, J. S. HAYNES, B. J. THACKER and S. U. EMERSON (1997) A novel virus in swine is closely related to the human hepatitis E virus. *Proc. Natl. Acad. Sci. USA* **94**: 9860–9865.

MURPHY, A. M., G. S. GROHMANN, P. J. CHRISTOPHER, W. A. LOPEZ, G. R. DAVEY and R. H. MILLSOM (1979) An Australia-wide outbreak of gastroenteritis from oysters caused by Norwalk virus. *Med. J. Aust.* **2**: 329–333.

NILSSON, M., K. O. HEDLUND, M. THORHAGEN, G. LARSON, K. JOHANSEN, A. EKSPONG and L. SVENSSON (2003) Evolution of human calicivirus RNA *in vivo*: accumulation of

© Woodhead Publishing Limited, 2011

mutations in the protruding P2 domain of the capsid leads to structural changes and possibly a new phenotype. *J. Virol.* **77**: 13117–13124.

OLIVER, S. L., C. A. BATTEN, Y. DENG, M. ELSCHNER, P. OTTO, A. CHARPILIENNE, I. N. CLARKE, J. C. BRIDGER and P. R. LAMBDEN (2006) Genotype 1 and genotype 2 bovine noroviruses are antigenically distinct but share a cross-reactive epitope with human noroviruses. *J. Clin. Microbiol.* **44**: 992–998.

PEBODY, R. G., T. LEINO, P. RUUTU, L. KINNUNEN, I. DAVIDKIN, H. NOHYNEK and P. LEINIKKI (1998) Foodborne outbreaks of hepatitis A in a low endemic country: an emerging problem? *Epidemiol. Infect.* **120**: 55–59.

PERRETT, K. and G. KUDESIA (1995) Gastroenteritis associated with oysters. *Commun. Dis. Rep CDR Rev.* **5**: R153–R154.

PINTO, R. M., M. I. COSTAFREDA and A. BOSCH (2009) Risk assessment in shellfish-borne outbreaks of hepatitis A. *Appl. Environ. Microbiol.* **75**: 7350–7355.

POWER, U. F. and J. K. COLLINS (1989) Differential depuration of poliovirus, *Escherichia coli*, and a coliphage by the common mussel, Mytilus edulis. *Appl. Environ. Microbiol.* **55**: 1386–1390.

POWER, U. F. and J. K. COLLINS (1990) Tissue distribution of a coliphage and Escherichia coli in mussels after contamination and depuration. *Appl. Environ. Microbiol.* **56**: 803–807.

PREVISANI, N., D. LAVANCHY and G. SIEGL (2004) Hepatitis A, pp. 1–30. In I.K. Musbawer (ed.), *Viral Hepatitis Molecular Biology, Diagnosis, Epidemiology and Control (Perspecives in medical Virology)*. Elsevier, Amsterdam.

RICHARDS, G. P. (1985) Outbreaks of shellfish-associated enteric virus illness in the United States: requisite for development of viral guidelines. *J. Food Prot.* **48**: 823.

RICHARDS, G. P. (1988) Microbial purification of shellfish a review of depuration and relaying. *J. Food Prot.* **51**: 218–251.

ROSENBLUM, L. S., I. R. MIRKIN, D. T. ALLEN, S. SAFFORD and S. C. HADLER (1990) A multifocal outbreak of hepatitis A traced to commercially distributed lettuce. *Am. J. Public Health* **80**: 1075–1079.

SANCHEZ, G., R. M. PINTO, H. VANACLOCHA and A. BOSCH (2002) Molecular characterization of hepatitis A virus isolates from a transcontinental shellfish-borne outbreak. *J. Clin. Microbiol.* **40**: 4148–4155.

SASAKI, Y., A. KAI, Y. HAYASHI, T. SHINKAI, Y. NOGUCHI, M. HASEGAWA, K. SADAMASU, K. MORI, Y. TABEI, M. NAGASHIMA, S. MOROZUMI and T. YAMAMOTO (2006) Multiple viral infections and genomic divergence among noroviruses during an outbreak of acute gastroenteritis. *J. Clin. Microbiol.* **44**: 790–797.

SCHWAB, K. J., F. H. NEILL, M. K. ESTES, T. G. METCALF and R. L. ATMAR (1998) Distribution of norwalk virus within shellfish following bioaccumulation and subsequent depuration by detection using RT-PCR. *J. Food Prot.* **61**: 1674–1680.

SJOGREN, M. H., H. TANNO, O. FAY, S. SILEONI, B. D. COHEN, D. S. BURKE and R. J. FEIGHNY (1987) Hepatitis A virus in stool during clinical relapse. *Ann. Intern. Med.* **106**: 221–226.

SOBSEY, M. D. and L. A. JAYKUS (1991) Human enteric viruses and depuration of bivalve mollusks, pp. 71–114. In W. S. Otwell, G. E. Rodrick and R. E. Martin (eds), *Molluscan Shellfish Depuration*. CRC Press, Florida.

STEELE, M. and J. ODUMERU (2004) Irrigation water as source of foodborne pathogens on fruit and vegetables. *J. Food Prot.* **67**: 2839–2849.

STRAUB, T. M., B. K. HONER ZU, P. OROSZ-COGHLAN, A. DOHNALKOVA, B. K. MAYER, R. A. BARTHOLOMEW, C. O. VALDEZ, C. J. BRUCKNER-LEA, C. P. GERBA, M. ABBASZADEGAN

© Woodhead Publishing Limited, 2011

and C. A. NICKERSON (2007) *In vitro* cell culture infectivity assay for human noroviruses. *Emerg. Infect. Dis.* **13**: 396–403.

TAN, M. and X. JIANG (2005) Norovirus and its histo-blood group antigen receptors: an answer to a historical puzzle. *Trends Microbiol.* **13**: 285–293.

TAUXE, R. V. (2002) Emerging foodborne pathogens. *Int. J. Food Microbiol.* **78**: 31–41.

TEI, S., N. KITAJIMA, K. TAKAHASHI and S. MISHIRO (2003) Zoonotic transmission of hepatitis E virus from deer to human beings. *Lancet* **362**: 371–373.

TEUNIS, P. F., C. L. MOE, P. LIU, E. MILLER, L. LINDESMITH, R. S. BARIC, P. J. LE and R. L. CALDERON (2008) Norwalk virus: How infectious is it? *J. Med. Virol.* **80**: 1468–1476.

TIAN, P., A. H. BATES, H. M. JENSEN and R. E. MANDRELL (2006) Norovirus binds to blood group A-like antigens in oyster gastrointestinal cells. *Lett. Appl. Microbiol.* **43**: 645–651.

TIERNEY, J. T., R. SULLIVAN and E. P. LARKIN (1977) Persistence of poliovirus 1 in soil and on vegetables grown in soil previously flooded with inoculated sewage sludge or effluent. *Appl. Environ. Microbiol.* **33**: 109–113.

TSAI, Y. L., B. TRAN, L. R. SANGERMANO and C. J. PALMER (1994) Detection of poliovirus, hepatitis A virus, and rotavirus from sewage and ocean water by triplex reverse transcriptase PCR. *Appl. Environ. Microbiol.* **60**: 2400–2407.

VASICKOVA, P., L. DVORSKA, A. LORENCOVA and I. PAVLIK (2005) Viruses as a cause of foodborne diseases: a review of the literature. *Vet. Med.* **50**: 89–104.

VERHOEF, L., H. VENNEMA, P. W. VAN, D. LEES, H. BOSHUIZEN, K. HENSHILWOOD and M. KOOPMANS (2010) Use of norovirus genotype profiles to differentiate origins of foodborne outbreaks. *Emerg. Infect. Dis.* **16**: 617–624.

VINJE, J., H. VENNEMA, L. MAUNULA, C. H. VON BONSDORFF, M. HOEHNE, E. SCHREIER, A. RICHARDS, J. GREEN, D. BROWN, S. S. BEARD, S. S. MONROE, B. E. DE, L. SVENSSON and M. P. KOOPMANS (2003) International collaborative study to compare reverse transcriptase PCR assays for detection and genotyping of noroviruses. *J. Clin. Microbiol.* **41**: 1423–1433.

WIDDOWSON, M. A., S. S. MONROE and R. I. GLASS (2005) Are noroviruses emerging? *Emerg. Infect. Dis.* **11**: 735–737.

WORM, H. C., W. H. VAN DER POEL and G. BRANDSTATTER (2002) Hepatitis E: an overview. *Microbes. Infect.* **4**: 657–666.

YAZAKI, Y., H. MIZUO, M. TAKAHASHI, T. NISHIZAWA, N. SASAKI, Y. GOTANDA and H. OKAMOTO (2003) Sporadic acute or fulminant hepatitis E in Hokkaido, Japan, may be food-borne, as suggested by the presence of hepatitis E virus in pig liver as food. *J. Gen. Virol.* **84**: 2351–2357.

YOTSUYANAGI, H., K. KOIKE, K. YASUDA, K. MORIYA, Y. SHINTANI, H. FUJIE, K. KUROKAWA and S. LINO (1996) Prolonged fecal excretion of hepatitis A virus in adult patients with hepatitis A as determined by polymerase chain reaction. *Hepatol.* **24**: 10–13.

ZHENG, D. P., T. ANDO, R. L. FANKHAUSER, R. S. BEARD, R. I. GLASS and S. S. MONROE (2006) Norovirus classification and proposed strain nomenclature. *Virol.* **346**: 312–323.

© Woodhead Publishing Limited, 2011

7

Ranking hazards in the food chain

F. Butler, University College Dublin, Ireland

Abstract: Ranking hazards that are likely to occur in the food chain is a key step in driving risk management actions and in prioritising resources for control of the hazards of most concern. A hazard is a biological, chemical, or physical substance that can cause harm to human health. Chain mapping and hazard identification are essential prerequisites before risk ranking can be undertaken. These steps have become more complex as food chains become more and more complex and increasingly global in their nature. It is only relatively recently that formal risk ranking techniques have emerged to rank microbial and chemical hazards in food. In the first instance, techniques have been developed to rank these hazards separately; however, there have been recent developments to rank both microbiological and chemical hazards in a unified framework. As with quantitative risk assessments, risk ranking techniques require good sources of data to minimise subjectivity of the outcomes.

Key words: risk ranking, microbial hazards, chemical hazards, hazard identification.

7.1 Introduction

This chapter outlines available and emerging techniques to identify and rank hazards in food and feed chains. Such hazards may be of chemical or biological origin, and may be added naturally, inadvertently or maliciously. Ranking hazards in the food chain is a key step to drive risk management actions in terms of prioritising resources to control the hazards of most concern. Well-publicised contamination incidents associated with food have occurred globally in the last few years. These incidents have, at times, undermined the public's confidence in

© Woodhead Publishing Limited, 2011

the safety of food and in the regulatory agencies' ability to police the global food chain. Lessons have been learnt from these incidents, in terms of the factors that contribute to the occurrence of crises, how they could have been prevented and how the contamination incident was subsequently managed. Ensuring the safety of food requires a consistent, global response with consistent standards based on sound science, and robust controls are required to ensure consumer health and consumer confidence is adequately protected. Food chains are all the links or steps in the production, processing and distribution of foods from primary food production to point of consumption. If feed chains are included, which they should be, all of the steps in the production, processing and distribution of the raw materials used to produce animal feed are also included in the overall feed/food chain associated with any given food. Food chains are also dynamic, as food processors change sources of ingredients depending on availability and economic considerations.

Even the most cursory examination of the contamination events that have occurred in recent years, reveals the enormous range of potential hazards that can be present in food. As defined by Codex (Codex Alimentarius, 1999), a hazard is a biological, chemical, or physical substance that can cause harm to human health, at levels that are reasonably likely to occur if not controlled. The European Union operates a rapid alert system for food and feed (RASFF). The RASFF was put in place to provide regulatory authorities within and beyond the European Union with a mechanism to exchange information about measures taken responding to serious risks detected in relation to food or feed (Europa, 2010). Accessing the database associated with RASSF (Europa, 2010), shows that on an ongoing weekly basis, contamination incidents occur in products destined for the European Union. Faced with the multitude of possible hazards that can enter food products, the food industry, regulatory authorities and relevant stakeholders need some mechanism to prioritise hazards in order to allocate resources that are always finite to control the most critical hazards. As stated by Taylor (2002), 'If the primary objective of the food safety system is to reduce the burden of disease, success requires risk-based resource allocation. The food safety system must make the best possible use of its resources to reduce the disease burden. This means focusing government effort on the greatest risks and the greatest opportunities to reduce risk, wherever they may arise.' Regulatory authorities require risk ranking methodologies to select hazard–food product combinations that pose the greatest risk and consequently require the highest level of regulatory focus to reduce the risk. This initial risk ranking may be followed by a full-scale quantitative risk assessment of the high ranking hazard–food product combinations to quantify the risk in terms of potential human illness. The disadvantage of a full-scale quantitative risk assessment is that it is time consuming and expensive (Evers and Chardon, 2010). Ideally, risk ranking techniques should not be as time consuming; however, it must always be recognised that, as a consequence, risk ranking outcomes may not be as accurate as the results obtained from quantitative risk assessment studies.

© Woodhead Publishing Limited, 2011

Food companies also require risk ranking techniques to prioritise hazard–food product combinations that arise in the portfolio of products they produce or to prioritise the hazards that might arise in a particular product that the company is manufacturing. Under the Codex Alimentarius HACCP programme (Codex Alimentarius, 1999), Principle One is to conduct a hazard analysis. As part of this step, a company 'should list all of the hazards that may be reasonably expected to occur at each step from primary production, processing, manufacture, and distribution until the point of consumption' (Codex Alimentarius, 1999). Once the hazards are identified, a company should 'conduct a hazard analysis to identify for the HACCP plan which hazards are of such a nature that their elimination or reduction to acceptable levels is essential to the production of a safe food' (Codex Alimentarius, 1999). This hazard analysis required under HACCP is essentially a risk ranking process and in many cases this first step of HACCP is conducted qualitatively rather than using formal risk ranking approaches. One simple approach, under Codex Alimentarius HACCP recommendations, is to use a risk assessment grid (Table 7.1) to evaluate hazards. The relative risk of a hazard is assessed by comparing severity (low, medium, high) and likelihood of occurrence (remote, low, medium, high). It is clear that a hazard with a rating of H/H has a high severity, is highly likely to occur, and should be addressed, and that one rated L/R may need little if any resource allocation. However, there is potential for confusion with regard to the midrange ratings. To alleviate some of the confusion in using the grid system, some experts recommend using a numerical scale to characterise food hazards. For example, severity and likelihood of occurrence are each ranked using a 10-point ranking scale and the individual scores multiplied. Again the system is arbitrary and there is potential for confusion with regard to the midrange ratings. For that reason there has been much recent interest in developing risk ranking criteria for hazards that are more quantitative and remove the potential confusion and subjectivity associated with risk assessment grid type approaches. Generally, the approaches that have been adopted rank either microbiological hazards or chemical hazards. However, a number of approaches that combine both microbiological hazards and chemical hazards into a common ranking have been developed. Physical hazards have generally not been considered in risk ranking criteria. Physical hazards cannot be ignored; however, they are generally few in number for a given food product, so they can be identified and their risk managed without the necessity for formal risk ranking techniques.

Table 7.1 Codex Alimentarius grid for ranking food hazards

Severity	Likelihood of occurrence			
High	H/R	H/L	H/M	H/H
Medium	M/R	M/L	M/M	M/H
Low	L/R	L/L	L/M	L/H
	Remote	Low	Medium	High

© Woodhead Publishing Limited, 2011

7.2 Hazard identification in the food/feed chain

7.2.1 Chain mapping

The first step in the identification and assessment of hazards is to map the food/feed chain to allow a systematic review of the information to be collected (Cnossen *et al.*, 2009). Mapping should go into sufficient detail to allow the subsequent hazard identification step to be properly carried out. However, it is difficult to be prescriptive as to the actual extent of the detail required. Detailed knowledge is needed about the technical aspects of the food chain, such as raw materials, products, processes, procedures, storage, transport and food contact materials, in order to map the respective food/feed chain and to identify and select relevant hazards, as well as their entry points. The actual level of detail required will depend on the complexity of the chain. The minimum level of detail should allow drawing a logical, consistent chain map which accurately represents the reality of the food/feed chain. Due regard must be given to the fact that food/feed chains are dynamic so, for example, as manufacturers switch suppliers of ingredients, that event potentially alters the chain and may introduce the potential for new hazards. Once the chain map is complete, there should be a formal process to confirm that the chain map accurately describes the processing operation under consideration at all stages and at all times of operation. The latter point is particularly important if there is evidence that the chain varies depending, for example, on time of year. So, for example, a fruit processor may source apples from the northern hemisphere at one time of the year and from the southern hemisphere later on in the year, thus altering the chain and potentially the hazards associated with the chain.

7.2.2 Hazard identification

As part of this step, a processor should list all of the hazards that may be reasonably expected to occur at each step from primary production, processing, manufacture, and distribution until the point of consumption (Codex Alimentarius, 1999). For animal and fish products, the feed chain is often a source of many hazards that may end up in the food product, so it is important that feed chains have been mapped and considered where relevant. The completed chain map forms the basis of a structured consideration of what hazards may enter the chain. Cnossen *et al.* (2009) set out the following criteria for the identification of hazards:

- they must have been shown to be directly or indirectly associated with the food chain;
- or have the potential to become hazards in the food chain (even if they are not or are not yet considered as such currently).

Where available, hazards should be identified using documented evidence (e.g. existing studies in the literature documenting the occurrence of the hazard in the food stuff under consideration) or expert opinion. Some 'outside the box thinking' may be required to identify unforeseen hazards (e.g. for bulk shipped

© Woodhead Publishing Limited, 2011

product, contamination by fuel oil entering the hold of the ship). While identifying the hazards, it is useful to classify them under the broad categories proposed by Codex Alimentarius – biological, chemical or physical agents (Codex Alimentarius, 1999) in advance of undertaking a formal ranking of the contaminants.

7.3 Risk ranking

7.3.1 Risk ranking of microbiological hazards

Recent efforts to develop risk ranking tools for microbiological hazards have primarily concentrated on systematically assessing the level of risk associated with pathogenic microorganisms in foods. Two broad approaches can be identified (Newsome *et al.*, 2009), models based on epidemiological data or models based on the established principles of food safety risk assessment which couple the probability of exposure to a hazard, the magnitude of the hazard in a food when present and the probability and severity of the outcomes that might arise. 'Risk Ranger' (Ross and Summer, 2002) was one of the first published models based on these principles of food safety risk assessment. It is a tool developed within Microsoft Excel that ranks the relative risk of various product/pathogen combinations. Users are asked to answer the following series of questions (a menu of potential answers is provided) in relation to food safety and factors contributing to food safety risk:

- hazard severity
- susceptibility of target population
- frequency of consumption
- proportion of population consuming the product
- size of population
- proportion of product contaminated
- effect of processing on eliminating the hazard
- potential for recontamination after processing
- the increase required from level present at processing to reach an infectious dose
- effectiveness of the post-processing control system
- effect of meal preparation.

The output of the model is presented in a number of ways including the predicted illnesses per annum in the population of interest and a relative risk ranking. 'The "Risk Ranking" value is scaled logarithmically between 0 and 100, where 0 represents no risk, and 100 represents the opposite extreme where every member of the population eats a meal that contains a lethal dose of the hazard every day' (Ross and Summer, 2002). Risk Ranger has been used in a number of published studies including Sumner *et al.* (2004), Sumner *et al.* (2005), Pointon *et al.* (2006), Mataragas *et al.* (2008). An alternative approach was developed by McNab (2003) who proposed a risk score for food–hazard–location combinations as follows:

© Woodhead Publishing Limited, 2011

$$\text{Risk 'score'} = \text{Pa} \times \text{Pb} \times \text{Pc} \times \text{Ia} \times \text{Ib} \times \text{Ic}$$

where Pa is a scale of consumption score (based on amount consumed per person per day), Pb is the 'proportion' contaminated with this hazard at this location, Pc is the 'proportion' that lead to consumer exposure to that hazard, Ia is the number of those exposed, the 'proportion' of consumers that become ill, Ib is the severity of illness among consumers who become ill, and Ic is the difficulty to reduce or limit impact.

In that study, McNab gave some indicative data for VTEC in ground beef. Recently, Evers and Chardon (2010) published a new 'Swift Quantitative Microbiological Risk Assessment (sQMRA)' tool. Developed in Microsoft Excel, The sQMRA tool consists of, firstly, an exposure assessment tool, which starts at the retail level and considers possible cross-contamination and the effect of cooking of the product in the home. Arising from the exposure assessment, the sQMRA tool calculates the risk of illness using information supplied on the probability of infection and the probability of illness. The output of the model is probability of illness for the product–pathogen combination being considered which then allows product–pathogen combinations to be ranked (Evers and Chardon, 2010). In the study illustrative calculations were made with the sQMRA-tool for *Campylobacter* spp. and *Salmonella* spp. in chicken fillet, filet americain and eggs. The results showed that the only public health relevant pathogen–food product combinations were *Campylobacter* in chicken fillet and *Salmonella* in table eggs.

A number of models for ranking of pathogenic microorganisms based on epidemiological data have been developed. One example is the Foodborne Illness Risk Ranking Model (FIRRM), developed by the Food Safety Research Consortium (Batz *et al.*, 2004). The study covered bacteria, parasites and viruses. This model used US surveillance data on pathogen illnesses and then attributed these illnesses back to food of origin. The model ranked pathogen–food combinations according to five measures of public health impact: estimated number of cases, hospitalisations, deaths, economic impact and loss of quality adjusted life years (QALYs) (Batz *et al.*, 2004). The study demonstrated that risk ranking outcomes depend on whether pathogens or pathogen–product combinations were ranked. In the study, the top-ranked pathogen was Norovirus, but *Salmonella* in egg dishes was the most significant pathogen–food combination. A separate Dutch study (Kemmeren *et al.*, 2006) also used disease burden as the primary criterion for the prioritisation of foodborne pathogens. To calculate the disease burden, an epidemiological approach was used based on the surveillance of food-related illness and deaths in the Dutch population. The public health impact was combined into a single measure, the disability adjusted life year (DALY). DALYs are the sum of years of life lost and years lived with disability (weighted by a factor between 0 and 1 for the severity of the disability). While, in principle, risk ranking models based on epidemiological data have advantages over models based on the principles of food safety risk assessment, many countries do not have the necessary epidemiological data

© Woodhead Publishing Limited, 2011

available to allow this approach to be adopted. In this case, models based on accepted risk assessment approaches will have to be used.

7.3.2 Risk ranking of chemical hazards

It is arguable that risk ranking of chemical hazards poses a greater challenge than ranking microbiological hazards as the potential number of chemical hazards in a foodstuff is far higher than the potential microbiological hazards. Accordingly, the hazard identification stage in the overall risk ranking process for chemical hazards requires more effort than the equivalent stage for microbiological hazards. The majority of chemical risk ranking models for foodstuffs take into account that the risk posed by a chemical in a foodstuff is a function of both its hazard and exposure. A considerable amount of work has been done in risk ranking of chemicals for environmental applications (see, for example, the review by Swanson and Socha, 1997). Some of the early published risk ranking of chemicals in foods ranked pesticide risks in foods (Low *et al.*, 2004; Calliera *et al.*, 2006). Low *et al.* (2004) calculated a worst-case scenario risk for each pesticide–food combination using the following risk ranking:

$$\text{Risk} = \text{Hazard} \times \text{Exposure}$$

In that study, hazard was defined as an intrinsic toxicological property of the pesticide that is inversely related to the acceptable daily intake (ADI). Worst-case exposure was calculated as food consumption × the highest concentration of the contaminant detected in the food. Calliera *et al.* (2006) formulated a similar chronic indicator, HAPERITIFchr, based on the ratio between the estimated daily intake and the acceptable daily intake (ADI). Cnossen *et al.* (2009) took a similar approach to these two studies. Risk ranking criteria were based on a risk-based advisory value (normally the ADI when available), which gives an estimate of the amount of a chemical substance that may be ingested over a lifetime without appreciable risk to health. This risk-based advisory value was coupled with an exposure estimate to give a risk ranking score. However, in addition to calculating a worst case exposure, the authors recommended recording a distribution for the concentration of the contamination detected in the foodstuff. The risk ranking was then categorised by some appropriate statistical measure of the concentration detected (e.g. mean, 95% quartile). This resulted in several ranking schemes depending on what statistical measure of the concentration detected was used. This approach was adopted as a single, possibly high profile contamination event could well distort the ranking if worst-case exposure alone was recorded.

The Veterinary Residue Committee in the UK developed an alternative 'Matrix Ranking' for prioritising substances recommended for surveillance under their annual Non-Statutory Surveillance Scheme (Veterinary Residues Committee, 2008). The Matrix Ranking scheme for veterinary residues calculates a 'substance total score' as follows:

© Woodhead Publishing Limited, 2011

$$\text{Substance total score} = (A + B) \times (C + D + E) \times F$$

where: A is nature of the hazard (scale 0–6), B is potency of the substance (scale 0–3), C is exposure 1 (scale 0–3), which represents the proportion of the whole population's diet that might come from animals that have been treated with a particular veterinary product, D is exposure 2 (scale 0–3) which represents the frequency of dosing with a particular substance, E is high exposure groups (scale 0–3) where there is evidence of high exposure groups, and F is evidence of detectable residues (scale 1–4).

The Veterinary Residue Committee has been operating Matrix Ranking for several years now and it represents a good working example where a formal risk ranking protocol is being used to underpin risk management decisions at a national level.

Recently, risk ranking approaches have been developed that can evaluate both microbiological and chemical hazards in a combined framework. Newsome *et al.* (2009) reported on the outcomes of a two-year grant awarded by the US Food and Drug Administration to the Institute of Food Technologists to develop a risk ranking framework to evaluate potential high-threat microbiological agents, toxins, and chemicals in food. The framework combines expert opinion with exposure and hazard related risk criteria for specific hazard food combinations. The final risk ranking value is as annual pseudo-disability adjusted life years (pDALY) which is a harmonisation of the very different dose–response relationships observed for chemicals and microbes (Newsome *et al.*, 2009). This study represents one of the first to attempts to rank both microbiological and chemical hazards in a combined framework. This approach is likely to be further developed if the global food industry is to objectively identify hazard–food product combinations that pose the greatest risk and consequently require the highest level of regulatory focus to manage that risk.

7.4 Conclusions

It is only relatively recently that formal risk ranking techniques emerged to rank microbial and chemical hazards in food. As has been demonstrated in this chapter, considerable progress has been made in the last ten years in developing risk ranking techniques for both microbial and chemical hazards. In the first instance, the techniques have ranked these hazards separately; however, there have been recent developments to rank both microbiological and chemical hazards in a unified framework. As with quantitative risk assessments, risk ranking techniques require good sources of data to minimise subjectivity of the outcomes. Data availability is a key issue that is holding back the widespread adoption of the risk ranking approaches that have been developed. This is particularly true for the risk ranking approaches for pathogenic microorganisms based on epidemiological data. Very few countries in the world have the necessary illness/pathogen/food attribution data collected to allow these approaches to be adopted at a national level.

© Woodhead Publishing Limited, 2011

As with the more formal quantitative risk assessment approaches, risk ranking of hazards should, by its nature, underpin subsequent risk management activities. As a consequence, risk ranking techniques should be fit for purpose, objective and transparent. It must be clear to the risk managers the data limitations and the assumptions underpinning the risk ranking approach used. Finally, risk ranking must be iterative as new data and challenges emerge.

7.5 References

BATZ, B.B., S.A. HOFFMANN, A.J. KRUPNICK, J.G. MORRIS, D.M. SHERMAN, M.R. TAYLOR and J.S. TICK (2004) *Identifying the Most Significant Microbiological Foodborne Hazards to Public Health: A New Risk Ranking Model*. Food Safety Research Consortium Discussion Paper no. 1. Food Safety Research Consortium, Washington.

CALLIERA, M., A. FINIZIO, G. AZIMONTI, E. BENFENATI and M. TREVISAN (2006) Harmonised pesticide risk trend indicator for food (HAPERITIF): the methodological approach. *Pest Management Science*, 62(12):1168-76

CNOSSEN, H.J., M.A. WASSENS, H.L. HEERES and N.B. LUCAS LUIJCKX (2009) *Vulnerabilities in the food chain, a stakeholders' guide*. Sigma Chain Project. Available from: http://www.sigmachain.eu/uploads/dateien/fp6-518451_stakeholders_guide_on_vulnerabilities_web.pdf (accessed 30/8/2010).

CODEX ALIMENTARIUS (1999) *Basic Texts on Food Hygiene*, 2nd edition, Codex Alimentarius Commission, Rome.

EUROPA (2010) *Rapid Alert System for Food and Feed*, European Commission, DG Health and Consumers. Available from: http://ec.europa.eu/food/food/rapidalert/index_en.htm (accessed 9/9/2010).

EVERS, E.G. and J.E. CHARDON (2010) A swift Quantitative Microbiological Risk Assessment (sQMRA) tool. *Food Control*. 21: 319–330.

KEMMEREN, J.M., M.-J.J. MANGEN, Y.T.H.P. VAN DUYNHOVEN and A.H. HAVELAAR (2006) *Priority Setting of Foodborne Pathogens: Disease Burden and Costs of Selected Enteric Pathogens*. RIVM report 330080001/2006. RIVM, Bilthoven.

LOW, F., H.M. LIN, J.A. GERRARD, P.J. CRESSEY and I.C. SHAW (2004) Ranking the risk of pesticide dietary intake. *Pest Management Science*, 60(9): 842–848.

MATARAGAS, M., P.N. SKANDAMIS and E.H. DROSINOS (2008) Risk profiles of pork and poultry meat and risk ratings of various pathogen/product combinations. *International Journal of Food Microbiology*, 126: 1–12.

McNAB, B. (2003) Food Safety Universe Database: A Semi-Quantitative Risk Assessment Tool. Ontario Ministry of Agriculture and Food. Ontario.

NEWSOME, R., N. TRAN, G.M. PAOLI, L.A. JAYKUS, B. TOMPKIN,M. MILIOTIS, T. RUTHMAN, E. HARTNETT, F.F. BUSTA, B. PETERSEN, F. SHANK, J. MCENTIRE, J. HOTCHKISS, M. WAGNER and D.W. SCHAFFNER (2009) Development of a risk-ranking framework to evaluate potential high-threat microorganisms, toxins, and chemicals in food. *Journal of Food Science*, 74: R39–R45.

POINTON, A., I. JENSON, D. JORDAN, P.VANDERLINDE, J. SLADE and J. SUMNER (2006) A risk profile of the Australian red meat industry: approach and management. *Food Control*, 17: 712–718.

ROSS, T. and J. SUMNER (2002) A simple, spreadsheet-based, food safety risk assessment tool. *International Journal of Food Microbiology*, 77: 39–53.

© Woodhead Publishing Limited, 2011

SUMNER, J., T. ROSS and L. ABABOUCH (2004) *Application of Risk Assessment in the Fish Industry*. FAO Fisheries Technical Paper 442, FAO, Rome.

SUMNER, J., T. ROSS, I. JENSON and A. POINTON (2005) A risk microbiological profile of the Australian red meat industry: risk ratings of hazard-product pairings. *International Journal of Food Microbiology* 105: 221–232.

SWANSON, M.B. and A.C. SOCHA (1997) *Chemical Ranking and Scoring: Guidelines for relative assessments of chemicals.* Setac Press, Florida.

TAYLOR, M.R. (2002) *Reforming Food Safety: A Model for the Future*. Resources for the Future, Washington. Available from: http://www.rff.org/rff/documents/RFF-IB-02-02.pdf (accessed 9/9/2010).

VETERINARY RESIDUES COMMITTEE (2008) Annual Report on Surveillance for Veterinary Residues in Food in the UK 2008. Veterinary Residues Committee, London.

© Woodhead Publishing Limited, 2011

8

Continuous temperature monitoring along the chilled food supply chain

M. Eden, TTZ Bremerhaven, Germany, V. Raab and J. Kreyenschmidt, University of Bonn, Germany and T. Hafliðason, G. Ólafsdóttir and S. G. Bogason, University of Iceland, Iceland

Abstract: This chapter addresses the need for continuous temperature monitoring along the entire food supply chain to ensure food quality and safety. In the first part, the needs and expectations as well as barriers of supply chain actors regarding novel technologies are discussed. In the second part, novel technologies for the continuous control of temperature conditions in chilled food chains are presented. These include temperature monitoring systems based on radio frequency identification (RFID) and smart labels, such as time temperature indicators (TTIs). Finally, solutions are presented which show how temperature monitoring systems can be linked with product characteristics to improve food quality and safety.

Key words: temperature monitoring, chain information management, food quality and safety, smart label, shelf life prediction.

8.1 Introduction

Growing international trade presents a major food quality and safety problem (Kuo and Chen, 2010; Orriss and Whitehead, 2000). Within these globalized food chains, temperature control is an important factor, as temperature during distribution and storage plays a key role in maintenance of food quality and safety (McMeekin *et al.*, 2008; Smolander *et al.*, 2004; Taoukis *et al.*, 1999). It largely influences microbial growth and associated risk of illness (Giannakourou *et al.*, 2005; Olafsdottir *et al.*, 2006; Zwietering *et al.*, 1991).

© Woodhead Publishing Limited, 2011

Incorrect storage conditions in the chain with regard to temperature lead to reduced shelf life and thus to substantial economic losses (Taoukis *et al.*, 1991; Fu and Labuza, 1992). To date, major problems regarding incorrect temperature conditions have been discovered at those points where products are transferred from one actor to another. Identification and control of these weaknesses and the improvement of continuous temperature monitoring in the cold chain are essential for the delivery of high quality and safe products (Olsson, 2004; Raab *et al.*, 2008).

Rising consumer awareness is forcing supply chain actors to implement efficient management systems (Kuo and Chen, 2010). A prerequisite for efficient cold chain management is a temperature monitoring system which controls conditions along the entire chilled food supply chain (Kreyenschmidt *et al.*, 2010; Kuo and Chen, 2010; Montanari, 2008). Currently, temperature monitoring is often only performed by measurement of environmental temperature instead of product core temperature (Kreyenschmidt *et al.*, 2010). Some logistics firms have already started to implement novel temperature monitoring solutions, but almost no inter-organizational exchange of the information and data collected takes place (Raab *et al.*, 2011). Although various temperature monitoring systems are applied in food distribution, there is still a need for new tools which facilitate temperature control and management. Tools are required, in particular, for electronic recording of full product temperature history, from production to consumer, in order to facilitate application of predictive models to estimate shelf life and pathogen growth. Such systems, which apply wireless temperature recording and data storage technologies, can provide supply chain actors with information about temperature history in real time, thus allowing improved transparency, more efficient supply chain management and waste reduction (Olafsdottir *et al.*, 2010).

A wide spectrum of different temperature measurement devices has been developed which aim to provide comprehensive temperature monitoring during the whole supply chain (Kreyenschmidt, 2008). Furthermore, researchers are working on holistic systems which ensure the tracking and tracing of goods from farm to fork (Chill-on project, 2010).

The effective implementation of such temperature control systems over the entire supply chain is currently uncommon. Although, logistic providers have already started to implement novel temperature monitoring solutions, the inter-organizational information exchange of these collected data is mostly not done. Furthermore, there is no combination of temperature data with data on product characteristics with regard to safety and quality aspects.

8.2 Needs of food supply chain stakeholders

Despite research efforts in the area of electronic traceability and information systems and development of tools to monitor quality (Olafsdottir *et al.*, 2004), the implementation of these novel technologies has been slow in the food supply chain.

© Woodhead Publishing Limited, 2011

Based on a survey of the opinions of stakeholders in the fish supply chain regarding their need for methods and dedicated tools for selected quality-related attributes such as odour or colour, or tools for the detection of bacteria, the respondents were more positive towards generic tools to monitor the overall concepts of freshness and quality, indicating that they wanted simple tools and were not knowledgeable regarding the details of complex spoilage changes related to loss of freshness or quality (Jørgensen *et al.*, 2003). Recent research activities funded by the EC have been aimed at implementing novel technologies to facilitate the electronic monitoring of temperature and obtaining real-time information on the remaining shelf life of products (Chill-on project, 2010). The assumption is that the transparency of the supply chain would be enhanced by the interoperability of real-time temperature monitoring. Verification of the traceability of products is generally considered to be a valid quality and safety indicator for consumers, but their perception of food quality, safety, and traceability appear to be interlinked and thus these concepts may all influence their purchase decisions (Rijswijk and Frewer, 2008).

The top-down approach of research or government when promoting new monitoring procedures or technologies is often not regarded positively by industry. Although innovative technologies may have promising features, their implementation may have various drawbacks. Focus groups and questionnaire surveys based on interviews have therefore been utilized to investigate the views of key commercial players in the fish supply chain and thus to identify the main drivers and barriers for implementing electronic information systems, including traceability and real-time temperature monitoring systems (Chill-on project, unpublished results).

The preliminary findings indicated that trust was considered necessary for business and therefore in place throughout the chain. However, some stated that they were not willing to share all information. In fact, when a question about potential mistrust was posed, it was generally agreed that mistrust can occur anywhere in the chain.

The majority of the respondents agreed with the statement that sharing of real-time temperature data between actors would enhance trust and commitment in the fish supply chain. However, those who did not agree stated that temperature was already controlled and monitored well by handheld devices or data loggers. The main concern was that sharing of electronic data would be too costly, extra work, 'annoyance', and with no added benefits. Proof and verification of critical temperature conditions was, however, considered important in the case of a breakdown in the cold chain. It was also acknowledged that it would be easier to determine where the break in the chain had occurred if sharing of electronic temperature data were in place. For the customer, verification of GPS location and proof of quality of the transport service were considered as benefits.

Regulations were mentioned most often as the key driver for implementing an electronically based information system. Consumer values were ranked in second place, where quality and shelf life of products were the key attributes. Thirdly, economic factors, such as cost-benefit ratio and improved supply chain

© Woodhead Publishing Limited, 2011

management, were considered important. Environmental issues were ranked last as important drivers for implementing new technologies. However, it was noted that the four choices given were highly dependent on each other. Environmental issues were considered as marketing tools and important to enhance companies' image and to address consumers' environmental awareness and sustainability demands.

According to Doluschitz *et al.* (2010), clear statements on costs, benefits and sustainability of developed IT systems are considered a prerequisite for acceptance in industry. The preliminary findings from the survey verified that cost was the main barrier for installing an electronically based system that is capable of sharing real-time information on traceability data, GPS location, temperature, and shelf life. The cost-benefit ratio was emphasized, as well as barriers regarding installation and operation cost and the short-term issue that it would be time consuming to have such a system implemented (Chill-on project, unpublished results).

In focus group discussions with fish supply chain actors in Europe, the view was expressed that the industry was not interested in solutions that were imposed upon them and made reference to the fact that the fish business is a very highly regulated industry, which is burdened by costly audits. They would therefore have to see the concrete benefits of implementing an electronic traceability system. Arguments on cost sharing among companies at different links of the supply chain have been reported and are based on the perceived benefits of traceability systems for different types of companies (Mai, 2009).

According to experts in the field of food risk management in Europe, effective food and ingredient traceability systems have the potential to improve food safety; however, further improvements regarding harmonization of practices and pan-global legislation are needed (Kher *et al.*, 2010). Other factors may also be barriers for the sharing of electronic temperature information as has been reported in the case of ERP implementation. The barriers were not technology-related issues such as technological complexity, compatibility, standardization, etc., but mostly concerned with organization and human-related issues, such as resistance to change, organizational culture, incompatible business processes, project mismanagement and top management commitment (Helo *et al.*, 2008).

8.3 Smart labels as temperature monitoring systems

8.3.1 Radio frequency identification (RFID)

In recent years, substantial technological progress has been achieved through the development and use in practice of wireless sensor networks combined with temperature sensors. The development of radio frequency identification (RFID) has particularly facilitated temperature control systems. Use of these tags is steadily increasing.

RFID is the generic name for technologies that use radio waves to identify items and gather data on items without human intervention or data entry (Wyld,

© Woodhead Publishing Limited, 2011

2006). In its most basic form, an RFID system has two components: tags and readers. RFID adds intelligence in order to minimize human intervention in the item identification process by using electronic tags. The tag contains unique identification information on the item to which it is attached; the reader emits and receives radio waves to read the information stored in the tag allowing processing of collected data (Roberts, 2006).

There are two main types of RFID tag, firstly the passive version with no internal battery and a short operating range, and secondly the active type which incorporates a battery and has a longer operating range (Roberts, 2006). Both can be equipped with additional sensors. Opera *et al.* (2009) describe temperature, humidity and gas sensor; Jederman *et al.* (2009) report semi-passive RFID loggers which allow spatial temperature profiling during transport of perishable foods. Temperature sensors combined with RFID offer the possibility to collect a full temperature history over the whole supply chain.

Commercially available active RFID data loggers are, for example, Temp Tale RF by Sensitech (Beverly, MA, USA), i-Q32T by Identec solutions (Lustenau, Austria), or the A927Z by CAEN (Viareggio, Italy). The semi-active solution Vario Sense® by KSW (Dresden, Germany), the Freshtime™ by Infratab (Oxnard, CA, USA), and the Log IC by American Thermal Instruments (Dayton, OH, USA) belong to the category of smart active labels. Here, an RFID inlet or inlay is connected with a power source embedded into a label. This offers the advantages of low-cost RFID inlets, flexible self-adhesive tags, use on smaller items, as well as the integration of shelf life algorithms. The disadvantage is that data acquisition and storage are limited (Kreyenschmidt *et al.*, 2005).

Besides the basic product identification potential, the RFID technology can enhance quality management, food logistic processes and supply chain management. Several authors emphasize the capability of this technology to identify, categorize, and manage product and information flow at important inspection and decision points (Kärkkäinen, 2003; Michael and McCathie, 2005; Dittmann, 2006; Jedermann *et al.*, 2006). It has the advantage that status information about the transport temperature of products can be collated in those areas where products are shipped and stored (Mousavi *et al.*, 2002; Kreyenschmidt *et al.*, 2005; Kerry *et al.*, 2006; Kumar and Budin, 2006; Bovenschulte *et al.*, 2007). The combination of this wireless technology with wireless communications systems, such as wireless wide area network (WWAN), wireless local area network (WLAN), and wireless sensor network (WSN) systems, can even enhance this potential area of application (Guo and Zhang, 2002; Wang *et al.*, 2006; Ruiz-Garcia *et al.*, 2007; Raab *et al.*, 2011).

One of the remaining obstacles for the general adoption of RFID is the cost of the individual tag. For cost reasons, RFID labels holding a battery are rarely applied to single product items but instead rather at pallet level. However, use of only few tags does not allow investigation of temperature distribution within larger transport units.

Another major obstacle is the lack of uniformity in global standards in the area of sensor technology, which has made implementation more difficult. Many

© Woodhead Publishing Limited, 2011

current commercial sensor solutions provide their own interfaces to communicate with their own tags (Lopez *et al.*, 2009).

Companies appear hesitant to adapt the solutions on a large scale because of the complexity of so many different standards, data formats (syntax, data structures, and algorithms), identification methods, and communication between RFID tags and RFID readers (Ruiz-Garcia and Lunadei, 2010). In addition to this, several authors point out that there are some limitations for temperature monitoring on the basis of RFID systems in meat supply chains, for example due to insufficient reliability, adverse impact on efficient operation in the presence of moisture, and the reflectivity of metals (Narsing, 2005; Labuza, 2006; Estrada-Flores and Tanner, 2008; Patti and Narsing, 2008).

8.3.2 Time temperature indicators (TTIs)

Besides RFID-based temperature monitoring systems, the use of time temperature indicators (TTIs) to control temperature conditions in chilled food supply chains from production to consumption has been discussed extensively in recent years, but implementation has only begun recently (Taoukis and Labuza, 1989; Kreyenschmidt *et al.*, 2005; Taoukis, 2008).

Time temperature indicators (TTIs) are simple and inexpensive devices, which reflect the time-temperature history of the product to which they are attached (Taoukis, 2001). The labels can indicate temperature abuse and show the accumulated time-temperature history from production to consumer. Various papers describe the application of TTIs as quality indicators and shelf life predictors (Giannakourou *et al.*, 2005; Kreyenschmidt *et al.*, 2010; Labuza and Fu, 1995; Nuin *et al.*, 2008; Smolander *et al.*, 2004; Taoukis *et al.*, 1999; Vaikousi *et al.*, 2009).

TTIs are able to check the integrity of the cold chain and indicate interruptions in the chill chain. The labels are user-friendly and cost-effective (Taoukis *et al.*, 1999), and can monitor, record and translate the overall effect of the temperature history of a food product on its quality and safety (Vaikousi *et al.*, 2009). TTI response can be correlated to a specific food product safety and quality status at any point in the distribution chain, thus providing an effective decision tool.

Since they give a visual indication of temperature history during distribution and storage, TTIs can be particularly useful for warning of temperature abuse for chilled food products. On the other hand, they are also used as 'freshness indicators' for estimating the remaining shelf life of perishable products. The responses of these labels are usually some distinct visual changes that are temperature-dependent, such as an increase in colour intensity and diffusion of a dye along a straight path (Yam *et al.*, 2005).

The TTI reaction is based on enzymatic, chemical, mechanical, electrochemical or microbiological reactions (Ellouze and Augustin, 2010; Kreyenschmidt *et al.*, 2010). Commercially available TTIs are, for example, Fresh-Check by LifeLines Technology, Inc. (Morris Plains, NJ, USA), which is based on a

© Woodhead Publishing Limited, 2011

polymerization reaction of diacetylenic monomers, leading to a gradual colour change. The TTI manufactured by 3M (St. Paul, MN, USA) uses a light transmission reaction where a tape layer dissolves, continuously revealing the colour underneath. The VITSAB TTI (Limhamn, Sweden) is based on an enzymatic reaction, leading to a pH-change in the reaction mixture and visualized through a colour change of a pH dye (Smolander *et al.*, 2004). The OnVuTM label developed by Freshpoint Quality Assurance Ltd, 8 Ramat-Gan, Israel) consists of a photochromic ink which is charged with UV light. A physical reaction leads to a time-temperature dependent back reaction in the form of a colour change (Eichen *et al.*, 2008; Kreyenschmidt *et al.*, 2010). All these commercial TTIs are suitable for monitoring the temperature history of chilled products (Smolander *et al.*, 2004).

Adaptation of TTIs to the specific spoilage process of a food requires knowledge of predictive food microbiology for each type of food and must be based on kinetic principles (Labuza and Fu, 1995; Taoukis *et al.*, 1999). TTIs are reliable indicators of end of shelf life for food products, if they have similar temperature sensitivities as the food deterioration mechanism (Taoukis *et al.*, 1991).

Combining TTI principles with the above described RFID technology would allow remote tracking of shelf life and quality status of food products (Estrada-Flores and Tanner, 2008). Such a combination of a chemically based TTI and a RFID chip is currently under development in the Chill-on project (www.chill-on.com).

8.4 Prediction of food quality and safety

One main aspect in the chilled food supply chain, besides the integration of temperature monitoring systems, is the linkage of product characteristic and temperature data. Research progress in the last years has shown that predictive food models can be possible solutions to link product quality with the temperature history of the product. Such models allow the prediction of food quality and remaining shelf life, based on microbiological growth depending on defined environmental factors. Generally, the prediction of quality deterioration in meat and fish is based on the growth of the specific spoilage organisms which are mostly responsible for spoilage. The number of microorganisms allows conclusions to be drawn on the quality status of the product. During the past years, several papers have been published on predictive food models (Koutsoumanis, 2001; Dalgaard *et al.*, 2002; Cayré *et al.*, 2003; Gospavic *et al.*, 2008; Kreyenschmidt *et al.*, 2010). Most of the models developed describe the growth of a specific spoilage organism depending on the temperature, as this is a main influence on food spoilage. However, a key element of the integration of such models and the calculation of the remaining shelf life is the continuous control of product temperature during the different stages of the supply chain, whereby aspects such as the positioning of the temperature device have to be

© Woodhead Publishing Limited, 2011

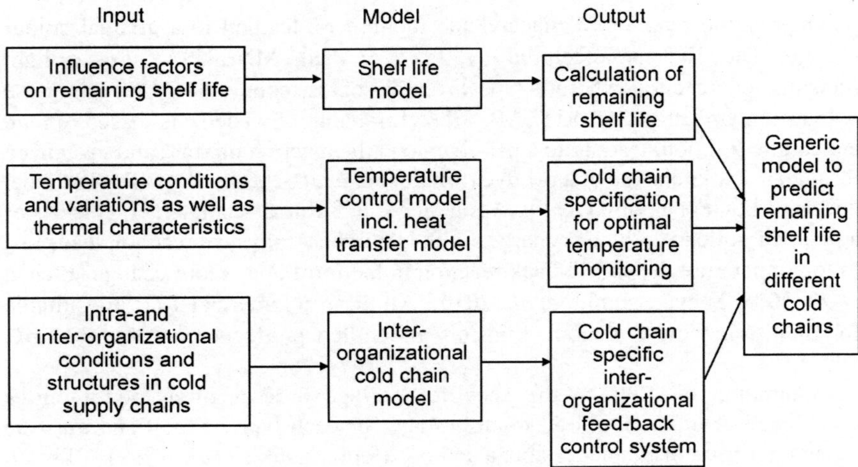

Fig. 8.1 An overview of different aspects which have to be considered in the integration and combination of shelf life models and temperature monitoring devices (Raab *et al.* 2008, Bruckner *et al.* 2009).

considered. In addition, the organization's structure and technical conditions in the supply chain have to be taken into account. In general, temperature conditions and variations, data on temperature history, system architecture regarding management of temperature, and quality data and microbiological data are needed for prediction of remaining shelf life in each step of the supply chain (Fig. 8.1).

As mentioned above, the actors in the food chain will only implement tools which guarantee an added value for their operations (Olafsdottir *et al.*, 2010). Such models, combined with temperature monitoring solutions, allow optimization of storage management. The well-established use of the first in – first out principle can be substituted by more quality-oriented procedures, such as least shelf life – first out. This will reduce waste and therefore costs, as described by Koutsoumanis *et al.* (2005) and Kreyenschmidt *et al.* (2010). A further benefit can be achieved by the combination and integration of the model into the risk assessment process.

While modelling of spoilage organisms allows prediction of shelf life and food quality, modelling of pathogen growth can be used to evaluate food safety. Quantitative Microbial Risk Assessment (QMRA) is a new method of risk analysis (Havelaar *et al.*, 2008). It aims at modelling pathogen growth and associated health risk based on dose-response relationship. Reliable QMRA systems, as presented by Janevska *et al.* (2010), are able to estimate risk to consumers resulting from contamination with pathogens. The main barrier for implementation of QMRA systems is seen in the time-consuming and expensive adaptation towards different food matrixes and pathogens (Evers and Chardon, 2010).

© Woodhead Publishing Limited, 2011

8.5 Information management to support inter-organizational cold chain management

The efficient and effective management of quality and temperature data can be highlighted as a major aspect in perishable food supply chains due to shelf life constraints (van der Vorst *et al.*, 2007). In particular, the concept of supply chain management is challenging and important in the field of perishable products due to short shelf life and therefore time requirements, a high number of product variants, strict traceability requirements, and the need for temperature control in the supply chain (Töyrylä, 1999; Kärkkainnen, 2003). A key element for optimal cold chain management and integrated disciplines such as supply chain management is a good data basis for operational and managerial decisions at important inspection and decision points within the supply chain.

Apart from technological innovations in the area of temperature monitoring, major technological progress has been made in the IT sector. IT has enabled coordination across organizational boundaries to achieve new levels of efficiency and productivity (Magretta, 1998) and opens up new possibilities for increasing value through better communication and information sharing in the supply chain. With electronic data interchange (EDI) and extensible markup language (XML), companies are able to share information seamlessly through integrated systems and the Internet (Laukkanen *et al.*, 2007). Recent studies show that EDI is still widely used (Narayanan *et al.*, 2009) in the food industry, but mostly for frequent and routine transactions such as despatching of invoices and purchasing orders, not for coordinating tasks such as transferring current schedules, production, and sales activities (Hill and Scudder, 2002). The XML business framework seems to be superior to the EDI and can carry more information (Nurmilaakso, 2008). There is still no single XML business framework that dominates. According to Nurmilaakso (2008), over 20 XML standards are in use, some are cross-industry while others are industry-specific. This has left many companies hesitant to take up a specific standard.

The value of information sharing in supply chains has been studied extensively in literature, especially in conjunction with sales and inventory information. However, very little focus has been placed in literature on other types of information shared in supply chains (Lee and Whang, 2000). The recent push for traceability in food supply chains and subsequent legislation have made several software suppliers take a holistic approach to the entire supply chain (Thompson *et al.*, 2005). Bollen *et al.* (2006) suggested that a very detailed traceability system would be able to track and locate a product at all points from harvest or slaughter to market and, by including temperature data, this could be used to predict quality at all times.

The development of holistic inter-organizational information management systems composed of a few stand-alone solutions offers the possibility for flexible systems which can be adapted to the needs of each user. On the one hand, it offers the basic functions of temperature monitoring in combination with advanced tools which permit the real-time collection of a full temperature

© Woodhead Publishing Limited, 2011

history within global supply chains. On the other hand, the collected temperature history makes it possible to add further components, e.g. heat transfer models, predictive models, and risk management tools which need such a data basis. The system integrates novel temperature monitoring and IT solutions and takes into account the needs of major actors and global supply chains, as well as the requirements of small and medium-sized companies, which are an important group in the food sector.

8.6 Conclusions

Food quality and safety along the chilled food supply chain can be ensured through various technologies and approaches. Some of the latest developments in the field are highlighted in this chapter.

Preliminary results from an industry survey identified barriers preventing the implementation of novel technologies for temperature monitoring and information management. Some industrial actors stated that there is a lack of trust in the chain and they are not willing to share all information, although they are aware of the added value it would bring. They are afraid that the novel tools are too costly and extra work. The lack of harmonization of standards, practices and legislation is a particular barrier for the implementation of RFID labels and information management software. If these technologies can only be applied to parts of the supply chain due to legal, organizational or regulatory aspects in some countries, their added value is lost. Costs are a barrier for the implementation of all novel technologies: Smart labels need to be purchased and adapted to the specific requirements; equipment and software has to be bought, and staff members have to be trained.

Despite these barriers, novel temperature monitoring tools and supply chain management software offer a wide range of advantages and make it easier to meet customer expectations and legal requirements.

The recording of the full temperature profile, and information sharing may enhance trust between supply chain actors. As quality is a value, enhanced product quality through efficient chill chain management may lead to an increase in market value. Last but not least, ensuring food safety along the entire supply chain may reduce cases of food-induced illness and the need for costly product recalls, which are also damaging for a company's reputation.

Overall, ensuring food safety, quality, and traceability from farm to fork will increase consumer trust.

8.7 References

BOLLEN, A.F., C.P. RIDEN, L.U. OPARA (2006) Traceability in postharvest quality management. *International Journal of Postharvest Technology and Innovation* 1(1): 93.

BOVENSCHULTE, M., P. GABRIEL, K. GAßNER, U. SEIDEL (2007) *RFID: Potenzial für*

© Woodhead Publishing Limited, 2011

Deutschland: Stand und Perspektiven von Anwendungen auf Basis der Radiofrequenz – Identifikation auf den nationalen und Internationalen Märkten. Bundesministerium für Wirtschaft und Technologie, Federal Ministry of Economics und Technology.

BRUCKNER, S., V. RAAB, J. KREYENSCHMIDT (2009) Concept for the implementation of a generic model for remaining shelf life prediction in meat supply chains. *6th International Conference Predictive Modeling in Foods, 8–12 September 2009, Washington DC.*

CAYRÉ, M.E., G. VIGNOLO, O. GARRO (2003) Modeling lactic acid bacteria growth in vacuum-packaged cooked meat emulsions stored at three temperatures. *Food Microbiology* 20(5): 561–566.

CHILL-ON PROJECT (2010) Developing and integrating novel technologies to improve safety, transparency and quality assurance of the chilled/frozen food supply chain. Funded by the European Commission in the 6th Framework Programme, Contract-No.: 016333. www.chill-on.com

DALGAARD, P., P. BUCH, S. SILBERG (2002) Seafood spoilage predictor – development and distribution of a product specific application software. *International Journal of Food Microbiology* 73: 343–349.

DITTMANN, L. (2006) *Der angemessene Grad an Visibilität in Logistik-Netzwerken: Die Auswirkungen von RFID.* Deutscher Universitäts-Verlag, Wiesbaden.

DOLUSCHITZ, R., B. ENGLER, C. HOFFMANN (2010) Quality assurance and traceability of foods of animal origin: major findings from the research project IT FoodTrace. *Journal für Verbraucherschutz und Lebensmittelsicherheit* 5: 11–19.

EICHEN, Y., D. HAARER, M. FEUERSTACK (2008) What it takes to make it work: the OnVu TTI. In *Proceedings of the Third International Workshop Cold-Chain-Management*, Bonn, Germany, p. 304.

ELLOUZE, M., J.-C. AUGUSTIN (2010) Applicability of biological time temperature integrators as quality and safety indicators for meat products. *International Journal of Food Microbiology* 138: 119–129.

ESTRADA-FLORES, S., D. TANNER (2008) RFID technologies for cold chain applications. Review article. Bulletin of the International Institute of Refrigeration, *Bulletin 2004–8. International Institute of Refrigeration* 2004(8): 4–8.

EVERS, E.G., J.E. CHARDON (2010) A swift quantitative microbial risk assessment (sQMRA) tool. *Food Control* 21: 319–330.

FU, B., T. LABUZA (1992) Considerations for the application of time-temperature integrators in food distribution. *Journal of Food Distribution Research* 23(1): 9–18.

GIANNAKOUROU, M.C., K. KOUSOUMANIS, G.J.E. NYCHAS, P.S. TAOUKIS (2005) Field evaluation of the application of time temperature integrators for monitoring fish quality in the chill chain. *International Journal of Food Microbiology* 102: 323–336.

GOSPAVIC, R., J. KREYENSCHMIDT, S. BRUCKNER, V. POPOV, N. HAQUE (2008) Mathematical modelling for predicting the growth of *Pseudomonas* spp. in poultry under variable temperature conditions. *International Journal of Food Microbiology* 127(3): 290–297.

GUO, L., Q. ZHANG (2002) A wireless LAN for collaborative off-road vehicle automation. *Proceedings of the automation technology for off-road equipment, July 26–27. Chicago, Illinois, USA*, 51–58.

HAVELAAR, A.H., E.G. EVERS, M.J. NAUTA (2008) Challenges of quantitative microbial risk assessment at EU level. *Trends in Food Science and Technology* 19: 26–33.

HELO, P., P. ANUSSORNNITISARN, K. PHUSAVAT (2008) Expectation and reality in ERP

© Woodhead Publishing Limited, 2011

implementation: consultant and solution provider perspective. *Industrial Management & Data Systems* 108(8): 1045–1059.

HILL, C., G.D. SCUDDER (2002) The use of electronic data interchange for supply chain coordination in the food industry. *Journal of Operations Management* 20(4): 375–387.

JANEVSKA, D.P., R. GOSPAVIC, E. PACHOLEWICZ, V. POPOV (2010) Application of a HACCP-QMRA approach for managing the impact of climate change on food quality and safety. *Food Research International*. Doi:10.1016/j.foodres.2010.01.025

JEDERMANN, R., C. BEHRENS, D. WESTPHAL, W. LANG (2006) Applying autonomous sensor systems in logistics – combining sensor networks, RFIDs and software agents. *Sensors and Actuators A: Physical* 132(1): 370–375.

JEDERMAN, R., L. RUIZ-GARCIA, W. LANG (2009) Spatial temperature profiling by semi-passive RFID loggers for perishable food transportation. *Computers and Electronics in Agriculture* 65: 145–154.

JØRGENSEN, B.M., J. OEHLENSCHLÄGER, G. ÓLAFSDÓTTIR, S.V. TRYGGVADÓTTIR, M. CARECHE, K. HEIA, P. NESVADBA, M.L. NUNES, B.M. POLI, C. DI NATALE, B. PÉREZ-VILLARREAL, H. BALLO, J.B. LUTEN, A. SMELT, W. DENTON, P. BOSSIER, T. HATTULA, G. ÅKESSON (2003) A study of the attitudes of the European fish sector towards quality monitoring and labelling. In J. B. Luten, J. Oehlenschlager, G. Olafsdottir (eds) (2003) *Quality of Fish from Catch to Consumer: Labelling, Monitoring and Traceability*. Wageningen Academic Publishers, The Netherlands.

KÄRKKÄINEN, M. (2003) Increasing efficiency in the supply chain for short shelf life goods using RFID tagging. *International Journal of Retail & Distribution Management* 10(10): 529–536.

KERRY, J., M. O'GRADY, S. HOGAN (2006) Past, current and potential utilisation of active and intelligent packaging systems for meat and muscle-based products: A review. *Meat Science* 74(1): 113–130.

KHER, S.V., L.J. FREWER, J.D. JONGE, M. WENTHOLT (2010) Experts' perspectives on the implentatation of traceability in Europe. *British Food Journal* 112 (3): 262–174.

KOUTSOUMANIS, K. (2001) Predictive modeling of the shelf life of fish under nonisothermal conditions. *Applied Environmental Microbiology* 67(4): 1821–1829.

KOUTSOUMANIS, K., P.S. TAOUKIS, G.-J.E. NYCHAS (2005) Development of a safety monitoring and assurance system for chilled food products. *International Journal of Food Microbiology* 100(1–3): 253–260.

KREYENSCHMIDT, J. (2008) Innovative Tools for supporting Cold-Chain-Management. In Kreyenschmidt, J., *Proceedings of the 3rd International Workshop 'Cold Chain-Management', 2–3 June 2008, Bonn.*

KREYENSCHMIDT, J., T. LETTMANN, B. PETERSEN (2005) Smart Labels – Frische angezeigt – neue Technologien verbessern die Kühlkettenlogistik und den Verbraucherschutz. *Fleischwirtschaft* 83(12): 66–72.

KREYENSCHMIDT, J., H. CHRISTIANSEN, A. HÜBNER, V. RAAB, B. PETERSEN (2010) A novel photochromic time-temperature indicator to support cold chain management. *International Journal of Food Science and Technology* 45: 208–215.

KUMAR, S., E. BUDIN (2006) Prevention and management of product recalls in the processed food industry: a case study based on an exporter's perspective. *Technovation* 26(5–6): 739–750.

KUO, J.-C., M.-C. CHEN (2010) Developing an advanced multi-temperature joint distribution system for the food cold chain. *Food Control* 21: 559–566.

LABUZA, T.P., B. FU (1995) Use of time/temperature integrators, predictive microbiology,

© Woodhead Publishing Limited, 2011

and related technologies for assessing the extent and impact of temperature abuse on meat and poultry products. *Journal of Food Safety* 15: 201–227.

LABUZA, T., J. KREYENSCHMIDT, B. PETERSEN (eds) (2006) Time-temperature integrators and the cold chain: What is next? *Cold Chain Management – Proceedings of the 2nd International Workshop, 8–9 May, Bonn*, Bonner Universitätsdruckerei: 43–52.

LAUKKANEN, S., S. SARPOLA, P. HALLIKAINEN (2007) Enterprise size matters: objectives and constraints of ERP adoption. *Journal of Enterprise Information Management* 20(3): 319–334.

LEE, H.L., S. WHANG (2000) Information sharing in a supply chain. *International Journal of Technology Management* 20(3/4): 373.

LÓPEZ, T.S., D.C. RANASINGHE, B. PATKAI, D. McFARLANE (2009) Taxonomy, technology and applications of smart objects. *Information Systems Frontiers*. DOI: 10.1007/s10796-009-9218-4.

MAGRETTA, J. (1998) The power of virtual integration: an interview with Dell Computer's Michael Dell. *Harvard Business Review*. 76(3): 72.

MAI, N., S.G. BOGASON, S. ARASON, S.V. ÁRNASON, T.G. MATTHÍASSON (2009) Benefit of traceability in fish supply chains – case studies. *British Food Journal* 112(9): 970–1002.

McMEEKIN, T., J. BOWMAN, O. McQUESTIN, L. MELLEFONT, T. ROSS, M. TAMPLIN (2008) The future of predictive microbiology: Strategic research, innovative applications and great expectations. *International Journal of Food Microbiology* 128: 2–9.

MICHAEL, K., L. McCATHIE (2005) The pros and cons of RFID in supply chain management. In Brookes, W. (ed.) *Proceedings of the International Conference on Mobile Business. 11–13 July 2005, IEEE Computer Society, Crowne Plaza Darling Harbour, Sydney, Australia.*

MONTANARI, R. (2008) Cold chain tracking: a managerial perspective. *Trends in Food Science & Technology* 19: 425–431.

MOUSAVI, A., M. SARHAVI, A. LENK, S. FAWCETT (2002) Tracking and traceability in the meat processing industry: a solution. *British Food Journal* 104(1): 7–19.

NARAYANAN, S., A.S. MARUCHECK, R.B. HANDFIELD (2009) Electronic data interchange: research review and future directions. *Decision Sciences* 40(1): 121–163.

NARSING, A. (2005) RFID and supply chain management: An assessment of its economic, technical and productive viability in global operations. *Journal of Applied Business Research* 12(5): 75–80.

NUIN, M., B. ALFARO, Z. CRUZ, N. ARGARATE, S. GEORGE, Y. LE MARC, J. OLLEY, C. PIN (2008) Modelling spoilage of fresh turbot and evaluation of a time-temperature integrator (TTI) label under fluctuating temperature. *International Journal of Food Microbiology* 127: 193–199.

NURMILAAKSO, J. (2008) EDI, XML and e-business frameworks: a survey. *Computers in Industry* 59(4): 370–379.

OLAFSDOTTIR G., P. NESVADBA, C. DI NATALE, C. CARECHE, J. OEHLENSCHLÄGER, S.V. TRYGGVADÓTTIR, R. SCHUBRING, M. KROEGER, K. HEI, M. ESAIASSEN, A. MACAGNANO, B.M. JØRGENSEN (2004) Multisensor for fish quality determination. *Trends in Food Science and Technology* 15: 86–93.

OLAFSDOTTIR, G., H.L. LAUZON, E. MARTINSDOTTIR, K. KRISTBERGSSON (2006) Influence of storage temperature on microbial spoilage characteristics of haddock fillets (*Melanogrammus aeglefinus*) evaluated by multivariate quality prediction. *International Journal of Food Microbiology* 111: 112–125.

OLAFSDOTTIR, G., S. BOGASON, C. COLMER, M. EDEN, T. HAFLIÐASON, M. KÜCK (2010)

© Woodhead Publishing Limited, 2011

Improved efficiency and real time temperature monitoring in the food supply chain. *1st IIR International Cold Chain and Sustainability Conferences*. Cambridge, 2010.

OLSSON, A. (2004) Temperature controlled supply chains call for improved knowledge and shared responsibility. In Aronsson, H. (ed.) *Conference Proceedings NOFOMA, 2004*: 569–582.

OPERA, A., J. COURBAT, N. BARSAN, D. BRIAND, N.F. DE ROOIJ, U. WEIMAR (2009) Temperature, humidity and gas sensors integrated on plastic foil for low power applications. *Sensors and Actuators B: Chemical* 140: 227–232

ORRISS, G.D., A.J. WHITEHEAD (2000) Hazards analysis and critical control point (HACCP) as a part of an overall quality assurance system in international food trade. *Food Control* 11: 345–351.

PATTI, A., A. NARSING (2008) Lean and RFID: friends or foes? *Journal of Business & Economics Research* 6(2): 83–90.

RAAB, V., S. BRUCKNER, E. BEIERLE, Y. KAMPMANN, B. PETERSEN, J. KREYENSCHMIDT (2008) Generic model for the prediction of remaining shelf life in support of cold chain management in pork and poultry supply chains. *Journal on Chain and Network Science* 8 (1): 59–73.

RAAB, V., B. PETERSEN, J. KREYENSCHMIDT (2011) Temperature monitoring in meat supply chains. *British Food Journal*. Accepted.

RIJSWIJK, W., L.J. FREWER (2008) Consumer perceptions of food quality and safety and their relation to traceability. *British Food Journal* 110 (10): 1034–1046

ROBERTS, C.M. (2006) Radio frequency identification (RFID). *Event (London)* 5: 18–26.

RUIZ-GARCIA, L., L. LUNADEI (2010) Sustainable Radio Frequency Identification Solutions. In: *Intech* (pp. 37–50). Croatia. Retrieved from http://sciyo.com/books/show/title/sustainable-radio-frequency-identification-solutions.

RUIZ-GARCIA, L., P. BARREIRO, J. RODRIGUEZ-BERMEJO, J. ROBLA (2007) Review. Monitoring the intermodal, refrigerated transport of fruit using sensor networks. *Spanish Journal of Agricultural Research* 5(2): 142–156.

SMOLANDER, M., H.-L. ALAKOMI, T. RITVANEN, J. VAINIOPÄÄ, R. AHVENAINEN (2004) Monitoring of the quality of modified atmosphere packaged broiler chicken cuts stored in different temperature conditions. A. Time-temperature indicators as quality-indicating tools. *Food Control* 15: 217–229.

TAOUKIS, P.S. (2001) Modelling the use of time-temperature indicators in distribution and stock rotation. In L.M.M Tijskens, M.L.A.T.M Hertog., B.M Nicolai. *Food Process Modelling*. Cambridge: Woodhead Publishing.

TAOUKIS, P.S. (2008) Application of time-temperature integrators for monitoring and management of perishable product quality in the cold chain. In: Kerry, J., P. Butler (Eds.) *Smart packaging technologies for fast moving consumer goods*, Chichester: Wiley, 61–74.

TAOUKIS, P.S., T.P. LABUZA (1989) Applicability of time-temperature indicators as shelf life monitors of food products. *Journal of Food Science* 54 (4): 783–788.

TAOUKIS, P.S., B. FU, T.P. LABUZA (1991) Time-temperature indicators. Food *Technology* 45 (10): 70–82

TAOUKIS, P.S., K. KOUTSOUMANIS, G.J.E. NYCHAS (1999) Use of time-temperature integrators and predictive modeling for shelf life control of chilled fish under dynamic storage conditions. *International Journal of Food Microbiology* 53: 21–31.

THOMPSON, M., G. SYLVIA, M.T. MORRISSY (2005) Seafood traceability in the United States: current trends, system design, and potential applications. *Comprehensive Reviews*

© Woodhead Publishing Limited, 2011

in Food Science and Food Safety 1: 1–7.

TÖYRYLÄ, I. (1999) *Realising the potential of traceability*, PhD Thesis, Helsinki, University of Technology, Espoo, Finland.

VAIKOUSI, H., C. G. BILIADERIS, K. P. KOUTSOUMANIS (2009) Application of microbial Time Temperature Indicator (TTI) for monitoring spoilage of modified atmosphere packaged minced meat. *International Journal of Food Microbiology* 133: 272–278.

VAN DER VORST, J. G. A. J., C. A.DA SILVA, J.H. TRIENEKENS (2007) Agro-industrial supply chain management: concepts and applications. Agricultural management, marketing and finance occasional paper, Food and Agricultural Organization of the United Nations.

WANG, N., N. ZHANG, M. WANG (2006) Review. Wireless sensors in agriculture and food industry – recent development and future perspective. *Computers and Electronics in Agriculture* 50(1): 1–14.

WYLD, D.C. (2006) RFID 101: the next big thing for management. *Management Research News* 29(4): 154–173.

YAM, K.L., P. T. TAKHISTOV, J. MILTZ (2005) Intelligent Packaging: Concept and Applications. *Journal of Food Science* 70 (1): 1–10

ZWIETERING, M.H., J.T. DE KOOS, B.E. HASENACK, J.C. DE WIT, K. VAN'T RIET (1991) Modeling of bacterial growth as a function of temperature. *Applied and Environmental Microbiology* 57: 1094–1101.

© Woodhead Publishing Limited, 2011

9

Emerging safety and quality issues of compound feed with implications for human foods

J. Hoorfar and I. Bang-Berthelsen, Technical University of Denmark, Denmark, F. T. Jones, University of Arkansas, USA, P. Häggblom, National Veterinary Institute, Sweden, G. Bruggeman, Vitamex Group, Belgium and J. Zentek, Free University of Berlin, Germany

Abstract: The increasing use of imported, low-cost feed raw materials from non-EU countries, the re-use of by-products from the biofuel industry and food production (i.e. alternative feed sources) as well as the emerging production of organic feeds and foods present new challenges for the safety and quality of the food derived from these feed materials. It is also important to recognize that to remain competitive, the European feed industry must apply its limited resources for control at critical points within the production chain. Furthermore, emerging technologies such as new methods of physical and chemical treatment and the increasing use of by-products of biotechnology processes deserve scientific attention. Finally, the increasing demand for organic animal derived foods has led to a rising demand for organic feed leading to the requirements for improved methods to distinguish organic from conventionally produced feed.

Key words: zoonoses, pathogens, salmonella, mycotoxins, biofuel, biodiesel, animal feed, organic feed, coccidiostats, sustainability.

© Woodhead Publishing Limited, 2011

9.1 Introduction

The European feed industry depends on the sustainable supply of cost-effective, high quality and safe feed ingredients. The term feed safety can be defined as the risk of health hazard to livestock animals or to the food produced hereby for human consumption. Feed quality is understood as the sum of nutritional and physiological values, supporting health and well-being of livestock animals and product quality.

Sustainability includes security and stability, cost-efficiency, production economics, consumer perception including animal welfare, and environmental aspects. European compound feeds largely comprise imported raw materials, which are at risk of containing unexpected microbial or chemical contaminants that could rapidly spread throughout the EU internationally. EC regulations require that products intended for animal feed may only be used if they are sound, genuine, merchantable and do not represent any danger to animal health, human health or the integrity of the environment (Directive 2002/32/EC) and finally to consider feed as integrated part of the food chain (Directive 2002/178/EC).

Within Europe, feed business operators are responsible for the implementation of measures and conditions necessary to control hazards and to ensure that feeds are hygienically safe for animal consumption (Directive 2005/183/EC). An understanding of how feed safety can be impaired has still not yet been achieved, as problems happen mostly through unexpected hazards. Risks for feed and food from animals can arise from unintended natural or anthropogenic contaminants, imprudent use of feed ingredients and additives, or overall impaired feed hygiene.

The European Commission has established maximum levels for a broad, but not exhaustive, spectrum of undesirable substances in feed including chemicals, toxic seeds and natural toxins (e.g. aflatoxins) (Directive 2002/32/EC). Risks can not only arise from unintended contaminants but also from negligent use of admitted substances such as certain feed additives. For example, the carry-over of anticoccidials into table eggs via cross contaminated layer feeds is still an issue in the EC and needs further follow-up and monitoring.

It has been known for many years that feed contains both spoilage and pathogenic microorganisms (Hinton, 1993). Therefore, improperly made, mishandled or misused feedstuffs can lead to spoilage but feedstuffs can *per se* turn into a vector for pathogens or their metabolic products. Here, mycotoxins are typical fungal products and important contaminants of feed. They have been studied in many projects funded by the EC. However, more work on sampling and methods is needed from the global and the European perspective towards feed.

9.2 Interaction between food and feed safety

While food and feed safety have traditionally been inextricably linked, recent crises concerning bovine spongiform encephalopathy (BSE), dioxins, foot-and-mouth disease (FMD), melamine, and other contaminants have strongly

© Woodhead Publishing Limited, 2011

emphasized the importance of this connection (Insall, 2009; Mantovani *et al.*, 2006). For the assessment of feed-borne risks, it is imperative that the toxicological hazards be characterized, that contamination pathways be identified and rates of transfer of the compound or its metabolites to foods of animal origin be determined (Mantovani *et al.*, 2006). However, toxicological hazards cannot be fully assessed until the physiology, i.e. absorption, metabolism, and persistence of feed contaminants, is known. Although these data are available for many problematic substances (Kan and Meijer, 2007), and contaminant exposure levels appear to be generally low in feedstuffs (as illustrated by the European RASFF (Rapid Alert System for Food and Feed) system) (Kleter *et al.*, 2009), only few specific physiological or concentration data or modelling approaches are available on most feed contaminants (van Raamsdonk *et al.*, 2009).

In addition, the effects emerging pathogenic microorganisms, mycotoxins, and naturally occurring contaminants have on animal physiology and product safety cannot be ignored (Kan and Meijer, 2007). A comprehensive assessment of feed contaminants requires an examination of the metabolic and biochemical impact of ingested compounds as well as exploring the interaction with diet composition, feeding technique, intestinal microbiota and histology, absorption, metabolism and excretion pathways. Calculation models for simulating the carry-over from feed to food need to be designed and have to be adjusted to the specific combination of contaminant animal-target product (van Raamsdonk *et al.*, 2009). To avoid animal experiments, advanced methods including *in vitro* and *ex vivo* models are necessary to use in order to study this interaction.

The scientific and technical knowledge necessary for the implementation of economic and effective control systems is still incomplete. Targeted and science-based sampling, tracing and analytical techniques are necessary to implement efficient management practices. Recently, the EFSA CONTAM panel evaluated more than 30 individual substances that are considered as undesirable contaminants of animal feed (http://www.efsa.europa.eu/en/panels/contam.htm). These include heavy metals, environmental pollutants, natural toxins and secondary metabolites from plants, including botanical impurities. In a recent technical report submitted to EFSA, information about current and new mycotoxins and natural plant toxins with relevance to the safety and quality of feed and food was summarized (Battilani *et al.*, 2009)

When assessing animal health effects and the risk to consumers resulting from contaminants in the feed or from residual amounts in animal-derived products, a large number of factors are important to consider. Hazard identification and exposure assessment in animals will depend on the variability of feed contamination, stability of the contaminant as well as feed composition and feeding regimes for different animals. Health effects in animals may include anti-nutritional effects inducing lower performance, specific toxicity affecting the animal or effects on the product. When assessing consumer risks resulting from undesirable substances in animal-derived food products, it is essential to analyse biotransformation, tissue disposition and, not least, consumer percep-

© Woodhead Publishing Limited, 2011

tion. These data have also to be linked to food consumption. In order to demonstrate emerging safety issues in feed, we use salmonellae, mycotoxins and cocciostats as 'case studies'.

9.3 Risks associated with *Salmonella*

In an opinion by the EFSA BIOHAZ panel, microbiological risks were assessed in feeding stuffs for food-producing animals (EFSA, 2008). Several zoonotic bacteria are present in animal feed; however, when assessing animal infections and the risk for animals and humans to become infected from animal-derived food, *Salmonella* contamination of feed is considered most relevant. Raw materials used in feed manufacturing are the main source for the introduction of *Salmonella* into the feed chain and the need for efficient monitoring and processing technologies to control *Salmonella* in feed is obvious (Jones, 2008a, 2008b). *Salmonella spp.* cause infectious disease in both humans and animals (Zhang *et al.*, 2008; Lan *et al.*, 2007) and contaminated feeds have been linked with salmonellosis in humans and animals (Davis *et al.*, 2003; Crump *et al.*, 2002; Mrema *et al.*, 2006). Indeed, Hald *et al.* (2006) concluded that if present in feedstuffs, more than 90% of *Salmonella* serotypes found in feedstuffs have the potential to infect humans via contaminated animal food products. Large outbreaks of feed-borne salmonellosis in pigs have demonstrated the economic dimension that can be caused by *Salmonella*-contaminated feeds (Österberg *et al.*, 2006). Feed is a particularly effective means of spreading *Salmonella* contamination because the organism can survive for long periods on a wide range of feed materials (Maciorowski *et al.*, 2007). Furthermore, every animal in a herd exposed to contaminated feed for extended periods of time will become a carrier. Finally, feed-borne contaminants can be disseminated through the food chain via litter, contaminated carcasses, animal manure or dust (Okelo *et al.*, 2006; Nowak *et al.*, 2007; Klerks *et al.*, 2006; Elizaquível and Aznar, 2008). Although feed ingredients can be the primary source of contamination in a feed mill, *Salmonella* contamination may also persist in oilseed plants or feed mills or other points in the processing chain, making these facilities important sources for permanent recontamination with *Salmonella* (Davies, 2005; Davies and Wray, 1997; Morita *et al.*, 2003, 2006).

Stationary contamination can often be linked with the use of coolers in feed mills. Improper equipment operation allows moisture to condense on interior surfaces, permitting *Salmonella* growth and feed contamination because no further heat treatment is done downstream (Davies and Hinton, 2000; Jones and Richardson, 2004)). One problem is that it occurs erratically and *Salmonella* is typically unevenly distributed within feeds (Maciorowski *et al.*, 2004) leading to a high degree of uncertainty in identifying relevant batches. The extent of *Salmonella* contamination of feeds and feed ingredients might be substantially underestimated because of inadequate sampling and detection methods (Davies, 2005; Durand *et al.*, 1990; Schluter *et al.*, 1994).

© Woodhead Publishing Limited, 2011

Indeed, *Salmonella* contamination of feed is associated with feed particles. One contaminated feed particle would be counted via traditional most probable number (MPN) procedures as one *Salmonella* colony forming unit (CFU). Yet, contaminated feed ingredients may contain *Salmonella* micro-colonies that comprise many CFUs. Hence, traditional quantification methods may dramatically underestimate contamination levels (Wales *et al.*, 2010). In addition, recent data suggest that certain feed ingredients (most notably palm kernel meal) bind to or protect *Salmonella*, so that the bacteria go undetected (Koyuncu and Häggblom, 2009).

Finally, the fact that heat-treated feeds may contain *Salmonella* that are sublethally injured further complicates procedural difficulties and may produce low estimates of cell numbers or false negative results (Wesche *et al.*, 2009). In one European research project (www.BIOTRACER.org) collaborations with global exporters of plant proteins used as raw material for feed production in Europe were established to develop a proactive strategy to prevent oil meal contamination.

However, for many feed ingredients there is a lack of understanding regarding *Salmonella* contamination. Palm kernel meal is an important raw material for concentrates used for dairy cows; however, current methods for *Salmonella* detection appear moderately successful to analyse this specific ingredient (Koyuncu and Häggblom, 2009). With regards to *Salmonella* it seems to be essential to focus on the development of time efficient and safe analytical methods allowing rapid characterization of the most important feed ingredients and to achieve a complete picture of the involved serotypes (Koyuncu *et al.*, 2010). The ultimate goal should be a full chain traceability of this important zoonotic agent.

Although *Salmonella* is an 'old' problem, i.e. it has long been associated with pig feed, there is an 'emerging' issue that is that palm kernel presents challenges to the detection of *Salmonella* that up until very recently has not been even considered.

9.4 Risks associated with mycotoxin

Mycotoxins (particularly aflatoxin) are toxic chemicals produced in low concentrations by moulds as they grow in or on feeds or foods. However, mycotoxins are – as *Salmonella* – distributed heterogeneously, and the detection of low – but effective – concentrations in large volumes of feed make sampling and detection methodologies challenging. Yet, mycotoxins can have serious consequences for human and animal health (Bryden, 2007). Since some mycotoxins in feeds can pass into foods destined for human consumption, it is essential to reduce their occurrence in animal feeds (Razzazi *et al.*, 2003). Indeed, the recalls of mycotoxin-contaminated feeds demonstrate the gravity of the situation. Although it is clear that addressing mycotoxin contamination as early as possible in the food chain is essential (Binder, 2007), critical steps in the tracing

© Woodhead Publishing Limited, 2011

of mycotoxins need to be identified. Rapid methods (including PCR (polymerase chain reaction) techniques and non-molecular based tests) for the identification of mycotoxigenic fungi have been developed (Schoen *et al.*, 2005), but toxin quantification, which is slow and costly, remains the gold standard for risk assessment.

Sampling, extraction and sample clean-up methods must be improved, especially for industrial purposes, requiring rapid and cost-effective methods. Furthermore, analytical methods for mycotoxin detection for feedstuffs must be validated to be acceptable by the legislative bodies. Finally, based on the progress in sampling and analytics, mathematical contamination models are needed to study future developments in emerging feed production chains requiring optimized sampling and sample management protocols. This issue has been thoroughly addressed in the BIOTRACER project (www.biotracer.org).

9.5 Coccidiostat carry-over as indicator of misuse of feed additives

Although the EU has authorized the use of a wide range of coccidiostats in feeds for broilers, turkeys, pullets and rabbits, none of these compounds are approved for used in commercial layers. Yet, coccidiostat residues in eggs at concentrations higher than the tolerance level established in Directive 2009/8/EC are reported (Cannavan *et al.*, 2000; Kennedy *et al.*, 1996, 1998; Mortier *et al.*, 2005; Yakkundi *et al.*, 2002). These unwanted residues in eggs come from accidental contamination of layer-diets in animal feed mills via carry-over from one batch of feed to another. In fact, in some feed mills, as much as 15% carry-over from one batch to another has been reported (Mortier *et al.*, 2005). Carry-over can also occur during transport of feed to the farm (Kan and Meijer, 2007). Also, some coccidiostats are strongly electrostatic and stick to equipment surfaces enhancing cross contamination (Yakkundi *et al.*, 2002). While the levels reported in eggs are low, studies have shown that when layers are exposed to coccidiostats at 5% treatment levels, residues appear in eggs up to 15 days later (Kennedy *et al.*, 1998). The reported incidence of coccidiostat residues in eggs in the EU seems to be low; however, changing production conditions and the consequences from ongoing globalization of production and trade indicates that more knowledge and improved control measures are needed. Uncertainty also exists concerning to what extent contamination occurs in imported products across Member States.

9.6 Tracing and tracking of contaminants in the feed chain

Hazard Analysis and Critical Control Point (HACCP) programmes have been instituted for the animal feed industry and are mandatory for feed producers in the European Union by the feed hygiene regulation (Directive (EC) 183/2005).

© Woodhead Publishing Limited, 2011

While programmes such as HACCP, Good Manufacturing Practices and risk analysis procedures are designed to prevent feed contamination (Cooke, 2002; Notermans and Beumer, 2003; Petri, 2009), such systems fail to address unexpected events such as manufacturing mistakes, cross contamination or deliberated contamination events.

The tracing and tracking components of feed and food safety systems address these unexpected events by allowing targeted recalls so that animal or human health consequences are minimized (Notermans, 2003). Indeed, total-chain traceability is a vital component of any food safety system because it allows the documentation of product history, identifies the location of products within the production chain at any given time and, thus, allows for prompt and targeted action when recalls are necessary. Proper traceability systems can also help create a feedback loop to improve product quality, conditions and delivery, thus optimizing related costs and their logistics. The European Union has mandated traceability throughout the food production chain in order to make use of the protection and benefits offered by such systems. Animal feed is rightfully included in the mandatory traceability standards. While traceability systems are vital to product safety and quality, traceability in animal feeds offer special challenges since feed formulations frequently change, feed manufacturers utilize a wide range of ingredients from a variety of ingredient suppliers from around the world, and a clear separation of raw material batches is impaired by many cross-points during raw materially import or production, transport, and (intermediate) storage. Nonetheless, traceability standards dictate that products are defined, lots are identified, the flow of materials is documented, data are managed correctly, and information can be retrieved efficiently; the latter involves a great deal of effort for feed operators (FAO, 1998; Directive (EC) 183/2005).

9.7 New feed sources as source of emerging risks

According to the European Feed Manufacturers Federation (FEFAC, 2008), the top three imported feed materials in the EU-27 are oil cakes and meals (27.3 million tons), feed cereals (10.0 million tons) and molasses (2.9 million tons). Collectively, they represent 89% of all imported feed materials in the EU-27. Although soybean and copra-palm kernel meals (self sufficiency of 3% and 0%, respectively, in EU-27) are particularly vital to Europés livestock industry, 10.0 million tons of cereals, corresponding to 14% of the feed consumption, are also imported. Thus, imported feed ingredients are crucial for European agriculture, and careful surveillance and control measures are necessary to ensure safety and quality of both feed and animal protein products for human consumption. There are a number of emerging issues connected to the consumption of the most important feed ingredient, soybean meal, including availability of GMO-free soya, contamination with *Salmonella*, and purity. Finally, the increasing global trade with soya is becoming an emerging production issue; i.e., concerns related

© Woodhead Publishing Limited, 2011

to sustainability, the environment and natural resources in the soya-producing countries. There is an increasing awareness regarding environmental consequences of the transfer of Brazilian rainforest to agricultural land. Biofuel production by-products can provide alternative protein sources to soya.

The European Commission has determined that biofuels have great potential in addressing EU energy needs and should be increased from the present level of 2% to the target level of 25% by 2030. This increase will likely commit about 18% of the total EU agricultural land to the production of biofuels (Coppola *et al.*, 2009). This situation will make the use of EU-produced DDGS (i.e. dried distillers grains with soluble) more attractive in the future and asks for improved quality control. After the EU biofuel Directive 2003/30/EC, the amount of by-products from biofuel production (DDGS) and condensed distillers soluble (CDS) increased from 0.5 million tons in 2005 to estimated 7.2 million tons in 2010 (van der Aar and Doppenberg, 2009). Ample availability and attractive prices will likely mean that by-products from the biofuel industry will, in large part, satisfy the demand for protein rich feedstuffs (PRF).

However, contaminants, such as pesticides and mycotoxins, can be concentrated in DDGS, and anti-nutritional factors can limit the use of these by-products in animal feed (Romer, 1984; Widyaratne, 2005). Questions have also been raised about the prevalence of zoonotic bacteria in by-products from the biofuel industry. Finally, it has been shown that DDGS can vary widely in nutrient level and level of anti-nutritional factors such as fibre and non-starch polysaccharides (Ortin and Yu, 2009; Stein and Shurson, 2009). Thus, the quality of food products from livestock animals may be either directly or indirectly affected by the use of DDGS as an ingredient in feeds. Several substrates, in Europe mainly wheat, in North and South America and Asia, mainly corn and sugar cane, are used for biofuel production. Depending on the area and geographical location and available logistics and infrastructure, dried or wet by-products are fed to livestock, especially ruminants, but also to pigs and poultry. The characteristics and especially the ileal digestibility of wheat DDGS are highly variable, probably due to the drying process and associated heating (Cozannet *et al.*, 2009).

Studies have shown that mycotoxins are concentrated up to three times in DDGS compared with the grain used alone (Wu and Munkvold, 2008). The losses to the US pig industry from weight gain reduction due to fumonisins in corn DDGS have been estimated at $9 million (Wu and Munkvold, 2008). Recently, concerns have also been raised about antibiotic residues in DDGS (Lundeen, 2008). In the US, antibiotics such as virginomycin, penicillin and erythromycin are added to fermentors to reduce bacteria which compete with the ethanol-producing yeast for sugar substrates and micronutrients. In the US, the FDA is concerned about potential animal and human health hazards associated with these antibiotic residues. These aspects have also to be taken into account in the EU.

Finally, special attention in terms of quality is needed for promising new technologies for production of raw materials for feed, like algae (Raja *et al.*,

© Woodhead Publishing Limited, 2011

2008). Critical parameters are production processes as well as downstream and residue processing.

9.8 Organic feedstuffs

A growing number of EU citizens are buying organic foods. Consequently, organic animal production has increased rapidly in recent years to keep up with the increasing demand. According to Eurostat, in 2005, 3.9% of the total agricultural area in the EU-25 (before the EU expansion to 27 members) was used for organic food production with the highest proportion of organic land in Austria (11%), Italy (8.4%), Czech Republic and Greece (both 7.2%). Many guidelines and restrictions apply to ingredients used in organic feeds and some difficulties arise when trying to ensure a well-balanced nutritious diet without the use of any supplements (Blair 2007, 2008). Because the quality and composition of organic livestock feeds is important to the production of organic meat and other animal products, these factors are strictly regulated (Council Regulation EC no. 834/2007). Yet, whether or not the products being marketed as 'organic' meet the organic standards and are fully compliant with all regulations is dependent on the integrity of the seller (Padel and Sundrum, 2006; Flaten and Lien, 2009). Clearly, improved methods by which organic feed ingredients can be objectively identified need to be settled.

9.9 Emerging production technologies

Improvements in feed manufacturing technology have dramatically increased production efficiencies in every commercial feed mill worldwide. Technology improvements have automated practically every process operating within feed manufacturing facilities (Koch, 2008). These improvements have also increased the quality of feeds produced while decreasing mistakes, labour costs and waste (Unger, 2006). However, some improvements have at times failed to consider contamination issues. In addition, concerns over microbial contamination, bovine spongiform encephalopathy (BSE) and other contaminants have enhanced the need for documentation, product tracking and integrated data analysis (Behnke, 2003). While advanced electronics are allowing feed manufacturing systems to have even more control over feed quality issues (Dunn, 2008), the impact of such systems on food safety and traceability must be examined.

An additional production trend is on-farm feed production, where home grown feedstuffs, mainly cereals and vegetable proteins, are processed and mixed with mineral supplements. While safe and high-quality feeds can certainly be produced on-farm, many producers do not realize the complexities involved in proper feed manufacturing procedures and do not realize that when they manufacture feeds they assume responsibility for feed quality control (Owsley and Van Dyke, 1992; Johnston and Hawton, 2008). Clearly, feeds

© Woodhead Publishing Limited, 2011

manufactured on-farm need to be examined for quality, the presence of undesirable substances, hygienic risks and carry-over of medications.

9.10 Conclusions

The main impact of a proactive approach proposed in the present review is improved quality, safety and sustainability of feed, especially new and upcoming feed ingredients, before they become normalized in their use in the EU. It also responds to industrial needs by introducing new feed processes and technologies. Finally, it can restrict the production of waste, specifically by promoting clean technologies and products which can be recycled and re-used. A full integration of the research elements with clear focus on targeted dissemination of the outcome is necessary if consumer trust is to be maintained in the future.

9.11 Acknowledgements

This work was supported in part by BIOTRACER Integrated Project under the EU FP6 programme (Contract no. 036272), and in part by the Danish Institute for Science, Technology and innovation (START grant no. 274-09-0388) and by the Flemish Institute for Science and Technology (IWT-Flanders) (SAFEFEED, grant no. 060369 and Eureka E!3934 project).

9.12 References

BATTILANI, P., L. G. COSTA, A. DOSSENA, M. L. GULLINO, R. MARCHELLI, G. GALAVERNA, A. PIETRI, C. DALL'ASTA, P. GIORNI, D. SPADARO and A. GUALLA (2009) Scientific information on mycotoxins and natural plant toxicants. CFP/EFSA/CONTAM/2008/1, Scientific Technical Report submitted to EFSA, Accepted for Publication on 23 November 2009.

BEHNKE, K. C. (2003) Emerging technologies in feed processing, pp. 1–8. *Proc. Eastern Nutrition Conference, Ottawa, CANADA*. www.dsm.com/en_US/downloads/dnpus/enc_03_15.pdf

BINDER, E. M. (2007) Managing the risk of mycotoxins in modern feed production. *Animal Feed Sci. Technol.* **133:** 149–166.

BLAIR, R. (2007) *Nutrition and feeding of organic pigs*. CABI, Wallingford.

BLAIR, R. (2008) *Nutrition and feeding of organic poultry*. CABI, Wallingford.

BRYDEN, W. L. (2007) Mycotoxins in the food chain: human health implications. *Asia Pac. J. Clin. Nutr.* **16 (Suppl 1)**: 95–101.

CANNAVAN, A., G. BALL and D. G. KENNEDY (2000) Nicarbazin contamination in feeds as a cause of residues in eggs. *Food Addit. Contam.* **17**: 829–836.

COOKE, B. C. (2002) The industrial production of safe animal feeds in Europe. In F. J. M. Smulders and J. D. Collins (eds), *Food safety assurance and veterinary public*

© Woodhead Publishing Limited, 2011

health, volume 1. Food safety assurance in the pre-harvest phase. Wageningen Academic Publishers, Wageningen, The Netherlands, pp. 71–86.

COPPOLA, F., S. BASTIANONI, H. and ØSTERGÅRD (2009) Sustainability of bioethanol production from wheat with recycled residues as evaluated by energy assessment. *Biomass and Bioenergy* **33**: 1626–1642.

COZANNET, P., Y. PRIMOT, J-P. MÉTAYER, C. GADY, M. LESSIRE, P. A. GERAERT, L. LE TUTOUR, F. SKIBA and J. NOBLET (2009) Wheat dried distiller grains with solubles for pigs. *INRA Productions Animales* **22**: 11–16.

CRUMP, J., P. M. GRIFFIN and F. J. ANGULO (2002) Bacterial contamination of animal feed and its relationship to human foodborne illness. *Clin Infect Dis.* **35**: 859–865.

DAVIES, R. H. (2005) Pathogen populations on poultry farms. In G.C. Mead (ed.), *Food safety control in the poultry industry*, CRC, Press, Boca Raton, FL, pp. 101–135.

DAVIES, R. H. and M. H. HINTON (2000) *Salmonella* in animal feed. In C. Wray and A. Wray (ed.), Salmonella *in domestic animals*. CABI, Wallingford, pp. 285–300.

DAVIES, R. H. and C. WRAY (1997) Distribution of *Salmonella* contamination in ten animal feedmills. *Veterinary Microbiology* **51**: 159–169.

DAVIS, M. A., D. D. HANCOCK, D. H. RICE, D. R. CALL, R. DIGIACOMO, M. SANADPOUR and T. E. BESSER (2003) Feedstuffs as a vehicle of cattle exposure to *Escherichia coli* O157:H7 and *Salmonella enterica*. *Veterinary Microbiology* **95**: 199–210.

DUNN, N. (2008) A revolution in feed quality control. *Feed Tech* **12(6)**: 2–5.

DURAND, A. M., W. H. GIESEKE, M. L. BARNARD, M. L. VAN DER WALT AND H. C. STERN (1990) *Salmonella* isolated from feeds and feed ingredients during the period 1982–1988: Animal and public health implications. *Onderstepoort J. Vet. Res.* **57**: 175–181.

EFSA (2008) Scientific Opinion of the Panel on Biological Hazards on a request from the Health and Consumer Protection, Directorate General, European Commission on Microbiological Risk Assessment in feedingstuffs for food producing animals. *The EFSA Journal* **720**: 1–84.

ELIZAQUÍVEL P. and R. AZNAR (2008) A multiplex RTi-PCR reaction for simultaneous detection of *Escherichia coli* O157:H7, *Salmonella* spp. and *Staphylococcus aureus* on fresh, minimally processed vegetables. *Food Microbiology* **25**: 705–713.

FAO (1998) Animal feeding and food safety. Expert Consultation, FAO Food and Nutrition paper no. 69, M-82, http://www.fao.org/docrep/w8901e/w8901e00.htm#Contents

FEFAC (EUROPEAN FEED MANUFACTURERS FEDERATION) (2008) *Food and Feed Statistical Yearbook 2008*. http://www.fefac.org/statistics.aspx?EntryID=629.

FLATEN, O. and G. LIEN (2009) Organic dairy farming in Norway under the 100% organically produced feed requirement. *Livestock Science* **126**: 28–37.

HALD, T., A. WINDSTRAND, T. BRONDSTED and D. M. A. L. F. WONG (2006) Human health impact of *Salmonella* contamination in imported soybean products: a semiquantitative risk assessment. *Foodborne Pathogens and Disease* **3(4)**: 422–431.

HINTON, M. (1993) Spoilage and pathogenic microorganisms in animal feed. *International Biodeterioration & Biodegradation* **32**: 67–74.

INSALL, L. (2009) Food industry guidance on supply of by-products/surplus food for feed use. Presentation at EU Feed and Food Safety Workshop, Brussels, 18 February 2009. http://www.copa-cogeca.eu/img/user/File/ADA_2009/ADA(09)1254EN%5B1%5D.pdf

JOHNSTON, L. J. and J. D. HAWTON (2008) Quality Control of On-Farm Swine Feed Manufacturing. Minnesota Cooperative Extension Pub no WW-05639.

© Woodhead Publishing Limited, 2011

JONES, F. T. (2008a) Quality control in feed manufacturing. *Feedstuffs* **80(38)**: 74–79.

JONES, F. T. (2008b) Control of toxic substances. *Feedstuffs* **80(38)**: 80–85.

JONES, F. T. and K. E. RICHARDSON (2004) *Salmonella* in commercially manufactured feeds. *Poultry Science* **83**: 384–391.

KAN, C. A. and G. A. L. MEIJER (2007) The risk of contamination of food with toxic substances present in animal feed. *Animal Feed Science and Technology* **133**: 84–108.

KENNEDY, D. G., W. J. BLANCHFLOWER, P. J. HUGHES and J. McCAUGHEY (1996) The incidence and cause of lasalocid residues in eggs in Northern Ireland. *Food Addit. Contam.* **13**: 787–794.

KENNEDY, D. G., P. J. HUGHES and W. J. BLANCHFLOWER (1998) Ionophore residues in eggs in Northern Ireland: incidence and cause. *Food Addit. and Contam.* **15**: 535–541.

KLERKS, M. M., A. H. C VAN BRUGGEN, C. ZIJLSTRA and M. DONNIKOV (2006) Comparison of methods of extracting *Salmonella enterica* Serovar Enteritidis DNA from environmental substrates and quantification of organisms by using a general internal procedural control. *Applied and Environmental Microbiology* **72**: 3879–3886.

KLETER, G. A., A. PRANDINI, L. FILIPPI and H. J. MARVIN (2009) Identification of potentially emerging food safety issues by analysis of reports published by the European Community's Rapid Alert System for Food and Feed (RASFF) during a four-year period. *Food Chem. Toxicol.* **47**: 932–950.

KOCH, K. (2008) Feed mill efficiency. Proc 16th Annual ASA-IM SEA Feed Technology and Nutrition Workshop. www.asasea.com/download_doc.php?file=FTNW08-Koch_Feedmill.pdf

KOYUNCU, S. and P. HÄGGBLOM (2009) A comparative study of cultural methods for the detection of *Salmonella* in feed and feed ingredients. *BMC Veterinary Research* **5**: 6.

KOYUNCU, S., G. ANDERSSON and P. HÄGGBLOM (2010) Accuracy and sensitivity of commercial PCR-based methods for detection of Salmonella in feed. *Appl. Environ. Microbiol.* doi:10.1128/AEM.02714-09.

LAN, R., G. STEVENSON, K. DONOHOE, L. WARD and P. R. REEVES (2007) Molecular markers with potential to replace phage typing for *Salmonella enterica* serovar typhimurium. *Journal of Microbiological Methods* **68**: 145–156.

LUNDEEN, T. (2008) DDGS use prompts questions. *Feedstuffs* **80(5)**: 5

MACIOROWSKI, K. G., F. T. JONES, S. D. PILLAI and S.C. RICKE (2004) Incidence, sources, and control of foodborne *Salmonella* spp. in poultry feeds. *World's Poultry Science Journal* **60**: 446–457.

MACIOROWSKI, K. G., P. HERRERA, F. T. JONES, S. D. PILLAI and S. C. RICKE (2007) Effects on poultry and livestock of feed contamination with bacteria and fungi. *Animal Feed Science and Technology* **133**: 109–136.

MANTOVANI, A., F. MARANGHI, I. PURIFICATO and A. MACRÌ (2006) Assessment of feed additives and contaminants: An essential component of food safety. *Ann Ist Super Sanità* **42(4)**: 432–437.

MORITA, T., Y. MURAYAMA and T. LIDA (2003) *Salmonella* contamination in materials for oil meal and oil meal manufacturing plant. *Jpn. J. Food Microbio.* **20(3)**: 117–122.

MORITA, T., H. KITAZAWA, T. LIDA and S. KAMATA (2006) Prevention of *Salmonella* cross-contamination in an oilmeal manufacturing plant. *J Appl. Microbiol.* **101**: 464–473.

MORTIER, L., A. C. HUET, E. DAESELEIRE, G. HUYGHEBAERT, T. FODEY, C. ELLIOTT, P. DELAHAUT and C. V. PETEGHEM (2005) Deposition and depletion of five anticoccidials in eggs.

© Woodhead Publishing Limited, 2011

J. Agric. Food Chem. **53**: 7142–7149.

MREMA, N., S. MPUCHANE and B. A. GASHE (2006) Prevalence of *Salmonella* in raw minced meat, raw fresh sausages and raw burger patties from retail outlets in Gaborone, Botswana. *Food Control* **17**: 207–212.

NOTERMANS, S. (2003) Food authenticity and traceability ensuring the safety of animal feed. Conference Proc. 'Ensuring the safety of animal feed', pp. 1–45. http://www.foodmicro.nl/ensuringsafety.pdf

NOTERMANS, S. and H. BEUMER (2003) Safety and traceability of animal feed. In M. Lees (ed.), *Food authenticity and traceability*, Woodhead Publishing, Cambridge, pp. 518–553.

NOWAK, B., T. VON MUFFLING, S. CHAUNCHOM and J. HARTUNG (2007) *Salmonella* contamination in pigs at slaughter and on the farm: a field study using an antibody ELISA test and a PCR technique. *Int. J. Food Microbiol.* **115**: 259–267.

OKELO, P. O., D. D. WAGNER, L. E. CARR, F. W. WHEATON, L. W. DOUGLAS and S. W. JOSEPH (2006) Optimization of extrusion conditions for elimination of mesophilic bacteria during thermal processing of animal feed mash. *Animal Feed Sci. Technol.* **129**: 116–137.

ORTIN, W. G. N. and P. YU (2009) Nutrient variation and availability of wheat DDGS, corn DDGS and blend DDGS from bioethanol plants. *J. Sci. Food Agric.* **89**: 1754–1761.

ÖSTERBERG, J., I. VÅGSHOLM, S. BOQVIST and S. S. LEWERIN (2006) Feed borne outbreak of *Salmonella cubana* in Swedish pig farms: risk factors and factors affecting the restriction period in infected farms. *Acta Vet. Scand.* **47**: 13–21.

OWSLEY, W. F. and N. J. VAN DYKE (1992) Controlling quality of farm mixed swine feed. Alabama Cooperative Extension Service pub no ANR-637.

PADEL, S. and A. SUNDRUM (2006) How can we achieve 100% organic diets for pigs and poultry? *Aspects of Applied Biology* **79**: 237–241.

PETRI, A. (2009) Quality management in European feed production. Proc.17th Annual ASAIM SEA Feed Technology and Nutrition Workshop, pp. 1–9, www.asasea.com/download_doc.php?file=FTNW09-Petri_EU.pdf

RAJA, R., S. HEMAISWARYA, N. ASHOK KUMAR, S. SRIDHAR and R. RENGASAMY (2008) A perspective on the biotechnological potential of microalgae. *Crit. Rev. Microbiol.* **34(2)**: 77–88.

RAZZAZI, E., J. BOEHM, A. ADLER and J. ZENTEK (2003) Fusarientoxine und ihre Bedeutung in der Nutztierfütterung: eine Übersicht. *Wien. Tieraertl. Msschr.* **90**: 202–210.

ROMER, T. (1984) Detecting mycotoxins in corn and corn-milling products, *Feedstuffs* **56(37)**: 22–23.

SCHLUTER, H., C. BEYER, I. HAGELSCHURER, L. GEUE and P. HAGELSCHURER (1994) Epidemiological studies on *Salmonella* infections in poultry flocks. *Tierartzliche Umschau* **49**: 400–410.

SCHOEN, C. D., M. SZEMES, P. J. M. BONANTS, A. SPEKSNIJDER, M. M. KLERKS, P. H. J. F. VAN DER BOORGERT., C. WAALWICK, J. M. VAN DER WOLF and C. ZÜLSTRA (2005) Novel molecular and biochemical techniques for quality control and monitoring in the agrofood production chain. In A. van Amerongen, D. Barug and M. Lauwaars (eds) *Rapid methods for biological and chemical contaminants in food and feed.* Wageningen Academic Publishers, Wageningen, The Netherlands, pp. 151–176.

STEIN, H. H. and G. C. SHURSON (2009) Board-invited review: the use and application of distillers dried grains with solubles in swine diets. *J Anim Sci* **87**: 1292–1303.

UNGER, W. (2006) The benefits of feedmill automation. *Feed processing and quality*

© Woodhead Publishing Limited, 2011

control, ASA Technical Report Series. www.asasea.com/download_doc.php?file=ASA-TR-Processing.pdf

VAN DER AAR, P. J. and J. DOPPENBERG (2009) *Biofuels: Consequences for feed formulation*, Schothorst Feed Research. http://www.eaap.org/Barcelona/Papers/published/03_van%20der%20Aar%20.pdf

VAN RAAMSDONK, L. W. D., J. C. H. V. EIJKEREN, G. A. L. MEIJER, M. RENNEN, M. J. ZEILMAKER, L. A. P. HOOGENBOOM and M. MENGELERS (2009) Compliance of feed limits, does not mean compliance of food limits. *Biotechnol. Agron. Soc. Environ.* **13**: 51–57.

WALES, A. D, V. M. ALLEN and R. H. DAVIES (2010) Chemical treatment of animal feed and water for the control of Salmonella. *Foodborne Pathogens and Disease* **7(1)**: 3–15.

WESCHE A. M., J. B. GURTHLER, B. P. MARKS and E. T. RYSER (2009) Stress, sublethal injury, resuscitation, and virulence of bacterial foodborne pathogens. *J. Food Prot.* **72(5)**: 1121–1138.

WIDYARATNE, G. P. (2005) Characterization and improvement of the nutritional value of ethanol by-products for swine. A thesis submitted in partial fulfilment of the requirements for the degree of Master of Science in the Department of Animal and Poultry Science University of Saskatchewan Saskatoon, Saskatchewan, Canada.

WU, F. and G. P. MUNKVOLD (2008) Mycotoxins in ethanol co-products: modeling economic impacts on the livestock industry and management strategies. (Biofuels and biobased products.). *J. Agric. Food Chem.* **56(11)**: 3900–3911.

YAKKUNDI, S., A. CANNAVAN, P. B. YOUNG, C. T. ELLIOTT and D. G. KENNEDY (2002) Halofuginone contamination in feeds as a cause of residues in eggs. *Analyt. Chim. Acta* **473**: 177–182.

ZHANG, X. L., V. T. JEZA and Q. PAN (2008) *Salmonella typhi*: from a human pathogen to a vaccine vector. *Cellular and Molecular Immunology* **5**: 91–97.

© Woodhead Publishing Limited, 2011

10

Improving microbial safety in the beef production chain

C. M. Burgess and G. Duffy, Teagasc, Ireland

Abstract: Foodborne disease of zoonotic origin is a considerable public health concern worldwide. Animals may host human pathogens in their gut which may then be excreted in faeces, resulting in direct or indirect contamination. Foods of animal origin, including beef and beef products, are therefore a major source of foodborne illness and are targets for interventions to minimise pathogen exposure. This chapter reviews the main pathogens associated with beef and the control measures which can be taken at each step in the beef production chain to improve microbial beef safety.

Key words: *E. coli* O157, *Salmonella, Listeria monocytogenes, Campylobacter,* beef.

10.1 Introduction

Bacterial foodborne zoonotic diseases are of major public health concern worldwide, as well as leading to significant economic losses for the agri-food sector. In the United States alone, foodborne illness from pathogens is responsible for 76 million cases of illness each year (Mead *et al.*, 1999), with annual cost of infection from the four most common bacterial foodborne agents being estimated at almost seven billion dollars (Allos *et al.*, 2004). Animals can harbour many pathogens in their gut, which are shed in faecal material, posing a source of direct and indirect contamination. Thus, foods of animal origin and especially meat are a major vehicle in the transmission of foodborne zoonotic pathogens (Adak *et al.*, 2005; Gillespie *et al.*, 2005; Mead *et al.*, 2006; CDC, 2006; Gallay *et al.*, 2008). Indeed, there is a high level of consumer concern about meat safety, and in particular about beef safety, resulting from many high-

© Woodhead Publishing Limited, 2011

profile beef safety scares such as bovine spongiform encephalopathy and *E. coli* O157 outbreaks (Sayed *et al.*, 2007), as well as concerns about the long-term health effects from exposure to residues of veterinary drugs and environmental toxins through the beef supply chain (Wilson *et al.*, 2003). The level of microbial illness which is attributable to beef varies widely across countries, depending on epidemiological factors, the per capita consumption of beef products consumed and the cooking and consumption habits of the country (Rhoades *et al.*, 2009). The beef production chain is quite complex and involves a number of steps. Measures to minimise pathogen transfer can be taken at different points. This chapter reviews the main pathogens associated with beef and the control measures which can be taken at each step in the chain to improve microbial beef safety.

10.2 Beef production

A diagram illustrating the main points of the beef production chain is outlined in Fig. 10.1.

Beef encompasses meat from bovine adult or near adult animals, while the meat from calves is referred to as veal. Animals are generally raised specifically for beef or veal production within certain herds. Dairy cows may also enter the beef production chain when they are no longer required for milk production. Different systems exist for rearing animals prior to slaughter. In countries such as the United States the feedlot system is used extensively where animals are kept in large pens and fed a controlled, mainly grain-based diet (Huntington, 1997). Other major beef-producing nations such as Ireland and Brazil use pasture-based systems for beef production (Crosson *et al.*, 2006; Ferraz and Felício, 2010). The production system and age of the animal can have a big impact on the carriage and shedding of pathogens by cattle (Rhoades *et al.*, 2009).

10.3 Pathogens associated with beef

There are many zoonotic pathogens associated with beef. The most predominantly associated are briefly described here.

10.3.1 *Escherichia coli* O157

E. coli O157 is a member of verocytotoxigenic *E. coli* (VTEC) which can cause a potentially fatal human illness, where symptoms include diarrhoea, haemorrhagic colitis and haemolytic uraemic syndrome. While the incidence of VTEC is low in comparison to other foodborne pathogens, 0.7 per 100 000 in the EU in 2008 (EFSA, 2010), the major disease burden lies in the low infectious dose and the severity of illness, which can arise from infection. While many

© Woodhead Publishing Limited, 2011

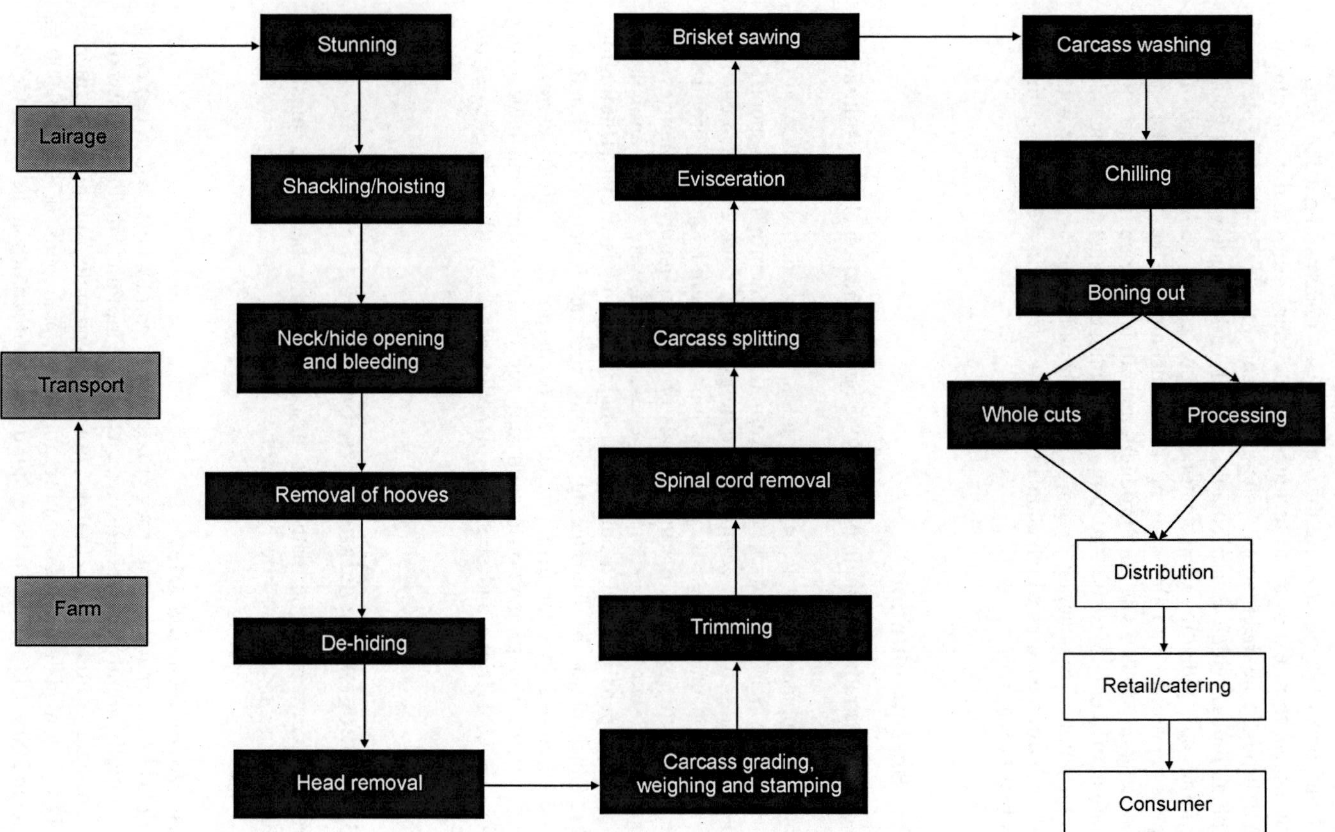

Fig. 10.1 The beef production chain.

© Woodhead Publishing Limited, 2011

VTEC serotypes have been associated with human illness, the majority of reported outbreaks of VTEC infection in humans have been linked to *E. coli* O157:H7. A number of factors contribute to the pathogenicity of *E. coli* O157, including the ability to produce one or both verocytotoxins (VT1 and VT2). Not all VTEC are equally pathogenic for humans, and those VTEC which are of most concern generally have the ability to adhere to and colonise the human large intestine forming attaching and effacing lesions (encoded by the locus of enterocyte effacement (LEE)). Other effectors thought to be involved in pathogenesis are located outside the LEE (Karmali *et al.*, 2010). Ruminant animals including cattle are the main reservoir of *E. coli* O157 and foods of bovine origin, especially undercooked beef products and unpasteurised milk, are major sources of human infection with bovine-derived products linked to approximately 75% of *E. coli* O157 outbreaks (Callaway *et al.*, 2009).

10.3.2 *Salmonella* spp.

Of the foodborne pathogens, *Salmonella* had the highest incidence of laboratory-confirmed infections in ten states of the US in 2009 (CDC, 2010). A similar trend has been witnessed in Europe with it being the most frequently reported cause of foodborne outbreaks (EFSA, 2010). *Salmonella* is divided into two species: *S. enterica* and *S. bongori*, with the former further divided into six subspecies. Most *Salmonella* belongs to the subspecies *S. enterica* subsp. *enterica* and over 2500 serovars have been identified. The most predominant serovars associated with the disease are region dependent with *S.* Typhimurium being the predominant serovar in the US, while in Europe it is *S.* Enteriditis and in many parts of Asia it is *S.* Choleraesuis (Foley *et al.*, 2008). Salmonellosis usually causes self-limiting gastroenteritis but occasionally infection may be more serious and potentially life threatening. The most common reservoir of *Salmonella* is the gastrointestinal tract of poultry, and pigs have also been shown to be frequent carriers (Davies *et al.*, 2004; Duffy *et al.*, 2010). However, beef also plays a role with beef products being linked to outbreaks of salmonellosis (Smelser, 2004). The most common serovars found in cattle in Europe are *S.* Typhimurium and *S.* Dublin (EFSA, 2010). *S.* Typhimurium is one of the most common *Salmonella* serovars associated with human infection and, therefore, beef can be an important vector for transmission of this infectious agent.

10.3.3 *Listeria monocytogenes*

L. monocytogenes causes wide-ranging symptoms, from mild influenza-like symptoms to meningitis, septicaemia and in pregnant women may lead to foetal infection and/or spontaneous abortion. While the disease rate is relatively low (0.3 cases per 100 000 in the EU in 2008; EFSA, 2010), the public health concern arises from the severity of the disease and its high fatality rate (17.5%; CDC, 2010). *L. monocytogenes* is ubiquitous in the environment and is a particular problem in ready-to-eat foods due to its persistence in food processing

© Woodhead Publishing Limited, 2011

environments and its ability to grow at refrigeration temperatures. Beef has been linked to outbreaks of listeriosis with some cases resulting in death (Smith *et al.*, 2010).

10.3.4 *Campylobacter* spp.

Campylobacter spp. are the number one cause of gastrointestinal illness of foodborne origin in many countries in the developed world with a prevalence rate of 40.7 per 100 000 in the EU in 2008 (EFSA, 2010). Campylobacteriosis is a self-limiting disease with symptoms including fever, vomiting, headaches, watery or bloody diarrhoea and abdominal pain and has also been identified as a risk factor in the development of inflammatory bowel disease (Garcia Rodriguez *et al.*, 2006). *Campylobacter* is commonly isolated from poultry and pigs, with poultry being generally recognised as the primary source of infection (Humphrey *et al.*, 2007). However, *Campylobacter* has recently been shown to have a high rate of carriage in cattle (Chatre *et al.*, 2010), and the consumption of undercooked beef has been shown to be a significant risk factor for *Campylobacter* infection (Gallay *et al.*, 2008). Typing of *Campylobacter* from different sources has shown clustering of isolates by host source (Miller *et al.*, 2006; Pittenger *et al.*, 2009). However, a recent study of Irish *C. jejuni* isolates from poultry, beef and clinical sources has shown clinical isolates to cluster with isolates from both chicken and beef, highlighting the potential role for beef in the transmission of this pathogen to humans (Burgess *et al.*, 2009).

10.4 Primary production

10.4.1 Pathogens in the farm environment

When pathogens are shed in cattle faeces they can persist in the underlying soil and grass, surviving for extended periods ranging from several weeks to many months (Duffy, 2003; Semenov *et al.*, 2009). This provides an important transmission route for pathogens within herds, farms, the fresh food chain, water courses and the wider environment. The prevalence of *E.coli* O157 in calf and cattle faeces ranges from 0 to 60% (Rhoades *et al.*, 2009). Cattle typically shed *E. coli* O157 at levels of less than 10^2 CFU/g of faeces but some cattle can shed the organism at levels greater than 10^4 CFU/g of faeces (Jacob *et al.*, 2010). High concentrations of *E. coli* O157 in bovine faeces may result from colonisation in the terminal rectum (Naylor *et al.*, 2005). *Salmonella* has been shown to colonise the bovine intestine (Morgan *et al.*, 2007) and the prevalence in cattle faeces ranges from 0 to 62% (Rhoades *et al.*, 2009). In cattle, *C. jejuni* colonises the proximal small intestine and may be found throughout the intestinal tract (Inglis *et al.*, 2005). Prevalence in cattle faeces ranges widely with studies finding from 16.5 to 94% (Chatre *et al.*, 2010; Krueger *et al.*, 2008). Few studies have examined the prevalence of *L. monocytogenes* in cattle faeces but those that have been have undertaken have shown it to range from 4.8 to 29.4%

© Woodhead Publishing Limited, 2011

(Nightingale *et al.*, 2004; Vilar *et al.*, 2010; Madden *et al.*, 2006; Fox *et al.*, 2009). Greater information would be required if quantitative risk assessments are to be undertaken at this primary production stage of the chain.

10.4.2 Control at primary production

Intervention strategies at primary production can be divided into broad categories of farm management practices which reduce bovine exposure to pathogens in the environment, and targeted intervention strategies such as vaccination and antimicrobial treatments, which prevent or limit pathogen colonisation and carriage (LeJeune and Wetzel, 2007).

Farm management practices

Biosecurity incorporates management procedures which are put in place on farms to prevent the introduction of an infectious agent to a herd, while biocontainment is the series of management procedures that help prevent the spread of an infectious agent within a herd or in products which leave the farm. General measures to control the spread of pathogens on the farm are outlined in Farm Quality Assurance Schemes and Good Agricultural Practices (Baghasa, 2008). These guidelines cover general issues such as management and housing of stock, pest control, management of feed and water supply, management of animal waste and disease prevention and control. The introduction of full on-farm HACCP systems which attempt to identify, monitor and control all hazards on the farm is hugely challenging due to the diverse nature of the farm environment. However, on-farm HACCP has been shown to work when it is kept very simple in terms of the key elements to control, the corrective actions to take and the records to keep. Good farm hygiene has been shown to reduce occurrence of *L. monocytogenes* (Fox *et al.*, 2009; Nightingale *et al.*, 2005) and *E. coli* O157 (Ellis-Iversen *et al.*, 2008).

10.4.3 Targeted interventions

A major difficulty with targeted treatments to reduce pathogen colonisation and carriage is that many animals are asymptomatic carriers of the pathogen and only shed them sporadically. Equally there can be a large variation in the species, serogroups and subtypes of a particular pathogen carried by cattle, thereby requiring a broad acting agent. However, it cannot be so broad that it impacts on the rest of the microflora in the animal gut which may have implications for rumen metabolism and digestion. Any proposed intervention strategies must also be economically feasible as a treatment for all the cattle in a given herd or farm. Targeted interventions may not completely eliminate pathogens but they should significantly reduce the burden of pathogen being shed. Some of these interventions which have been proposed for the beef production industry are briefly reviewed below.

© Woodhead Publishing Limited, 2011

Vaccination
Vaccination has been investigated as a way of promoting the animal's immune system to reduce pathogen loads by producing antigens against particular pathogens. They also have the advantage that many farmers are already familiar with vaccine administration and it is relatively easy to incorporate into existing farm management. They are, for example, already widely used to control *Salmonella* in poultry (EFSA, 2007).

Vaccines which have been developed for cattle have generally targeted *E. coli* O157. One study examined using the cell binding domain of intimin or a truncated EHEC factor for adherence (Efa-1) as potential vaccines. Both induced humoral immunity in calves but did not protect against colonisation by *E. coli* O157 upon subsequent challenge (van Diemen *et al.*, 2007). A recent study showed that inoculation with a recombinant *S.* Dublin strain expressing *E. coli* O157 intimin reduced enteric colonisation and faecal shedding of *E. coli* O157 in calves, although it did not correspond with the presence of intimin-specific IgA in faeces (Khare *et al.*, 2010). Another study examined vaccines containing *E. coli* O157 type III secreted proteins and it was found that faecal shedding and duration of shedding was reduced (Potter *et al.*, 2004) but a subsequent field trial in feedlot cattle showed no significant association between pen prevalence of faecal *E. coli* O157 and vaccination (Van Donkersgoed *et al.*, 2005). Thomson *et al.* (2009) have examined the effect of a siderophore receptor and porin (SRP) proteins-based vaccine on *E. coli* O157 faecal shedding in a large feedlot trial and found reduced shedding and the prevalence to be lower in vaccinated animals.

Vaccines against *Salmonella* have shown that a DNA adenine methylase deficient *S.* Typhimurium vaccine provided cross protection against *S.* Newport and *S.* Dublin with vaccinated calves exhibiting reduced faecal shedding and colonisation (Mohler *et al.*, 2006, 2008). Conversely, a study using a commercially available *S.* enterica subunit vaccine showed no evidence of reduced shedding of *S.* enterica in cows (Heider *et al.*, 2008). Such studies clearly demonstrate that while vaccination has clear potential many hurdles remain to be overcome.

Some studies have examined anti-*Campylobacter fetus* vaccines in bovine herds, but this focuses on its role as a sexually transmitted disease which may affect fertility as opposed to food safety (Cobo *et al.*, 2004). To the authors' knowledge, no studies on using vaccination as an intervention for the control of *Listeria* in cattle have been published.

Direct fed microbials
This intervention covers both probiotics and competitive exclusion. Direct fed microbials (DFM) have been shown to be beneficial for many different aspects of beef production such as increased daily gain and improved efficiency (Krehbiel *et al.*, 2003), as well as potential for food safety. DFM are believed to function by a number of mechanisms such as competitive exclusion, immune modulation or bactericidal activity by the production of particular metabolites

© Woodhead Publishing Limited, 2011

(Krehbiel *et al.*, 2003). Like other intervention strategies in cattle, the research has focused mainly on *E. coli* O157 and less so on *Salmonella*. One study examined the use of a *Bacillus subtilis* DFM but showed no significant difference in faecal or hide prevalence of *E. coli* O157 or on the number of animals shedding the pathogen at high levels (Arthur *et al.*, 2010). A number of studies have focused on *Lactobacillus acidophilus*. A study using a DFM comprised of *Lb. acidophilus* (LA51) and *Propionibacterium freudenreichii* (PF 24) showed a significant reduction in faecal shedding of *E. coli* O157 but not *Salmonella* in naturally infected feedlot cattle (Tabe *et al.*, 2008). Another *Lb. acidophilus* study showed significant reduction of both *E. coli* O157 and *Salmonella* on hide (Stephens *et al.*, 2007). Selection of strains for inclusion in a DFM appears to be important, as highlighted when the beneficial effect of NP51 on the faecal prevalence of *E. coli* O157 was reduced when fed with another strain of *Lactobacillus* (Younts-Dahl *et al.*, 2004). Like vaccination studies, DFM trials have mixed results regarding pathogen reduction which may be as a result of study design or mode of action but promising results warrant further investigation.

Bacteriophage

Bacteriophages are viruses that infect and kill bacterial cells by reproducing in the bacteria, disrupting metabolic pathways and causing cell lysis. They specifically target bacterial cells and not animal cells and therefore, have potential as a pathogen reduction strategy. Because some bacteriophage have a narrow host range they may be potentially useful for the selective elimination of pathogens in a mixed microbial population such as that in the bovine rumen.

Like other intervention strategies in beef the work has predominantly focused on *E. coli* O157. One study targeting *E. coli* O157 applied two bacteriophages orally and rectally to sheep and cattle, respectively. While no difference in intestinal numbers of *E. coli* O157 was observed when administered orally, the prevalence was reduced when applied rectally in cattle although the infection was not cleared (Sheng *et al.*, 2006). A more recent study used encapsulation for delivery of anti *E. coli* O157 bacteriophage and while the delivery was found to be successful, it did not reduce shedding in comparison to controls (Stanford *et al.*, 2010). Rozema *et al.* (2009) also saw no significant difference in the number of *E. coli* O157 positive samples among treatments applied both orally and rectally with control animals. Conversely, a study undertaken in calves using oral inoculation of anti *E. coli* O157 bacteriophage was shown to reduce the duration of shedding of the pathogen. A commercial bacteriophage produced by OmniLytics Inc has been licensed in the US for hide administration to minimise contamination by *E. coli* O157.

The use of bacteriophage for reducing pathogen numbers in the beef production chain has been investigated as a strategy for a wider range of pathogens. A three bacteriophage cocktail was used on inoculated meat samples and was shown to reduce *E. coli* O157 numbers from 3 $\log_{10}$ CFU/ml to undetectable levels after two hours (O'Flynn *et al.*, 2004). The combined use of nisin and a

© Woodhead Publishing Limited, 2011

listeriophage was found to be successful in reducing cell numbers of *L. monocytogenes* and preventing regrowth in a broth system, but this combined effect was not evident when examined on vacuum packed fresh beef (Dykes and Moorhead, 2002). A study which examined the effect of two bacteriophages on *Samonella* and *C. jejuni* on cooked and raw meat showed significant reduction in pathogen numbers, in particular when there was a high host cell density. However, the authors point out that some of the experimental conditions were not particularly realistic for application to foods (Bigwood *et al.*, 2008). While the use of bacteriophage is promising there is no doubt that further research is required on areas such as appropriate selection of bacteriophages, modes of administration, maintaining bacteriophage viability and the emergence of bacteriophage resistant bacteria (Johnson *et al.*, 2008).

Bacteriocins and antimicrobial peptides

Bacteriocins are ribosomally synthesised antimicrobial peptides or proteins produced by bacteria that kill or inhibit the growth of other bacteria, either in the same species (narrow spectrum), or across genera (broad spectrum) (Cotter *et al.*, 2005). Bacteriocins can be administered to animals by mixing dried or wet cultures with feed or drinking water.

A number of studies have looked at the potential role of colicins (bacteriocins produced by *E. coli*) to reduce *E. coli* O157 in cattle populations (Diez-Gonzalez, 2007; Schamberger *et al.*, 2004). In calves fed with a colicin-producing *E. coli*, levels of *E. coli* O157:H7 were reduced by 1.8 $\log_{10}$ CFU/g over 24 days (Schamberger *et al.*, 2004). A controlled approach based on the expression of colicins by bacteria which are normally present in the animal rumen would decrease the chances of transfering colicin genes to other potentially pathogenic *E. coli* strains, such as *E. coli* O157:H7 (McCormick *et al.*, 1999). Coinoculation of an *E. coli* strain which possessed bacteriocin like activity was shown to reduce shedding of *Salmonella* in experimentally inoculated calves (Patton *et al.*, 2009).

Bacteriocins have also been employed later in the production chain with their application on beef and beef products. They may be incorporated into the food as an ingredient in a purified/semi purified form or as a protective culture where the bacteriocin producing bacteria produce the bacteriocin *in situ* in the food. The majority of studies investigating bacteriocins in meat products have focused on *L. monocytogenes*, due to its ability to grow at refrigeration temperatures. One study used different bacteriocins either alone or in combination in a meat slurry and showed reductions in *L. monocytogenes* of up to $\log_{10}$ 5.2 CFU/g when used singly and $\log_{10}$ 9.9 CFU/g when used in combination (Vignolo *et al.*, 2000). Azuma *et al.* (2006) showed that while piscicocin did not completely eliminate *L. monocytogenes* in a minced beef and pork mixture it did reduce the numbers and could possibly be used in combination with other treatments. Another study examined the use of various lactic acid bacteria and some were shown to reduce the recovery of *C. jejuni* from vacuum packaged beef (Jones *et al.*, 2009). The authors of this study emphasised the importance of demonstration of

© Woodhead Publishing Limited, 2011

antimicrobial activity in the environment proposed for application. One of the most commonly used bacteriocins in food is nisin which has US Food and Drug Administration approval. Although more commonly used in dairy products, it has also been employed in beef studies (Ariyapitipun *et al.*, 2000; Barboza De Martinez *et al.*, 2002; Zhang and Mustapha, 1999).

10.5 Slaughter and processing

Pathogens present in intestines, stomach contents, oral cavity or oesophagus of cattle may be transferred to carcasses during slaughter and dressing operations in particular at the eviseration stage. It is also generally accepted that the amount of faecal material adherent to cattle hides significantly influences the levels of microbial contamination on derived carcasses. A number of studies on the prevalence of *E. coli* O157 have shown clear evidence that hides are the primary source of contamination on beef carcasses (Koohmaraie *et al.*, 2005; Nou *et al.*, 2003; Barkocy-Gallagher *et al.*, 2003). The prevalence of *E. coli* O157 varies widely with some studies showing a prevalence as low as 4.7% (Elder *et al.*, 2000) while another study reported a prevalence of 75.7% on hide of feedlot cattle (Arthur *et al.*, 2004). Factors which may influence prevalence include the sampling site choice (Reid *et al.*, 2002; Kalchayanand *et al.*, 2009), feeding systems and season (Rhoades *et al.*, 2009). The concentration of *E. coli* O157 on the hides is typically around 100 CFU/cm^2 (Arthur *et al.*, 2004; O'Brien *et al.*, 2005).

A seasonal variation has also been shown in relation to *Salmonella* prevalence on hides varying from 27.7% in the winter to 97.7% in the autumn (Barkocy-Gallagher *et al.*, 2003). Broadly, it has been observed that *Salmonella* prevalence on hides is comparable with that of *E. coli* O157 (Rhoades *et al.*, 2009). Very few studies are available on the prevalence of *Listeria* and *Campylobacter* on bovine hide. Rivera-Betancourt *et al.* (2004) reported the prevalence of *Listeria* spp. on hides to range from 37.7% in one plant to 75.5% in another and the prevalence of *L. monocytogenes* to range from 0.8 to 18.7%. A study on the effect of transport on pathogen prevalence showed that for *Campylobacter* the hide contamination level decreased from 25 to 13% for feedlot cattle but remained unchanged for adult animals at 1 to 2% (Beach *et al.*, 2002).

10.5.1 Control measures

The risk of faecal contamination on the carcass at slaughter can be reduced by specific procedures including 'rodding' (a technique used to separate the oesophagus from the trachea and diaphragm). Bagging and tying of the bung can also help prevent contamination of the carcass. Removal of hides should be carried out in a manner that avoids contact between the hide and the carcass. This can be achieved by a number of measures including the use of hide pulling

© Woodhead Publishing Limited, 2011

equipment and using clean equipment (immersion of knives in water at 82 °C) for the dehiding operation.

Carcass chilling has been reported to reduce microbial contamination on carcasses. McEvoy *et al.* (2004) reported a reduction in prevalence on carcasses after chilling for 24 hours. Similarly, Gill *et al.* (1996) reported a reduction in coliforms and *E. coli* of between 0.5 $\log_{10}$ units and 2 $\log_{10}$ units on carcasses following cooling processes. Sheridan (2000) noted that carcass contamination may increase, decrease or remain unchanged following chilling and depended on parameters such as temperature, air speed and relative humidity.

The prevalence rates for *Salmonella* on pork carcasses have been reported to drop from 15.2% on pre-chill carcass to 3.1% on post chill carcasses (Duggan *et al.*, 2010). However, it should be noted that there may be methodology difficulties with recovery of bacteria from chilled carcasses as the bacteria may be sub-lethally injured by the combination of chilling and drying rendering them non-culturable and thus underestimating the number of bacteria on the chilled carcass. Specific interventions as outlined below may also be applied where legislation permits.

Many different hide and carcass decontamination strategies have been trialled with varying degrees of success. These studies often examine general microflora as a measure of carcass hygiene, though some studies have also examined the effect of interventions on the prevalence of particular pathogens, such as *E. coli* O157. Some of the hide decontamination strategies which have been trialled are summarised in Table 10.1. Carcass decontamination treatments which may be used, including washes, antimicrobial treatments, trimming, vacuuming and pasteurisation, have been reviewed elsewhere (Gill, 2009) and will not be discussed further here.

Organic acids

Organic acid washes are the most frequently used decontaminants in the meat processing industry and acids such as lactic and acetic acid are widely used in the USA to decontaminate carcasses at concentrations of up to 2.5% (USDA-FSIS, 2004). Like many pathogen reduction interventions, the results achieved have varied between studies. Some studies in beef have shown that organic acid rinses show little (less than one $\log_{10}$ CFU reduction) or no pathogen reduction (Brackett *et al.*, 1994; Fratamico *et al.*, 1996; Podolak *et al.*, 1996). Other studies on beef have shown pathogen reduction of greater than 1 $\log_{10}$ CFU reduction. Greer and Dilts (1992) reported a 1 to 1.2 $\log_{10}$ CFU reduction in numbers of *L. monocytogenes* and *C. jejuni* following use of organic acids. *E. coli* O157 levels have been reportedly reduced by 0.9 to 1.1 $\log_{10}$ CFU (Heller *et al.*, 2007) and up to 1.84 $\log_{10}$ CFU (Raftari *et al.*, 2009). Acidified sodium chlorite, which is a combination of any acid generally regarded as safe and sodium chlorite in aqueous solution, is approved by the US Department of Agriculture as a direct food additive for decontamination of red meat carcasses (CFR, 1998). This treatment has been shown to reduce *L. monocytogenes* numbers by greater than 1 $\log_{10}$ CFU/g and has a greater effect when used after a

© Woodhead Publishing Limited, 2011

Table 10.1 Examples of hide decontamination strategies in the beef chain

Strategy	Target pathogens	Reduction (log_{10} CFU/% prevalence*)	Reference
Preslaughter wash (3 min)	*E. coli* O157:H7	1.92	Byrne *et al.* (2000)
Preslaughter wash	*S.* Typhimurium		Mies *et al.* (2004)
with lactic acid (2–6%)		1.3–5.1	
with acetic acid (2–6%)		2.4–4.8	
with chlorine (100-400 ppm)		0.6–1.3	
with ethanol (70–90%)		5.0–5.5	
Steam condensing of hide at subatmospheric pressure	*E. coli* O157:H7	1.94–5.99	McEvoy *et al.* (2001)
Chemical dehairing of hide	*E. coli* O157:H7, *S.* Typhimurium	>4.6	Castillo *et al.* (1998)
Chemical dehairing of hide	*E. coli* O157:H7	49*	Nou *et al.* (2003)
Ozonated water hide wash	*E. coli* O157:H7	58*	Bosilevac *et al.* (2005b)
Electrolysed oxidising water hide wash		47*	
Online hide cabinet with sodium hydroxide wash and a chlorinated water rinse	*E. coli* O157:H7	27*	Bosilevac *et al.* (2005a)
Water wash followed by cetylpyridinium chloride treatment (1%)	*E. coli* O157:H7	22* (hide) 20* (carcass)	Bosilevac *et al.* (2004)
Microbial immobilisation on hair using Shellac in ethanol	*E. coli* O157:H7	72.7*	Antic *et al.* (2010)
Commercial bacteriophage	*E. coli* O157:H7 *Salmonella*	Not available	OmniLytics, Inc.

© Woodhead Publishing Limited, 2011

hot water wash (Ozdemir *et al.*, 2006). Similar results were obtained when it was tested against *E. coli* O157 and *S.* Typhimurium with 1.5–2.5 $\log_{10}$ CFU/g reductions reported in beef trim and ground beef (Harris *et al.*, 2006).

10.6 Storage and distribution

It is well recognised that pathogens can persist and even proliferate during storage and distribution depending on the temperature and other environmental conditions. Numerous studies have examined the prevalence of *E. coli* O157 in retail minced beef and raw beef burgers internationally. In Europe prevalence in different countries has ranged from 0% to 2.8% (Chapman *et al.*, 2000, 2001; Cagney *et al.*, 2004; Vernozy-Rozand *et al.*, 2002; Tutenel *et al.*, 2003; Conedera *et al.*, 2004; Dontorou *et al.*, 2003). Prevalence studies in retail ground beef in North America have detected VTEC in 9 to 36.4% of samples, with *E. coli* O157 in 0 to 3.7% of samples tested (Doyle and Schoeni, 1987; Read *et al.*, 1990; Samadpour *et al.*, 2006; Tarr *et al.*, 1999). Similarly, in South America *E. coli* O157 prevalence in minced beef was found to range from 0 to 3.8% (Silveira *et al.*, 1999; Chinen *et al.*, 2001). *Salmonella* has been detected in minced beef at a prevalence of 0 to 4.2% in studies conducted internationally (Ghafir *et al.*, 2005; Rose *et al.*, 2002; Samadpour *et al.*, 2006; Wong *et al.*, 2007). Prevalence of *L. monocytogenes* was found to vary widely at this stage in the beef production chain with levels internationally ranging from 3.5 to 52% (Fantelli and Stephan, 2001; Samadpour *et al.*, 2006; Bohaychuk *et al.*, 2006). The prevalence of *Campylobacter* spp. in retail beef has been found to range from 2.4 to 14.1% (Sammarco *et al.*, 2010; Rahimi *et al.*, 2010; Medeiros *et al.*, 2008; Little *et al.*, 2008; Whyte *et al.*, 2004).

10.6.1 Control measures

Similarly to earlier steps in the beef production chain, measures can be taken to minimise pathogen contamination of products or proliferation during storage of the product. Key elements to this include maintaining the appropriate temperature conditions during transport and storage and preventing contamination from external sources by storing and covering appropriately. Good personal hygiene also plays a key role.

Consumers are increasingly demanding minimally processed food with lower amounts of chemical additives. This may have a negative impact on the safety of foods and therefore, there is a drive to find more natural agents for ensuring food safety. As mentioned earlier, some anti-pathogen interventions such as the use of bacteriophage and bacteriocins have been employed in beef products with varying success. Essential oils, obtained from plant materials, are often used in foods as flavourings but their antimicrobial properties have also attracted attention. Eugenol was shown to reduce *L. monocytogenes* numbers by 0.6–1.77 $\log_{10}$ CFU/g in cooked beef (Hao *et al.*, 1998), while reduction of *Salmonella*

© Woodhead Publishing Limited, 2011

and *L. monocytogenes* was in the range of 1–3 $\log_{10}$ CFU/g when oregano oil was used on beef fillets (Skandamis *et al.*, 2002; Tsigarida *et al.*, 2000).

Packaging systems

Packaging plays an important role, both in minimising microbial proliferation during storage and also in preventing contamination of the product. Various forms of flexible and rigid packaging exist. Rigid packaging included glass, metal containers and rigid plastic and these may be used either alone or in combination with other preservation methods to ensure the safety of products. Flexible packaging includes controlled atmosphere packaging, vacuum packaging, modified atmosphere packaging, active packaging and edible films (Cutter, 2002). The range of packaging options and their benefits are too broad to consider here but some examples will be given. Most studies consider the overall microbiological profile and the effect on shelf life as opposed to that of specific pathogens when examining packaging options. However, some studies have specifically examined pathogen survival in beef products following packaging. Edible chitosan films have been shown to control *L. monocytogenes* on ready-to-eat roast beef (Beverly *et al.*, 2008). Another study which looked at alginate-based films incorporating essential oils for preservation of whole beef muscle saw significant reductions in *S.* Typhimurium and *E. coli* O157 (Oussalah *et al.*, 2006). An active polythene film which contained an anti-listerial bacteriocin was shown to reduce *L. monocytogenes* numbers in ground beef by about one $\log_{10}$ CFU/g (Mauriello *et al.*, 2004). Modified atmosphere packaging was also been shown to be beneficial in reducing numbers of *E. coli* O157 and *Salmonella* in comparison to controls following storage (Brooks *et al.*, 2008). Another study which examined the use of vacuum packaging and 100% carbon dioxide found that there was no reduction in the numbers of *Salmonella* and *E. coli* O157 on beef primal cuts during storage, which contradicted earlier studies (Dykes *et al.*, 2001).

A key contributor to the spoilage of beef is a breakdown in the chill chain during distribution. Spoilage prediction and the application of an optimised quality and safety assurance scheme for chilled storage and distribution of beef and beef products would be greatly aided by the continuous monitoring of temperature during distribution and storage. Time temperature indicators (TTIs) can provide a solution as they can be used to record and translate, at a unit level, the overall effect of the temperature history of a food product on its quality and safety (Ellouze and Augustin, 2010). An effective TTI must indicate a clear, continuous and irreversible reaction in response to changes in temperature. TTIs are devices which contain a thermally labile substance which can be biological (microbiological or enzymatic), chemical or physical. Biological TTIs are the best studied. Studies have demonstrated the use of TTIs as quality and safety indicators in beef. A study undertaken by Ellouze and Augustin (2010) showed that a microbiological TTI reduced consumer exposure to high concentrations of indigenous microflora in beef. Another study in modified atmosphere packaged beef also demonstrated that the end point of a TTI based on the growth and

© Woodhead Publishing Limited, 2011

metabolic activity of a *Lb. sakei* strain was very close to the end of the product's sensorial shelf life (Vaikousi *et al.*, 2009). Such studies show the potential use of TTIs for monitoring the quality status of beef during distribution and storage and their use as a management tool.

Thermal inactivation

The person in charge of food preparation, whether it is in the home, a restaurant or catering facility, is the last line of defence when it comes to preventing foodborne infection. One of the easiest ways to ensure that beef products are pathogen free is thorough cooking of the products. Thermal inactivation of pathogens varies with the food matrix and product constituents, for example *Salmonella* has been shown to have a D value of 2.7 and 1.2 minutes at 60 ºC in whole beef muscle and ground beef samples, respectively (Mogollón *et al.*, 2009). It is therefore necessary to establish these values for individual product and pathogen combinations before guidance on cooking can be given. Following cooking guidelines, such as those provided by health promotion and regulatory agencies worldwide, can help prevent viable pathogens being consumed. This underlines the key role that education, both of the consumer and those working in food processing and food preparation, plays in preventing foodborne infectious disease.

10.7 Conclusions

Undoubtedly, great progress has been made in recent years in improving the safety of beef and beef products. Interventions at all stages of the beef production chain which have been found to minimise the pathogen load of the final product and increased consumer awareness of the importance of appropriate hygiene and storage measures will together help minimise the role played by beef in the transmission of key foodborne pathogens. Sound scientific data of the pathogen load at all stages of the chain allows for quantitative risk assessments to be undertaken and identification of key control areas for the different pathogens. Nonetheless, the human and economic losses associated with foodborne illness continue to rise, indicating that lapses in procedure still occur and that bacteria possess the ability to adapt to evade the interventions put in place. The emergence of pathogens with greater resistance to food processing stresses is of great concern as it increases their likelihood of reaching the consumer. This is a situation which needs to be constantly monitored. Furthermore, while many interventions have shown some promise, there remains much variation between studies, and many interventions have been investigated in very limited trials. In order to increase confidence in such interventions larger trials, in real conditions, need to be undertaken in a standardised manner. Any successful anti-pathogen intervention will have to be economical and easy to use without having any other deleterious effects on the final product. It is vital that a whole chain approach is taken starting at primary production right through to

© Woodhead Publishing Limited, 2011

consumer. Steps taken at each sector will help to contribute to greater enhancement of the microbial safety of beef and beef products. It is an area where research will without doubt continue to focus on.

10.8 References

ADAK, G. K., S. M. MEAKINS, H. YIP, B. A. LOPMAN and S. J. O'BRIEN (2005) Disease risks from foods, England and Wales, 1996–2000. *Emerg Infect Dis* **11**: 365–372.

ALLOS, B. M., M. R. MOORE, P. M. GRIFFIN and R. V. TAUXE (2004) Surveillance for sporadic foodborne disease in the 21st century: the FoodNet perspective. *Clin Infect Dis* **38 Suppl 3**: S115–120.

ANTIC, D., B. BLAGOJEVIC, M. DUCIC, R. MITROVIC, I. NASTASIJEVIC and S. BUNCIC (2010) Treatment of cattle hides with Shellac-in-ethanol solution to reduce bacterial transferability – a preliminary study. *Meat Sci* **85**: 77–81.

ARIYAPITIPUN, T., A. MUSTAPHA and A. D. CLARKE (2000) Survival of *Listeria monocytogenes* Scott A on vacuum-packaged raw beef treated with polylactic acid, lactic acid and nisin. *J Food Prot* **63**: 131–136.

ARTHUR, T. M., J. M. BOSILEVAC, X. NOU, S. D. SHACKELFORD, T. L. WHEELER, M. P. KENT, D. JARONI, B. PAULING, D. M. ALLEN and M. KOOHMARAIE (2004) *Escherichia coli* O157 prevalence and enumeration of aerobic bacteria, *Enterobacteriaceae*, and *Escherichia coli* O157 at various steps in commercial beef processing plants. *J Food Prot* **67**: 658–665.

ARTHUR, T. M., J. M. BOSILEVAC, N. KALCHAYANAND, J. E. WELLS, S. D. SHACKELFORD, T. L. WHEELER and M. KOOHMARAIE (2010) Evaluation of a direct-fed microbial product effect on the prevalence and load of *Escherichia coli* O157:H7 in feedlot cattle. *J Food Prot* **73**: 366–371.

AZUMA, T., D. K. BAGENDA, T. YAMAMOTO, Y. KAWAI and K. YAMAZAKI (2006) Inhibition of *Listeria monocytogenes* by freeze-dried Piscicocin CS526 fermentate in food. *Lett Appl Microbiol* **44**: 138–144.

BAGHASA, H (2008) European System Related to Good Agricultural Practice (EUREPGAP). Policy Brief No. 26. http://ageconsearch.umn.edu/bitstream/48602/2/26_pol_brf_eurepgap_hb_en.pdf

BARBOZA DE MARTINEZ, Y., K. FERRER and E. MARQUEZ SALAS (2002) Combined effects of lactic acid and nisin solution in reducing levels of microbiological contamination in red meat carcasses. *J Food Prot* **65**: 1780–1783.

BARKOCY-GALLAGHER, G. A., T. M. ARTHUR, M. RIVERA-BETANCOURT, X. NOU, S. D. SHACKELFORD, T. L. WHEELER and M. KOOHMARAIE (2003) Seasonal prevalence of Shiga toxin-producing *Escherichia coli,* including O157:H7 and non-O157 serotypes, and *Salmonella* in commercial beef processing plants. *J Food Prot* **66**: 1978–1986.

BEACH, J. C., E. A. MURANO and G. R. ACUFF (2002) Prevalence of *Salmonella* and *Campylobacter* in beef cattle from transport to slaughter. *J Food Prot* **65**: 1687–1693.

BEVERLY, R. L., M. E. JANES, W. PRINYAWIWATKUL and H. K. NO (2008) Edible chitosan films on ready-to-eat roast beef for the control of *Listeria monocytogenes*. *Food Microbiol* **25**: 534–537.

BIGWOOD, T., J. A. HUDSON, C. BILLINGTON, G. V. CAREY-SMITH and J. A. HEINEMANN (2008)

© Woodhead Publishing Limited, 2011

Phage inactivation of foodborne pathogens on cooked and raw meat. *Food Microbiology* **25**: 400–406.

BOHAYCHUK, V. M., G. E. GENSLER, R. K. KING, K. I. MANNINEN, O. SORENSEN, J. T. WU, M. E. STILES and L. M. McMULLEN (2006) Occurrence of pathogens in raw and ready-to-eat meat and poultry products collected from the retail marketplace in Edmonton, Alberta, Canada. *J Food Prot* **69**: 2176–2182.

BOSILEVAC, J. M., T. M. ARTHUR, T. L. WHEELER, S. D. SHACKELFORD, M. ROSSMAN, J. O. REAGAN and M. KOOHMARAIE (2004) Prevalence of *Escherichia coli* O157 and levels of aerobic bacteria and *Enterobacteriaceae* are reduced when hides are washed and treated with cetylpyridinium chloride at a commercial beef processing plant. *J Food Prot* **67**: 646–650.

BOSILEVAC, J. M., X. NOU, M. S. OSBORN, D. M. ALLEN and M. KOOHMARAIE (2005a) Development and evaluation of an on-line hide decontamination procedure for use in a commercial beef processing plant. *J Food Prot* **68**: 265–272.

BOSILEVAC, J. M., S. D. SHACKELFORD, D. M. BRICHTA and M. KOOHMARAIE (2005b) Efficacy of ozonated and electrolyzed oxidative waters to decontaminate hides of cattle before slaughter. *J Food Prot* **68**: 1393–1398.

BRACKETT, R. E., Y. Y. HAO and M. P. DOYLE (1994) Ineffectiveness of hot acid sprays to decontaminate *Escherichia coli* O157:H7 on beef. *J Food Prot* **57**: 198–203.

BROOKS, J. C., M. ALVARADO, T. P. STEPHENS, J. D. KELLERMEIER, A. W. TITTOR, M. F. MILLER and M. M. BRASHEARS (2008) Spoilage and safety characteristics of ground beef packaged in traditional and modified atmosphere packages. *J Food Prot* **71**: 293–301.

BURGESS, C. M., C. CAGNEY, D. McNULTY, S. HUYNH, K. SCANLAN, A. CARROLL, E. MCNAMARA, C. T. PARKER and G. DUFFY (2009) Comparative genomic indexing of Gram negative pathogens. In Conference Proceedings of: Food processing stresses and associated genomics of Gram negative foodborne bacteria. Dublin, Ireland, pp. 25–27.

BYRNE, C. M., D. J. BOLTON, J. J. SHERIDAN, D. A. McDOWELL and I. S. BLAIR (2000) The effects of preslaughter washing on the reduction of *Escherichia coli* O157:H7 transfer from cattle hides to carcasses during slaughter. *Lett Appl Microbiol* **30**: 142–145.

CAGNEY, C., H. CROWLEY, G. DUFFY, J. J. SHERIDAN, S. O'BRIEN, E. CARNEY, W. ANDERSON, D. A. McDOWELL, I. S. BLAIR and R. H. BISHOP (2004) Prevalence and numbers of *Escherichia coli* O157:H7 in minced beef and beef burgers from butcher shops and supermarkets in the Republic of Ireland. *Food Microbiol* **21**: 203–212.

CALLAWAY, T. R., M. A. CARR, T. S. EDRINGTON, R. C. ANDERSON and D. J. NISBET (2009) Diet, *Escherichia coli* O157:H7, and cattle: a review after 10 years. *Curr Issues Mol Biol* **11**: 67–79.

CASTILLO, A., J. S. DICKSON, R. P. CLAYTON, L. M. LUCIA and G. R. ACUFF (1998) Chemical dehairing of bovine skin to reduce pathogenic bacteria and bacteria of fecal origin. *J Food Prot* **61**: 623–625.

CDC (2006) Multistate outbreak of *Salmonella typhimurium* infections associated with eating ground beef – United States, 2004. *MMWR Morb Mortal Wkly Rep* **55**: 180–182.

CDC (2010) Preliminary FoodNet data on the incidence of infection with pathogens transmitted commonly through food – 10 States, 2009. *MMWR Morb Mortal Wkly Rep* **59**: 418–422.

CFR (1998) Secondary direct food additives permitted in food for human consumption. In: Federal Register. pp. 11118–11119.

CHAPMAN, P. A., C. A. SIDDONS, A. T. CERDAN MALO and M. A. HARKIN (2000) A one year

© Woodhead Publishing Limited, 2011

study of *Escherichia coli* O157 in raw beef and lamb products. *Epidemiol Infect* **124**: 207–213.

CHAPMAN, P. A., A. T. CERDAN MALO, M. ELLIN, R. ASHTON and M. A. HARKIN (2001) *Escherichia coli* O157 in cattle and sheep at slaughter, on beef and lamb carcasses and in raw beef and lamb products in South Yorkshire, UK. *Int J Food Microbiol* **64**: 139–150.

CHATRE, P., M. HAENNI, D. MEUNIER, M. A. BOTREL, D. CALAVAS and J. Y. MADEC (2010) Prevalence and antimicrobial resistance of *Campylobacter jejuni* and *Campylobacter coli* isolated from cattle between 2002 and 2006 in France. *J Food Prot* **73**: 825–831.

CHINEN, I., J. D. TANARO, E. MILIWEBSKY, L. H. LOUND, G. CHILLEMI, S. LEDRI, A. BASCHKIER, M. SCARPIN, E. MANFREDI and M. RIVAS (2001) Isolation and characterization of *Escherichia coli* O157:H7 from retail meats in Argentina. *J Food Prot* **64**: 1346–1351.

COBO, E. R., C. MORSELLA, D. CANO, A. CIPOLLA and C. M. CAMPERO (2004) Immunization in heifers with dual vaccines containing *Tritrichomonas foetus* and *Campylobacter fetus* antigens using systemic and mucosal routes. *Theriogenology* **62**: 1367–1382.

CONEDERA, G., P. DALVIT, M. MARTINI, G. GALIERO, M. GRAMAGLIA, E. GOFFREDO, G. LOFFREDO, S. MORABITO, D. OTTAVIANI, F. PATERLINI, G. PEZZOTTI, M. PISANU, P. SEMPRINI and A. CAPRIOLI (2004) Verocytotoxin-producing *Escherichia coli* O157 in minced beef and dairy products in Italy. *Int J Food Microbiology* **96**: 67–73.

COTTER, P. D., C. HILL and R. P. ROSS (2005) Bacteriocins: developing innate immunity for food. *Nat Rev Microbiol* **3**: 777–788.

CROSSON, P., P. O'KIELY, F. P. O'MARA and M. WALLACE (2006) The development of a mathematical model to investigate Irish beef production systems. *Agr Syst* **89**: 349–370.

CUTTER, C. N. (2002) Microbial control by packaging: a review. *Crit Rev Food Sci Nutr* **42**: 151–161.

DAVIES, R., R. DALZIEL, J. GIBBENS, J. WILESMITH, J. RYAN, S. EVANS, C. BYRNE, G. PAIBA, S. PASCOE and C. TEALE (2004) National survey for *Salmonella* in pigs, cattle and sheep at slaughter in Great Britain (1999–2000). *J Appl Microbiol* **96**: 750–760.

DIEZ-GONZALEZ, F. (2007) Applications of bacteriocins in livestock. *Curr Issues Intest Microbiol* **8**: 15–24.

DONTOROU, C., C. PAPADOPOULOU, G. FILIOUSSIS, V. ECONOMOU, I. APOSTOLOU, G. ZAKKAS, A. SALAMOURA, A. KANSOUZIDOU and S. LEVIDIOTOU (2003) Isolation of *Escherichia coli* O157:H7 from foods in Greece. *Int J Food Microbiol* **82**: 273–279.

DOYLE, M. P. and J. L. SCHOENI (1987) Isolation of *Escherichia coli* O157:H7 from retail fresh meats and poultry. *Appl Environ Microbiol* **53**: 2394–2396.

DUFFY, G. (2003) Verocytoxigenic *Escherichia coli* in animal faeces, manures and slurries. *J Appl Microbiol* **94 Suppl**: 94S–103S.

DUFFY, G., F. BUTLER, D. PRENDERGAST, S. DUGGAN, U. GONZALES BARRON, N. LEONARD, C. MANNION, S. FANNING, R. H. MADDEN, S. SPENCE, P. NAUGHTON, D. EGAN and M. CORMICAN (2010) Salmonella *in pork on the island of Ireland: a microbial risk assessment*. Dublin, Teagasc.

DUGGAN, S.J., MANNION, C., PRENDERGAST, D.M., LEONARD, N., FANNING, S., GONZALES-BARRON, U., EGAN, J., BUTLER, F. and DUFFY, G. (2010) Tracking the *Salmonella* status of pigs and pork from lairage through the slaughter process in the Republic of Ireland. *J Food Prot* **73**: 2148–2160.

DYKES, G. A. and S. M. MOORHEAD (2002) Combined antimicrobial effect of nisin and a

© Woodhead Publishing Limited, 2011

listeriophage against *Listeria monocytogenes* in broth but not in buffer or on raw beef. *Int J Food Microbiol* **73**: 71–81.

DYKES, G. A., S. M. MOORHEAD and S. L. ROBERTS (2001) Survival of *Escherichia coli* O157:H7 and *Salmonella* on chill-stored vacuum or carbon dioxide packaged primal beef cuts. *Int J Food Microbioly* **64**: 401–405.

EFSA (2007) Opinion of the Scientific Panel on Animal Health and Welfare (AHAW) related with the vaccination against avian influenza of H5 and H7 subtypes in domestic poultry and captive birds. *EFSA Journal* 489.

EFSA (2010) Trends and sources of zoonoses and zoonotic agents and food-borne outbreaks in the European Union in 2008. *EFSA Journal* **8**: 1496.

ELDER, R. O., J. E. KEEN, G. R. SIRAGUSA, G. A. BARKOCY-GALLAGHER, M. KOOHMARAIE and W. W. LAEGREID (2000) Correlation of enterohemorrhagic *Escherichia coli* O157 prevalence in feces, hides, and carcasses of beef cattle during processing. *Proc Natl Acad Sci USA* **97**: 2999–3003.

ELLIS-IVERSEN, J., R. P. SMITH, S. VAN WINDEN, G. A. PAIBA, E. WATSON, L. C. SNOW and A. J. COOK (2008) Farm practices to control *E. coli* O157 in young cattle – a randomised controlled trial. *Vet Res* **39**: 3.

ELLOUZE, M. and J. C. AUGUSTIN (2010) Applicability of biological time temperature integrators as quality and safety indicators for meat products. *Int J Food Microbiol* **138**: 119–129.

FANTELLI, K. and R. STEPHAN (2001) Prevalence and characteristics of shigatoxin-producing *Escherichia coli* and *Listeria monocytogenes* strains isolated from minced meat in Switzerland. *Int J Food Microbiol* **70**: 63–69.

FERRAZ, J. B. S. and P. E. D. FELÍCIO (2010) Production systems – an example from Brazil. *Meat Sci* **84**: 238–243.

FOLEY, S. L., A. M. LYNNE and R. NAYAK (2008) *Salmonella* challenges: prevalence in swine and poultry and potential pathogenicity of such isolates. *J Anim Sci* **86**: E149–162.

FOX, E., T. O'MAHONY, M. CLANCY, R. DEMPSEY, M. O'BRIEN and K. JORDAN (2009) *Listeria monocytogenes* in the Irish dairy farm environment. *J Food Prot* **72**: 1450–1456.

FRATAMICO, P. M., F. J. SCHULTZ, R. C. BENEDICT and R. L. BUCHANAN (1996) Factors influencing attachment of *Escherichia coli* O157:H7 to beef tissues and removal using selected sanitizing rinses. *J Food Prot* **59**: 453–459.

GALLAY, A., V. BOUSQUET, V. SIRET, V. PROUZET-MAULEON, H. VALK, V. VAILLANT, F. SIMON, Y. LE STRAT, F. MEGRAUD and J. C. DESENCLOS (2008) Risk factors for acquiring sporadic *Campylobacte*r infection in France: results from a national case-control study. *J Infect Dis* **197**: 1477–1484.

GARCIA RODRIGUEZ, L. A., A. RUIGOMEZ and J. PANES (2006) Acute gastroenteritis is followed by an increased risk of inflammatory bowel disease. *Gastroenterology* **130**: 1588–1594.

GHAFIR, Y., B. CHINA, N. KORSAK, K. DIERICK, J. M. COLLARD, C. GODARD, L. DE ZUTTER and G. DAUBE (2005) Belgian surveillance plans to assess changes in *Salmonella* prevalence in meat at different production stages. *J Food Prot* **68**: 2269–2277.

GILL, C. O. (2009) Effects on the microbiological condition of product of decontaminating treatments routinely applied to carcasses at beef packing plants. *J Food Prot* **72**: 1790–1801.

GILL, C.O., McGINNIS, J.C. and BADONI, M. (1996). Use of total of *Escherichia coli* counts to assess the hygiene characteristics of a beef carcass dressing process. *Int J Food Microbiol* **31**: 181–196.

GILLESPIE, I. A., S. J. O'BRIEN, G. K. ADAK, T. CHEASTY and G. WILLSHAW (2005) Foodborne

© Woodhead Publishing Limited, 2011

general outbreaks of Shiga toxin-producing *Escherichia coli* O157 in England and Wales 1992–2002: where are the risks? *Epidemiol Infect* **133**: 803–808.

GREER, G. G. and B. D. DILTS (1992) Factors affecting the susceptibilitiy of meatborne pathogens and spoilage bacteria to organic acids. *Food Res Int.* **25**: 355–364.

HAO, Y. Y., R. E. BRACKETT and M. P. DOYLE (1998) Inhibition of *Listeria monocytogenes* and *Aeromonas hydrophila* by plant extracts in refrigerated cooked beef. *J Food Prot* **61**: 307–312.

HARRIS, K., M. F. MILLER, G. H. LONERAGAN and M. M. BRASHEARS (2006) Validation of the use of organic acids and acidified sodium chlorite to reduce *Escherichia coli* O157 and *Salmonella* Typhimurium in beef trim and ground beef in a simulated processing environment. *J Food Prot* **69**: 1802–1807.

HEIDER, L. C., R. W. MEIRING, A. E. HOET, W. A. GEBREYES, J. A. FUNK and T. E. WITTUM (2008) Evaluation of vaccination with a commercial subunit vaccine on shedding of *Salmonella enterica* in subclinically infected dairy cows. *J Am Vet Med Assoc* **233**: 466–469.

HELLER, C. E., J. A. SCANGA, J. N. SOFOS, K. E. BELK, W. WARREN-SERNA, G. R. BELLINGER, R. T. BACON, M. L. ROSSMAN and G. C. SMITH (2007) Decontamination of beef subprimal cuts intended for blade tenderization or moisture enhancement. *J Food Prot* **70**: 1174–1180.

HUMPHREY, T., S. O'BRIEN and M. MADSEN (2007) *Campylobacters* as zoonotic pathogens: a food production perspective. *Int J Food Microbiol* **117**: 237–257.

HUNTINGTON, G. B. (1997) Starch utilization by ruminants: from basics to the bunk. *J Anim Sci* **75**: 852–867.

INGLIS, G. D., L. D. KALISCHUK, H. W. BUSZ and J. P. KASTELIC (2005) Colonization of cattle intestines by *Campylobacter jejuni* and *Campylobacter lanienae*. *Appl Environ Microbiol* **71**: 5145–5153.

JACOB, M. E., D. G. RENTER and T. G. NAGARAJA (2010) Animal- and truckload-level associations between *Escherichia coli* O157:H7 in feces and on hides at harvest and contamination of preevisceration beef carcasses. *J Food Prot* **73**: 1030–1037.

JOHNSON, R. P., C. L. GYLES, W. E. HUFF, S. OJHA, G. R. HUFF, N. C. RATH and A. M. DONOGHUE (2008) Bacteriophages for prophylaxis and therapy in cattle, poultry and pigs. *Anim Health Res Rev* **9**: 201–215.

JONES, R. J., M. ZAGOREC, G. BRIGHTWELL and J. R. TAGG (2009) Inhibition by *Lactobacillus sakei* of other species in the flora of vacuum packaged raw meats during prolonged storage. *Food Microbiol* **26**: 876–881.

KALCHAYANAND, N., D. M. BRICHTA-HARHAY, T. M. ARTHUR, J. M. BOSILEVAC, M. N. GUERINI, T. L. WHEELER, S. D. SHACKELFORD and M. KOOHMARAIE (2009) Prevalence rates of *Escherichia coli* O157:H7 and *Salmonella* at different sampling sites on cattle hides at a feedlot and processing plant. *J Food Prot* **72**: 1267–1271.

KARMALI, M. A., V. GANNON and J. M. SARGEANT (2010) Verocytotoxin-producing *Escherichia coli* (VTEC). *Vet Microbiol* **140**: 360–370.

KHARE, S., W. ALALI, S. ZHANG, D. HUNTER, R. PUGH, F. C. FANG, S. J. LIBBY and L. G. ADAMS (2010) Vaccination with attenuated *Salmonella enterica* Dublin expressing *E. coli* O157:H7 outer membrane protein Intimin induces transient reduction of fecal shedding of *E. coli* O157:H7 in cattle. *BMC Vet Res* **6**: 35.

KOOHMARAIE, M., T. M. ARTHUR, J. M. BOSILEVAC, M. GUERINI, S. D. SHACKELFORD and T. L. WHEELER (2005) Post-harvest interventins to reduce/eliminate pathogens in beef. *Meat Sci* **71**: 79–91.

KREHBIEL, C. R., S. R. RUST, G. ZHANG and S. E. GILLILAND (2003) Bacterial direct-fed

© Woodhead Publishing Limited, 2011

microbials in ruminant diets: performance response and mode of action. *J. Anim Sci.* **81**: E120–132.

KRUEGER, N. A., R. C. ANDERSON, W. K. KRUEGER, W. J. HORNE, I. V. WESLEY, T. R. CALLAWAY, T. S. EDRINGTON, G. E. CARSTENS, R. B. HARVEY and D. J. NISBET (2008) Prevalence and concentration of *Campylobacter* in rumen contents and feces in pasture and feedlot-fed cattle. *Foodborne Pathog Dis* **5**: 571–577.

LEJEUNE, J. T. and A. N. WETZEL (2007) Preharvest control of *Escherichia coli* O157 in cattle. *J Anim Sci* **85**: E73–80.

LITTLE, C. L., J. F. RICHARDSON, R. J. OWEN, E. DE PINNA and E. J. THRELFALL (2008) *Campylobacter* and *Salmonella* in raw red meats in the United Kingdom: prevalence, characterization and antimicrobial resistance pattern, 2003–2005. *Food Microbiol* **25**: 538–543.

MADDEN, R. H., K. A. MURRAY and A. GILMOUR (2006) Carriage of four bacterial pathogens by beef cattle in Northern Ireland at time of slaughter. *Lett Appl Microbiol* **44**: 115–119.

MAURIELLO, G., D. ERCOLINI, A. LA STORIA, A. CASABURI and F. VILLANI (2004) Development of polythene films for food packaging activated with an antilisterial bacteriocin from *Lactobacillus curvatus* 32Y. *J Appl Microbiol* **97**: 314–322.

McCORMICK, J. K., T. R. KLAENHAMMER and M. E. STILES (1999) Colicin V can be produced by lactic acid bacteria. *Lett Appl Microbiol* **29**: 37–41.

McEVOY, J. M., A. M. DOHERTY, J. J. SHERIDAN, I. S. BLAIR and D. A. McDOWELL (2001) Use of steam condensing at subatmospheric pressures to reduce *Escherichia coli* O157:H7 numbers on bovine hide. *J Food Prot* **64**: 1655–1660.

McEVOY, J.M., SHERIDAN, J.J., BLAIR, I.S. and McDOWELL, D.A (2004) Microbial contamination on beef in relation to hygiene assessment based on criteria used in EU Decision 2001/471/EC. Int J Food Microbiol **92**:217–225

MEAD, P. S., L. SLUTSKER, V. DIETZ, L. F. McCAIG, J. S. BRESEE, C. SHAPIRO, P. M. GRIFFIN and R. V. TAUXE (1999) Food-related illness and death in the United States. *Emerg Infect Dis* **5**: 607–625.

MEAD, P. S., E. F. DUNNE, L. GRAVES, M. WIEDMANN, M. PATRICK, S. HUNTER, E. SALEHI, F. MOSTASHARI, A. CRAIG, P. MSHAR, T. BANNERMAN, B. D. SAUDERS, P. HAYES, W. DEWITT, P. SPARLING, P. GRIFFIN, D. MORSE, L. SLUTSKER and B. SWAMINATHAN (2006) Nationwide outbreak of listeriosis due to contaminated meat. *Epidemiol Infect* **134**: 744–751.

MEDEIROS, D. T., S. A. SATTAR, J. M. FARBER and C. D. CARRILLO (2008) Occurrence of *Campylobacter* spp. in raw and ready-to-eat foods and in a Canadian food service operation. *J Food Prot* **71**: 2087–2093.

MIES, P. D., B. R. COVINGTON, K. B. HARRIS, L. M. LUCIA, G. R. ACUFF and J. W. SAVELL (2004) Decontamination of cattle hides prior to slaughter using washes with and without antimicrobial agents. *J Food Prot* **67**: 579–582.

MILLER, W. G., M. D. ENGLEN, S. KATHARIOU, I. V. WESLEY, G. WANG, L. PITTENGER-ALLEY, R. M. SILETZ, W. MURAOKA, P. J. FEDORKA-CRAY and R. E. MANDRELL (2006) Identification of host-associated alleles by multilocus sequence typing of *Campylobacter coli* strains from food animals. *Microbiology* **152**: 245–255.

MOGOLLÓN, M. A., B. P. MARKS, A. M. BOOREN, A. ORTA-RAMIREZ and E. T. RYSER (2009) Effect of beef product physical structure on *Salmonella* thermal inactivation. *J Food Sci* **74**: M347–M351.

MOHLER, V. L., D. M. HEITHOFF, M. J. MAHAN, K. H. WALKER, M. A. HORNITZKY, C. S. McCONNELL, L. W. SHUM and J. K. HOUSE (2006) Cross-protective immunity in calves conferred by

© Woodhead Publishing Limited, 2011

a DNA adenine methylase deficient *Salmonella enterica* serovar Typhimurium vaccine. *Vaccine* **24**: 1339–1345.

MOHLER, V. L., D. M. HEITHOFF, M. J. MAHAN, K. H. WALKER, M. A. HORNITZKY, L. W. SHUM, K. J. MAKIN and J. K. HOUSE (2008) Cross-protective immunity conferred by a DNA adenine methylase deficient *Salmonella enterica* serovar Typhimurium vaccine in calves challenged with *Salmonella* serovar Newport. *Vaccine* **26**: 1751–1758.

MORGAN, E., A. J. BOWEN, S. C. CARNELL, T. S. WALLIS and M. P. STEVENS (2007) SiiE is secreted by the *Salmonella enterica* serovar Typhimurium pathogenicity island 4-encoded secretion system and contributes to intestinal colonization in cattle. *Infect Immun* **75**: 1524–1533.

NAYLOR, S. W., D. L. GALLY and J. C. LOW (2005) Enterohaemorrhagic *E. coli* in veterinary medicine. *Int J Med Microbiol* **295**: 419–441.

NIGHTINGALE, K. K., Y. H. SCHUKKEN, C. R. NIGHTINGALE, E. D. FORTES, A. J. HO, Z. HER, Y. T. GROHN, P. L. McDONOUGH and M. WIEDMANN (2004) Ecology and transmission of *Listeria monocytogenes* infecting ruminants and in the farm environment. *Appl Environ Microbiol* **70**: 4458–4467.

NIGHTINGALE, K. K., E. D. FORTES, A. J. HO, Y. H. SCHUKKEN, Y. T. GROHN and M. WIEDMANN (2005) Evaluation of farm management practices as risk factors for clinical listeriosis and fecal shedding of *Listeria monocytogenes* in ruminants. *J Am Vet Med Assoc* **227**: 1808–1814.

NOU, X., M. RIVERA-BETANCOURT, J. M. BOSILEVAC, T. L. WHEELER, S. D. SHACKELFORD, B. L. GWARTNEY, J. O. REAGAN and M. KOOHMARAIE (2003) Effect of chemical dehairing on the prevalence of *Escherichia coli* O157:H7 and the levels of aerobic bacteria and *enterobacteriaceae* on carcasses in a commercial beef processing plant. *J Food Prot* **66**: 2005–2009.

O'BRIEN, S. B., G. DUFFY, D. DALY, J. J. SHERIDAN, I. S. BLAIR and D. A. McDOWELL (2005) Detection and recovery rates achieved using direct plate and enrichment/ immunomagnetic separation methods for *Escherichia coli* O157:H7 in minced beef and on bovine hide. *Lett Appl Microbiol* **41**: 88–93.

O'FLYNN, G., R. P. ROSS, G. F. FITZGERALD and A. COFFEY (2004) Evaluation of a cocktail of three bacteriophages for biocontrol of *Escherichia coli* O157:H7. *Appl Environ Microbiol* **70**: 3417–3424.

OUSSALAH, M., S. CAILLET, S. SALMIERI, L. SAUCIER and M. LACROIX (2006) Antimicrobial effects of alginate-based film containing essential oils for the preservation of whole beef muscle. *J Food Prot* **69**: 2364–2369.

OZDEMIR, H., A. KOLUMAN and Y. YILDIRIM (2006) Effects of acidified sodium chlorite, cetylpyridinium chloride and hot water on populations of *Listeria monocytogenes* and *Staphylococcus aureus* on beef. *Lett Appl Microbiol* **43**: 168–173.

PATTON, T. G., V. K. SHARMA and S. A. CARLSON (2009) Evaluation of the control of pathogen load by an anti-*Salmonella* bacterium in a herd of cattle with persistent *Salmonella* infection. *Am J Vet Res* **70**: 92–98.

PITTENGER, L. G., M. D. ENGLEN, C. T. PARKER, J. G. FRYE, B. QUINONES, S. T. HORN, I. SON, P. J. FEDORKA-CRAY and M. A. HARRISON (2009) Genotyping *Campylobacter jejuni* by comparative genome indexing: an evaluation with pulsed-field gel electrophoresis and *flaA* SVR sequencing. *Foodborne Pathog Dis* **6**: 337–349.

PODOLAK, R. K., J. F. ZAYAS, C. L. KASTNER and D. Y. C. FUNG (1996) Inhibition of *Listeria monocytogenes* and *Escherichia coli* O157:H7 on beef by application of organic acids. *J Food Prot* **59**: 370–373.

POTTER, A. A., S. KLASHINSKY, Y. LI, E. FREY, H. TOWNSEND, D. ROGAN, G. ERICKSON, S.

© Woodhead Publishing Limited, 2011

HINKLEY, T. KLOPFENSTEIN, R. A. MOXLEY, D. R. SMITH and B. B. FINLAY (2004) Decreased shedding of *Escherichia coli* O157:H7 by cattle following vaccination with type III secreted proteins. *Vaccine* **22**: 362–369.

RAFTARI, M., F. A. JALILIAN, A. S. ABDULAMIR, R. SON, Z. SEKAWI and A. B. FATIMAH (2009) Effect of organic acids on *Escherichia coli* O157:H7 and *Staphylococcus aureus* contaminated meat. *Open Microbiol J* **3**: 121–127.

RAHIMI, E., M. AMERI and H. R. KAZEMEINI (2010) Prevalence and antimicrobial resistance of Campylobacter species isolated from raw camel, beef, lamb, and goat meat in Iran. *Foodborne Pathog Dis* **7**: 443–447.

READ, S. C., C. L. GYLES, R. C. CLARKE, H. LIOR and S. McEWEN (1990) Prevalence of verocytotoxigenic *Escherichia coli* in ground beef, pork, and chicken in southwestern Ontario. *Epidemiol Infect* **105**: 11–20.

REID, C. A., A. SMALL, S. M. AVERY and S. BUNCIC (2002) Presence of food-borne pathogens on cattle hides. *Food Control* **13**: 411–415.

RHOADES, J. R., G. DUFFY and K. KOUTSOUMANIS (2009) Prevalence and concentration of verocytotoxigenic *Escherichia coli*, *Salmonella enterica* and *Listeria monocytogenes* in the beef production chain: a review. *Food Microbiol* **26**: 357–376.

RIVERA-BETANCOURT, M., S. D. SHACKELFORD, T. M. ARTHUR, K. E. WESTMORELAND, G. BELLINGER, M. ROSSMAN, J. O. REAGAN and M. KOOHMARAIE (2004) Prevalence of *Escherichia coli* O157:H7, *Listeria monocytogenes*, and *Salmonella* in two geographically distant commercial beef processing plants in the United States. *J Food Prot* **67**: 295–302.

ROSE, B. E., W. E. HILL, R. UMHOLTZ, G. M. RANSOM and W. O. JAMES (2002) Testing for Salmonella in raw meat and poultry products collected at federally inspected establishments in the United States, 1998 through 2000. *J Food Prot* **65**: 937–947.

ROZEMA, E. A., T. P. STEPHENS, S. J. BACH, E. K. OKINE, R. P. JOHNSON, K. STANFORD and T. A. MCALLISTER (2009) Oral and rectal administration of bacteriophages for control of *Escherichia coli* O157:H7 in feedlot cattle. *J Food Prot* **72**: 241–250.

SAMADPOUR, M., M. W. BARBOUR, T. NGUYEN, T. M. CAO, F. BUCK, G. A. DEPAVIA, E. MAZENGIA, P. YANG, D. ALFI, M. LOPES and J. D. STOPFORTH (2006) Incidence of enterohemorrhagic *Escherichia coli*, *Escherichia coli* O157, *Salmonella*, and *Listeria monocytogenes* in retail fresh ground beef, sprouts, and mushrooms. *J Food Prot* **69**: 441–443.

SAMMARCO, M. L., G. RIPABELLI, I. FANELLI, G. M. GRASSO and M. TAMBURRO (2010) Prevalence and biomolecular characterization of *Campylobacter* spp. isolated from retail meat. *J Food Prot* **73**: 720–728.

SAYED, H. S., J. M. LEIGH and R. R. MICHAEL (2007) The effects of *E. coli* 0157:H7, FMD and BSE on Japanese retail beef prices: a historical decomposition. *Agribusiness* **23**: 131–147.

SCHAMBERGER, G. P., R. L. PHILLIPS, J. L. JACOBS and F. DIEZ-GONZALEZ (2004) Reduction of *Escherichia coli* O157:H7 populations in cattle by feeding colicin E7-producing *E. coli*. *Appl Environ Microbiol* **70**: 6053–6060.

SEMENOV, A. V., L. VAN OVERBEEK and A. H. C. VAN BRUGGEN (2009) Percolation and survival of *Escherichia coli* O157:H7 and *Salmonella* enterica Serovar Typhimurium in soil amended with contaminated dairy manure or slurry. *Appl. Environ. Microbiol.* **75**: 3206–3215.

SHENG, H., H. J. KNECHT, I. T. KUDVA and C. J. HOVDE (2006) Application of bacteriophages to control intestinal *Escherichia coli* O157:H7 levels in ruminants. *Appl Environ Microbiol* **72**: 5359–5366.

© Woodhead Publishing Limited, 2011

SHERIDAN, J.J. (2000) Monitoring CCPs in HACCP systems. In: *HACCP in the Meat Industry*. M. Brown (ed.). Boca Raton, FL: CRC Press, pp. 203–230.

SILVEIRA, N. F., N. SILVA, C. CONTRERAS, L. MIYAGUSKU, M. L. BACCIN, E. KOONO and N. J. BERAQUET (1999) Occurrence of *Escherichia coli* O157:H7 in hamburgers produced in Brazil. *J Food Prot* **62**: 1333–1335.

SKANDAMIS, P., E. TSIGARIDA and G.-J. E. NYCHAS (2002) The effect of oregano essential oil on survival/death of *Salmonella typhimurium* in meat stored at 5C under aerobic, VP/MAP conditions. *Food Microbiol* **19**: 97–103.

SMELSER, C. (2004) Outbreak of salmonellosis associated with beef jerky in New Mexico. *New Mexico Epidemiology Report* **2004**.

SMITH, B., J. T. LARSSON, M. LISBY, L. MÜLLER, S. B. MADSEN, J. ENGBERG, J. BANGSBORG, S. ETHELBERG and M. KEMP (2010) Outbreak of listeriosis caused by infected beef meat from a meals-on-wheels delivery in Denmark 2009. *Clin Microbiol Infec* doi: 10.111/j.1469-0691.2010.03200.x.

STANFORD, K., T. A. McALLISTER, Y. D. NIU, T. P. STEPHENS, A. MAZZOCCO, T. E. WADDELL and R.P. JOHNSON (2010) Oral delivery systems for encapsulated bacteriophages targeted at *Escherichia coli* O157:H7 in feedlot cattle. *J Food Prot* **73**: 1304–1312.

STEPHENS, T. P., G. H. LONERAGAN, E. KARUNASENA and M. M. BRASHEARS (2007) Reduction of *Escherichia coli* O157 and *Salmonella* in faeces and on hides of feedlot cattle using various doses of a direct-fed microbial. *J Food Prot* **70**: 2386–2391.

TABE, E. S., J. OLOYA, D. K. DOETKOTT, M. L. BAUER, P. S. GIBBS and M. L. KHAITSA (2008) Comparative effect of direct-fed microbials on fecal shedding of *Escherichia coli* O157:H7 and *Salmonella* in naturally infected feedlot cattle. *J Food Prot* **71**: 539–544.

TARR, P. I., N. T. TRAN and R. A. WILSON (1999) *Escherichia coli* O157:H7 in retail ground beef in Seattle: results of a one-year prospective study. *J Food Prot* **62**: 133–139.

THOMSON, D. U., G. H. LONERAGAN, A. B. THORNTON, K. F. LECHTENBERG, D. A. EMERY, D. T. BURKHARDT and T. G. NAGARAJA (2009) Use of a siderophore receptor and porin protein-based vaccine to control the burden of *Escherichia coli* O157:H7 in feedlot cattle. *Foodborne Pathog Dis* **6**: 871–877.

TSIGARIDA, E., P. SKANDAMIS and G.-J. E. NYCHAS (2000) Behaviour of *Listeria monocytogenes* and autochthonous flora on meat stored under aerobic, vacuum and modified atmosphere packaging conditions with or without the presence of oregano essential oil at 5 °C. *J Appl Microbiol* **89**: 901–909.

TUTENEL, A. V., D. PIERARD, J. VAN HOOF, M. CORNELIS and L. DE ZUTTER (2003) Isolation and molecular characterization of *Escherichia coli* O157 isolated from cattle, pigs and chickens at slaughter. *Int J Food Microbiol* **84**: 63–69.

USDA-FSIS (2004) Safe and suitable ingredients used in the production of meat and poultry products. In: FSIS Directive 7120.1 Amendment 6. F. S. a. I. S. United States Department of Agriculture (ed).

VAIKOUSI, H., C. G. BILIADERIS and K. P. KOUTSOUMANIS (2009) Applicability of a microbial time temperature indicator (TTI) for monitoring spoilage of modified atmosphere packed minced meat. *Int J Food Microbiol* **133**: 272–278.

VAN DIEMEN, P. M., F. DZIVA, A. ABU-MEDIAN, T. S. WALLIS, H. VAN DEN BOSCH, G. DOUGAN, N. CHANTER, G. FRANKEL and M. P. STEVENS (2007) Subunit vaccines based on intimin and Efa-1 polypeptides induce humoral immunity in cattle but do not protect against intestinal colonisation by enterohaemorrhagic *Escherichia coli* O157:H7 or O26:H. *Vet Immunol Immunopath* **116**: 47–58.

VAN DONKERSGOED, J., D. HANCOCK, D. ROGAN and A. A. POTTER (2005) *Escherichia coli*

© Woodhead Publishing Limited, 2011

O157:H7 vaccine field trial in 9 feedlots in Alberta and Saskatchewan. *Can Vet J* **46**: 724–728.

VERNOZY-ROZAND, C., S. RAY-GUENIOT, C. RAGOT, C. BAVAI, C. MAZUY, M. P. MONTET, J. BOUVET and Y. RICHARD (2002) Prevalence of *Escherichia coli* O157:H7 in industrial minced beef. *Lett Appl Microbiol* **35**: 7–11.

VIGNOLO, G., J. PALACIOS, M. E. FARIAS, F. SESMA, U. SCHILLINGER, W. H. HOLZAPFEL and OLIVER, G. (2000) Combined effect of bacteriocins on the survival of various *Listeria* species in broth and meat system. *Curr Microbiol* **41**: 410–416.

VILAR, M. J., F. J. PENA, I. PEREZ, F. J. DIEGUEZ, M. L. SANJUAN, J. L. RODRIGUEZ-OTERO and E. YUS (2010) Presence of *Listeria*, *Arcobacter*, and *Campylobacter* spp. in dairy farms in Spain. *Berl Munch Tierarztl Wochenschr* **123**: 58–62.

WHYTE, P., K. MCGILL, D. COWLEY, R. H. MADDEN, L. MORAN, P. SCATES, C. CARROLL, A. O'LEARY, S. FANNING, J. D. COLLINS, E. McNAMARA, J. E. MOORE and M. CORMICAN (2004) Occurrence of *Campylobacter* in retail foods in Ireland. *Int J Food Microbiol* **95**: 111–118.

WILSON, J. S., T. OTSUKI and B. MAJUMDSAR (2003) Balancing food safety and risk: do drug residue limits affect international trade in beef? *J Int Trade Econ Dev* **12**: 377–402.

WONG, T. L., C. NICOL, R. COOK and S. MACDIARMID (2007) *Salmonella* in uncooked retail meats in New Zealand. *J Food Prot* **70**: 1360–1365.

YOUNTS-DAHL, S. M., M. L. GALYEAN, G. H. LONERAGAN, N. A. ELAM and M. M. BRASHEARS (2004) Dietary supplementation with *Lactobacillus*- and *Propionibacterium*-based direct-fed microbials and prevalence of *Escherichia coli* O157 in beef feedlot cattle and on hides at harvest. *J Food Prot* **67**: 889–893.

ZHANG, G. and A. MUSTAPHA (1999) Reduction of *Listeria monocytogenes* and *Escherichia coli* O157:H7 numbers on vacuum-packaged fresh beef treated with nisin or nisin combined with EDTA. *J Food Prot* **62**: 1123–1127.

© Woodhead Publishing Limited, 2011

11

Animal welfare: an essential component in food safety and quality

L. A. Boyle and K. O'Driscoll, Teagasc, Ireland

Abstract: Animal welfare is considered an important attribute of an overall 'food quality concept', and there is a growing realisation of a link between animal welfare and food safety. Threats to animal welfare, including stress-induced immunosuppression and promotion of foodborne pathogens, and genetic modifications have the potential to compromise the healthiness and safety of food. The growth in the organic farming sector presents new animal welfare and food safety concerns with regard to animal health security, the prevalence of zoonotic diseases, and the presence of toxic residues in the food produced.

Key words: animal welfare, stress, immunosuppression, genetic modification, organic farming.

11.1 Introduction

Disagreement about the treatment of animals dates back at least to ancient Greece (Sorabji, 1993). However, the animal welfare debate as we know it can be traced back to the 1960s with the publication of Ruth Harrison's book *Animal Machines* (Harrison, 1964). This book focused on the welfare of animals in intensive production systems such as battery cages and veal crates. It caused enormous public debate that has since grown to encompass a much broader range of issues including transport, slaughter and genetics. Animal welfare is fundamentally an ethical issue, but ethical reasoning alone is insufficient to answer the welfare dilemmas raised by modern agricultural practices (Fraser, 2001). Animal welfare science has emerged in recent years to add a valuable empirical element to the debate. It has not only increased our understanding of

© Woodhead Publishing Limited, 2011

the link between biological systems and animal welfare (Dawkins, 2006), but has also contributed to improvements in housing, handling and management practices for farm animal species. Importantly, it has provided a scientific basis for some of the welfare legislation passed in recent years in the EU including the prohibition of tether and stall housing of gestating sows and of conventional battery cages for laying hens (SVC, 1997; EFSA, 2005). The growth in animal welfare legislation in the EU is also mirrored in other parts of the world such as the US, where some intensive farming practices have been prohibited in individual states (e.g. confinement of calves in veal crates and sows in gestation stalls banned in Arizona in 2006). Some of the most significant improvements to animal welfare in the US have been in the area of handling and slaughter techniques (Grandin, 2001). These were primarily driven by the relatively recent involvement of fast food companies in the auditing of plants for compliance with the American Meat Institute Guidelines (Grandin, 1997).

Most of the advances described above were prompted by vigorous campaigning by animal welfare advocates. In the future, change is more likely to be subtly driven by consumer expectations. Nowadays consumers are very well informed and have requirements that their food is produced with greater respect for animal welfare, for the environment, and for food safety. Indeed, animal welfare is considered an important attribute of an overall 'food quality concept' (Blokhuis *et al.*, 2008).

Traditionally, producers, processors and retailers of food of animal origin were more concerned with supply, price and competition, while they paid little attention to consumer expectations. However, there is a growing awareness, possibly initiated by the major retailers, that consumer requirements must be at the forefront of efforts to ensure the sustainability of agri-food chains (Blokhuis *et al.*, 2008). Indeed, retailers and producers are also recognising that efforts to meet consumer concerns and requirements relating to animal welfare represent a business opportunity, and might therefore be worthwhile paying attention to. Sossidou and Elson (2009) discuss that public concern for improvements to laying hen welfare and egg quality is an area of potential added value for producers, retailers, processors, manufacturers and producer cooperatives. The Freedom Food label developed by the Royal Society for the Prevention of Cruelty to Animals in the UK is the most successful example of a scheme that markets produce from farms that meet high animal welfare standards to an expanding customer base.

11.2 Animal health, animal welfare and food safety

In the food industry, there is already an appreciation that it is important to consider animal welfare because of related implications for food quality (e.g. conditions that negatively affect animal welfare such as fear can also increase the occurrence of tough or watery meat as well as the incidence of bruising, bone breakage, blood spots and abnormal eggshells; Jones, 1997; Faure *et al.*, 2003).

© Woodhead Publishing Limited, 2011

However, there is also growing realisation of a link between animal welfare and food safety. There are instances where this link brings animal welfare into conflict with food safety, such as in the case of organic farming, which will be discussed later in this chapter. However, consumers generally consider animal welfare in their food purchasing decisions because they intuitively perceive it to be linked in some way to healthy and safe food. Scientific data in support of this are scarce, not only because of a lack of research in this area, but also because the multi-dimensional nature of animal welfare makes it difficult to establish such links directly.

The health status of an animal is obviously a major contributor to its welfare state, and animals with pathology, whether from infectious disease, metabolic or other disorders, can have very poor welfare (Broom, 2006). Indeed, many endemic health problems (e.g. lameness and mastitis in high producing dairy cows; pathologies associated with selection for fast growth (ascites and tibial dyschondroplasia) in broilers) are among the most serious welfare problems (Broom, 2006). It is important to note that animal health and animal welfare are not synonymous in that an animal with poor welfare (e.g. a calf in distress because of separation from its mother) need not necessarily have a health problem. Nevertheless, chronically poor welfare resulting from a wide variety of different causes (e.g. overcrowding, isolation) often makes disease more likely because it suppresses immune functioning (Broom and Kirkden, 2004). Conversely, a high level of welfare, such as that provided by the social support of conspecifics, can help to protect individuals against disease (Sachser, 2001; Lutgendorf, 2001). These links between animal health and animal welfare facilitate a better appreciation of the link between animal welfare and food safety. In this regard, threats to animal health (and therefore animal welfare) have the potential to affect the healthiness and safety of food through the following channels:

- Stress-induced immunosuppression leading to infectious disease and consequently to antimicrobial usage, with food safety issues surrounding antibiotic resistance and antibiotic residues.
- Stress-induced promotion of foodborne pathogens such as *Escherichia coli* and *Salmonella*.
- Genetic modifications to animals focused on improving production efficiency at the expense of health and fitness, resulting in changes to the composition of food products associated with disease in humans.

11.2.1 Stress, immunosuppression and susceptibility to disease

An animal's response to stress is the main mechanism through which poor welfare is translated into poor health and consequently to food safety problems (Humphrey, 2006; Rostagno, 2009). Stress is a word used to describe experiences that are challenging psychologically (e.g. fear or anxiety), physically (usually with a strong psychological component, e.g. electric shock) or physiologically (e.g. heat stress). A stressor is the stimulus that causes the

© Woodhead Publishing Limited, 2011

stress, and in farm animals these include weaning, mixing, overcrowding, transport and restraint, among others. The welfare of an animal or the success of its attempts to cope with its environment is determined by the animal's ability to respond appropriately to a particular stressor (Broom, 1993). If a stressor is so severe or so prolonged that resources are diverted from normal biological functioning to an extreme extent then the animal enters a pre-pathological state (Moberg, 2000). In this state the animal is more susceptible to disease, impaired reproduction or inefficient metabolism and therefore impaired welfare (Moberg, 1987). The stress response starts with the perception of the stressor by the central nervous system (CNS) and the bi-directional communication of the stressor between the brain and the peripheral organs. Because of this cross-communication, stress can directly affect organ functioning, and physiological changes within the organs of the body can directly affect the CNS. These physiological changes are characterised by activation of the sympathetic–adrenal–medullary axis and the hypothalamic–pituitary–adrenal axis. Activation of these axes results in systemic elevations of the glucocorticoids (e.g. cortisol) and catecholamine stress hormones (adrenaline and noradrenaline) (reviewed by Llamas Moya *et al.*, 2006).

Many complex pathways modulate the effects that the hormonal products of the stress response have on the immune system and susceptibility to infectious disease, making it impossible to generalise the effects of stress. This also explains that while there are numerous examples of stressors affecting immunological function in farm animals (e.g. Ekkel *et al.*, 1995 and Hicks *et al.*, 1998; social stress in pigs) there are few clear examples of stress-induced immunosuppression leading to a greater incidence of disease or illness in farm animals. One of the exceptions is in the case of transport and the respiratory disease complex in cattle. Broom and Kirkden (2004) review how the myriad of environmental challenges associated with transport activate physiological coping mechanisms, modulate different aspects of the immune system and ultimately increase the incidence of respiratory disease. In any case, it is clear that stressors which elicit physiological coping mechanisms in animals and therefore cause poor welfare also alter their health status (Biondi and Zannino, 1997). Overstocking in most species of farm animals leads to welfare problems and increased susceptibility to disease. For example, Maes *et al.* (2000) found that a higher number of pigs per pen increased the risk of being seropositive for respiratory pathogens. Conversely, management systems associated with access to pasture and exercise for dairy cows and hence improvements to animal welfare are also linked to improvements in health (Regula *et al.*, 2004; Olmos *et al.*, 2009a,b).

The problem for human health arises because many of the diseases that cause suffering and welfare problems for farm animals such as pneumonia, diarrhoea and mastitis are also associated with a high level of antimicrobial usage (McEwen and Fedorka-Cray, 2002). Employing practices to improve animal welfare, such as reducing stocking densities, combined with good management practices could help to control the incidence of such diseases (McEwen and

© Woodhead Publishing Limited, 2011

Fedorka-Cray, 2002; Cagienard *et al.*, 2005). The result would not only be an improvement in animal health but also a significant reduction in the use of antimicrobials. There is some evidence to suggest that 'animal welfare friendly' farming practices are playing a part in contributing to reduced antibiotic resistance (Regula *et al.*, 2003; Schwaiger *et al.*, 2010). However, as de Passille and Rushen (2005) point out, it is not clear whether this is because of fewer health problems on these farms or because of a more prudent approach to antibiotic usage. The spread of resistant bacteria arising from antimicrobial use in farm animals is a significant human health concern (Khachatourians, 1998). It is also a concern for animal health but less is known about this problem. To slow the development of resistance, the EU and some other countries have restricted antimicrobial use in feed to promote growth and feed efficiency. The published evidence suggests that such bans have reduced overall antibiotic use in animals, although the impact, or lack of, on humans is the subject of much debate (Casewell *et al.*, 2003; Phillips, 2007). In any case, it appears that some growth promoters had health promotional or prophylactic effects (Wierup, 2001) which have not been replicated by improvements to aspects of animal husbandry (Casewell *et al.*, 2003). Some argue that their withdrawal has resulted in more health and welfare problems in animals resulting in more use of therapeutic drugs which are identical to those used in human medicine (reviewed by Casewell *et al.*, 2003). In any case, there is a wide spectrum of alternatives to growth promoting and prophylactic uses of antimicrobials in agriculture including improved management practices, wider use of vaccines, and introduction of probiotics (McEwen and Fedorka-Cray, 2002). These in combination with monitoring programmes, prudent use guidelines, and educational campaigns could not only protect animal welfare but provide approaches to minimise the further development of antimicrobial resistance.

11.2.2 Stress induced promotion of foodborne pathogens

Pathogens such as *Salmonella enterica, E. coli* O157:H7 and *Campylobacter* are a major threat to food safety and there is now a large body of evidence to support the link between poor welfare brought about by stress and the proliferation of these pathogens. In pigs, the stress associated with weaning, handling and mixing can result in an increase in colonisation and faecal shedding of both *Salmonella enterica* and *E. coli* O157:H7 (Jones *et al.*, 2001; Callaway *et al.*, 2006). While it is uncertain if cattle perceive dirty, muddy pens as a stressor, such conditions are equated with poor welfare (Whay *et al.*, 2003) and result in higher bacterial counts at the abattoir (McEvoy *et al.*, 2000). Forced moulting of poultry is an extreme practice of food deprivation so as to manipulate egg production in commercial egg and meat type birds that is widely employed in the US (Bell, 2003). The practice not only causes serious welfare problems for poultry (Duncan, 2001) but it also increases intestinal numbers of *S. enteritidis* and consequently transmission to infected hens via increased faecal shedding (Holt *et al.*, 1995). While forced moulting is not practised in the EU, feed is

© Woodhead Publishing Limited, 2011

routinely withdrawn from most farm animal species prior to slaughter in order to reduce the chances of carcass contamination by the gut contents. The resulting nutritional stress can have negative welfare implications (Terlouw *et al.*, 2008) and in poultry, has similar implications for enteric pathogens as forced moulting (Byrd *et al.*, 1998; increases in *Campylobacter* in broilers).

Rostagno (2009) reviews how stress indirectly encourages the proliferation of enteric pathogens by suppression of the immune system and by physiological changes in the gastrointestinal tract via the action of the stress hormones. However, enteric pathogens also appear to have evolved systems for directly sensing stress hormones (Freestone *et al.*, 2008). For example, *E. coli* O157:H7 and *Salmonella enterica serovar typhimurium* appear to be as responsive to stress hormones as are the cells of their host (Vlisidou *et al.*, 2004; Toscano *et al.*, 2007). Hence, even brief exposure of these pathogens to physiological concentrations of stress hormones can result in massive increases in growth and marked changes in expression of virulence factors such as adhesins and toxins (reviewed by Freestone and Lyte, 2010). The shedding of even low numbers of *E. coli* O157:H7 in the faeces of cattle and their entry into the food chain is a primary route to human infections (Tarr and Neill, 2001). Post-slaughter antimicrobial treatments and HACCP policies in slaughter plants can significantly reduce carcass contamination (Elder *et al.*, 2000). However, it is preferable to prevent the initial colonisation and proliferation of these fast-growing enteropathogens. Freestone and Lyte (2010) suggest that certain animal husbandry techniques associated with slaughter need to be modified so as to avoid stress hormone-induced promotion of undesirable bacterial species. Such changes would not only make meat microbiologically safer but could concurrently result in significant improvements to animal welfare prior to slaughter (Terlouw *et al.*, 2008).

11.2.3 The threat posed by genetic manipulation of farm animals to welfare, food safety and human health

Artificial selection is a process of genetic modification of farm animal species that has increased production levels of livestock species considerably. As feed costs are economically the most important costs, the breeding goal has traditionally been to select single production traits to create a population with high economic efficiency, i.e. high productivity combined with relatively low feed intake (Luiting, 1990). This means that animals have been bred for earlier and faster growth, efficient feed conversion and partitioning, and increased prolificacy. This approach has been hugely successful, meaning cheaper and more readily available animal products for consumers. However, it has not been without adverse consequences for animal health and welfare, genetic diversity and the environment (Fisher and Mellor, 2008). Indeed, it is responsible for some of the worst global animal welfare problems exemplified by dystocia in double muscled beef cattle, lameness, metabolic disorders, infertility and mastitis in dairy cows, porcine stress syndrome in pigs, osteoporosis and cloacal

© Woodhead Publishing Limited, 2011

prolapse in egg-laying hens, and skeletal and cardiovascular disorders in turkeys and broiler chickens (Broom, 1994; Phillips, 1997; Rauw *et al.*, 1998).

Dramatic changes to behavioural, physiological and immunological traits underlie these welfare problems in farm animals (Rauw *et al.*, 1998). Hence, it should not be surprising if there are associated consequences for food safety, although this issue has not received much research attention to date. The negative implications for human health of too much fat in the diet are, however, well known and there are some examples of how genetic selection in farm animals might have had an influence. For example, increased concentrations of insulin and glucagon in plasma and perhaps even insulin resistance are thought to drive the increased fat deposition seen in broilers selected for fast growth (Sinsigalli *et al.*, 1987). The result is that the proportion of fat in the carcasses of selected lines is higher (Emmerson, 1997). This effect has only partially been ameliorated by modified selection and nutritional schemes. Selection for high milk yield in dairy cows has also had an adverse effect on the fat composition of milk mediated by metabolic changes in these animals. In early lactation, cows can be in negative energy balance (NEB). This means that they utilise more energy to produce milk then they can ingest and therefore mobilise body reserves for milk production (Butler and Smith, 1989). This is a naturally occurring state during the first 2 weeks of lactation, and cows recover at a variable rate (Butler and Smith, 1989). However, cows selected for high milk production can have a significantly lower energy balance than non-selected cows well into lactation as they continue partitioning energy to milk production rather than to body reserves (Harrison *et al.*, 1990). These cows not only lose a significant amount of weight (Gallo *et al.*, 1996) and are at greater risk of production-related diseases (Ingvartsen *et al.*, 2003), but also produce a higher proportion of saturated fats in their milk (Stoop *et al.*, 2009). While the former are welfare concerns, the latter has negative implications for the healthiness of milk with up to 35% of the total saturated (or bad) fat in the human diet coming from bovine milk fat (Grummer, 1991).

Interestingly, factors unrelated to genetic selection, but beneficial from an animal welfare point of view can also improve the ratio of saturated to unsaturated fats in food (e.g. increased space; Simsek *et al.*, 2009). Good animal welfare practices such as increased space and access to pasture also promote a lower omega-6/omega-3 ratio in foods of animal origin (Enser *et al.*, 1998; Castellini *et al.* 2002). The enormous increase in the consumption of omega-6 polyunsaturated fatty acids (PUFA) in Western diets and a high omega-6/omega-3 ratio are inextricably linked to diseases such as coronary heart disease, hypertension and type 2 diabetes, which are of high prevalence in Western societies (Simopoulos, 2002). Hence, although not traditionally associated with food safety, changes to the composition of foods of animal origin arising from genetic selection, which are coupled with welfare problems, are potentially as threatening to human health as microbial or chemical contamination. Particularly considering that the effect extends to entire populations and not just to unlucky individuals or groups of individuals.

© Woodhead Publishing Limited, 2011

Given the mass consolidation of animal agriculture, breeding decisions and practices have the potential to rapidly improve the health and welfare of farm animals (Greger, 2010) and by implication, the quality and health of the food they produce. RobustMilk (www.robustmilk.eu) is one such example, where geneticists from six leading European agricultural research institutes, funded by the EU 7th Framework Programme have come together in a project with the aim of producing healthier milk from healthier cows (McParland and Berry, 2010). However, other examples of a concerted effort to re-focus breeding objectives with the aim of improving the welfare of farm animals are rare. On the contrary, in spite of lessons learned, most of the new reproduction techniques and increasingly complex means of altering the DNA of farm animal species still have the sole objective of enhancing performance. For example, animals have been transformed with growth-related genes, such as those for growth hormone (somatotropin), growth hormone releasing factor, and insulin-like growth factor (Pursel *et al.*, 2004; Adams and Briegel, 2005). Given the range of production related diseases associated with traditional techniques of genetic manipulation (Rauw *et al.*, 1998) it is likely that the adoption of such technologies to speed up genetic progress will lead to further unfavourable side effects and disorders with welfare implications (Van Reenan *et al.*, 2001; Greger, 2010). Kleter and Kuiper (2002) highlight the main food safety issues that they envisage for such animals. Nevertheless, the current state of scientific development is still some distance away from genetically engineered farm animals that are close to being commercially viable. In contrast, there is a range of experience already available for transgenic fish which are likely to be the first animals to be commercially available (Kaiser, 2005).

Fish in particular have been targeted for growth hormone (GH) gene transgenesis because they respond dramatically having undergone little or no selection for enhanced growth in comparison to livestock where gains arising from transgenesis have been more modest (Devlin *et al.*, 2009). Elevation of GH increases rates of protein synthesis and lipid mobilization, affecting not only growth, but also feed conversion efficiency, metabolic rate, body composition, head and body morphometrics, osmoregulation, age at maturity, swimming ability and aggressive behaviour, with other modifications particular to species and transgenic lines (Hallerman *et al.*, 2007). Although welfare issues posed by such morphological, physiological, and behavioural alternations in GH-transgenic fish have not been well characterised (Hallerman *et al.*, 2007) much of the available data indicates that they are less fit than non-transgenic fish (Dunham, 2003). One of the greatest future potential benefits of gene transfer in fish will be enhancement of disease resistance (Dunham, 2003) which not alone has beneficial welfare implications but implies a reduction in drug and antibiotic usage with consequent improvements to food safety. However, the production related health problems inherent to reduced fitness (Rauw *et al.*, 1998) could counteract any benefits arising from gene transfer to improve disease resistance. Other food safety issues posed by transgenic fish are discussed by Berkowitz and Kryspin-Sorensen (1994) and include the possible risks of consumption of

© Woodhead Publishing Limited, 2011

transgenes, their resulting protein, potential production of toxins by aquatic transgenic organisms, changes in the nutritional composition of foods, activation of viral sequences and allergenicity of transgenic products. Nevertheless, studies designed to directly assess these concerns are rare.

In spite of the imminent arrival of transgenic fish on supermarket shelves in some countries (Kaiser, 2005), there are huge gaps in our knowledge of the welfare and food safety implications of these modifications. This does not bode well for the other farm animal species. In analyses of concerns raised about transgenic animals, animal welfare is one of the main issues deemed as ethically significant (Sandøe and Holtug, 1993). Nevertheless, it is not considered as a key priority for strategic research but only in the context of consumer requirements in the EU's Farm Animal Breeding and Reproduction Technology Platform vision document for 2025. Encouragingly, it recognises that safe and wholesome food should be the first requirement, followed closely by improved product quality and that this is a 'significant shift from the food sector's position in the last century, when improving the level of production was the obvious goal of a [*sic*] production system'. However, as shown above there is a continuing worldwide trend among those concerned with genetic engineering to make animals grow even faster. This is difficult to reconcile with a genuine concern for animal welfare or safe and wholesome food of high quality. Genetically engineering farm animals for accelerated growth was deemed to be the least acceptable application of biotechnology in a consumer survey conducted in the US (Mench, 1999). A consumer backlash against biotechnology resulting from an application perceived to worsen the plight of animals in agriculture could not only adversely affect the public's view of medical applications of biotechnology and the science of genomics as a whole, but could seriously undermine confidence in the food system (PEW, 2006). In accordance with Broom (2009) we cannot depend upon the moral consciences of those who develop and use transgenic animals and so specific legislation is needed such that modifications of animals which are carried out for commercial purposes only but which result in poor welfare and by implication disregard the safety, wholesomeness and quality of food should not be permitted.

11.3 Organic farming, animal welfare, and food safety

The primary interest of consumers of organic produce relates to supposed beneficial effects on the environment and human health through the reduced use of chemicals (medication, pesticides and fertilisers). However, the public also perceives that organic farming results in improved animal welfare (Roitner-Schobesberger *et al.*, 2008) and these perceived animal welfare benefits are an important factor in the motivation of consumers to buy organic produce (Harper and Henson, 1998). Indeed, the majority of codes and regulations that deal with organic animal production systems include specific standards dealing with animal welfare. Such minimal organic standards and their implementation via certification

© Woodhead Publishing Limited, 2011

procedures are likely to provide several preconditions for good living conditions for farm animals. They primarily pertain to 'respect for aspects of the animals' innate behaviour' and 'meeting animals species specific behavioural needs'. This is mainly achieved by offering less restrictive environments (Hörning, 1998) and providing more variety in the diet (Leeb and Baumgartner, 2000). In organic farming, unrestricted environments are synonymous with loose housing, free-ranging, access to an outdoor area, straw bedding, or pasture. These do facilitate more behavioural freedom and should improve welfare but animals are also brought into closer contact with the environment (soil, water, wild fauna, etc.). This is a risk factor for animal health security, the prevalence of zoonotic diseases, and the presence of toxic residues in the food, which are three of the primary food safety concerns with organic farming (SAFO, 2004).

11.3.1 Animal health security

Closed herds and flocks and improved health security on farms are recommended practices of the European Organic Livestock Standards (EU Regulation 1804/1999). However, it can be difficult to reconcile this with organic practices (e.g. free ranging) that expose livestock to increased disease challenge. In addition, other bio-security precautions can be more difficult to implement on organic farms. For example, it can be difficult to source disease-free animals when buying in new stock because of the scarcity of organic stock (SAFO, 2004). Furthermore, as vaccine use is restricted, the risk of buying in animals with diseases of a sub-clinical or immuno-suppressive nature such as bovine virus diarrhoea (BVD) is higher (SAFO, 2004). Should these compromised animals contract other infections their welfare may be compromised as the use of medication is strictly controlled.

11.3.2 Zoonotic diseases

Zoonoses such as *Campylobacter* spp. and *Toxoplasma* gondi are important causes of human illness in both the developed and developing world and also appear to be more of a problem in organic compared to conventional systems because of greater exposure to the environment (Kijlstra and Bos, 2008). For example, Heuer *et al.* (2001) found that while only 49.2% of extensive indoor and 36.7% of conventional flocks on farms in Denmark had *Campylobacter* spp. isolates, 100% of organic farms tested positive. *Campylobacter* spp. infection has few clinical effects on poultry but practices to reduce its risk (e.g. improved hygiene practices and better farmer attention to the birds) (Huneau-Salaün *et al.*, 2007; McDowell *et al.*, 2008) are also associated with improved welfare. Furthermore, even though zoonotic agents may be more likely to be present on organic bird carcasses, the bacteria are less likely to be resistant to antibiotics compared with isolates from conventionally farmed birds, which has important implications for treatment (Miranda *et al.*, 2008). Organically farmed pigs are also more likely to be infected with *Toxoplasma* because of increased exposure

© Woodhead Publishing Limited, 2011

to the pathogen (Dubey, 2009). Moreover, unlike *Campylobacter* in poultry, infection with *Toxoplasma* can cause mortality in pigs, particularly in neonates, and thus can have a negative impact on the animals' welfare (Dubey, 2009). In general, however, there are significant welfare benefits for organically farmed pigs compared with intensively reared pigs not least because they are less susceptible to respiratory diseases (Leeb and Baumgartner, 2000).

11.3.3 Toxic residues

Organic food production prohibits the use of conventional synthetic pesticides and only a very limited selection of biological, naturally derived and simple chemical substances are permitted for use. These materials are in turn used infrequently meaning that organic foods are largely free of pesticides (Baker *et al.*, 2002; Benbrook *et al.*, 2008). However, the increased exposure of livestock in organic systems to the environment also means that they are more exposed to environmental contaminants. This explains why, for example, free-foraging laying hens produce eggs that have higher dioxin levels than conventionally farmed birds (de Vries *et al.*, 2006). Access to the outdoor run allows these birds to consume soil, insects and worms, all of which contain dioxins which are efficiently transferred to the egg yolk (de Vries *et al.*, 2006). Management measures that reduce dioxin levels in animal products (i.e. increasing flock size and reducing time spent outside; de Vries *et al.*, 2006) go against the principles of good welfare inherent to organic farming standards.

Supplementation of feed with trace elements (TE) (e.g. Cu, I, Mg) is common practice as TE deficiency is an important cause of poor performance, fertility and disease problems in farm animals. The amount of TE supplementation is well regulated, which is important to avoid deficiencies in animals but also because their deposition in edible tissues and animal products has food safety implications for consumers. Organic farming encourages the nutritional needs of the animals to be met using minerals and TE of natural origin (e.g. herbs and mixed forages) although supplementation is sometimes required for high producing animals (Coonan *et al.*, 2002). Organic forms of TE are thought to be more bioavailable than inorganic forms, and as a consequence, lower levels of supplementation could be sufficient to meet nutritional needs. However, higher bioavailability could lead to increased deposition in animal products, and thus increased carryover of potentially toxic elements to consumers (Mantovani *et al.*, 2010). Several studies have shown that animals supplemented with TEs from organic sources have greater TE concentrations in the carcass, compared with animals fed TEs from an inorganic source (e.g. Vignola *et al.* 2009). Although Kessler *et al.* (2003) found no difference in Zn levels in the carcasses of fattening bulls fed either an organic or inorganic source of Zn. Thus, risk assessment needs to be carried out on new organic sources for each TE, in order to determine whether lower supplementation limits are necessary.

© Woodhead Publishing Limited, 2011

11.4 Conclusions

Although there is a clear link between animal welfare and food safety, the relationship is not uni-directional. Systems and practices associated with good animal welfare can potentially have positive and negative effects on food safety. Nevertheless, many of the negative effects on food safety are associated with the increased contact with the environment afforded by free ranging and organic systems. Such systems enhance the behavioural freedom of farm animals and hence improve one aspect of their welfare. However, the perception that behavioural freedom is the only welfare aspect of importance to consumers of organic produce has meant that other aspects of animal welfare (freedom from hunger and thirst, freedom from discomfort, pain, disease and injury and freedom from fear and distress) have been largely ignored. If the high level of integrity enjoyed by the organic sector is to be maintained it is essential that these other aspects of animal welfare are also taken into consideration. Indeed clarification of what exactly consumers expect from 'organic' animal welfare is going to be an essential component of the resolution to the conflict between animal welfare objectives and the food safety objectives in organic farming.

In general, the better framework for animal health and welfare management offered by organic standards means that the food safety risks posed by agro-chemical and veterinary medications and the development of antibiotic resistant pathogenic microorganisms are considerably reduced compared to conventional systems (Leifert *et al.*, 2008). In conventional systems, where the vast majority of animals are still farmed, mental and physical stress combined with genetic manipulations contribute to a myriad of problems with food safety implications, examples of which were documented in this chapter. Conversely, the positive relationship between good animal welfare and safer and healthier food is almost unanimous. Where conflicts will arise, past experience indicates that zero tolerance approaches to food safety risks and indeed environmental threats mean that animal welfare considerations will be sacrificed. This approach will not fit well within the sustainable food production systems envisaged for the future. As international standards governing sustainable animal production systems begin to develop, enormous effort is clearly required to integrate standards for good animal welfare with food safety standards and to ensure that they are not mutually exclusive. One of the first steps will be to develop better ways of measuring, monitoring and marketing animal welfare. The EU-funded Welfare Quality® project is a good example of an improved system for on-farm welfare assessment for several species. It has also led to the development of accepted procedures for the standardised conversion of welfare measures into accessible and understandable product information. Such developments will not only contribute to the raising of animal welfare standards but will also help to ensure that all the stakeholder requirements are being met. Ultimately, the role that high welfare standards play in ensuring the safety and healthiness of food will have to be acknowledged. This realisation will not only improve the quality and safety

© Woodhead Publishing Limited, 2011

of the food we eat but will also play an important role in improving the welfare of millions of farm animals.

11.5 References

ADAMS, N. R. and J. R. BRIEGEL (2005) Multiple effects of an additional growth hormone gene in adult sheep. *J. Anim. Sci.* 83: 1868–1874.

BAKER, B. P., C.M. BENBROOK, E. GROTH 3RD and K. LUTZBENBROOK (2002) Pesticide residues in conventional, integrated pest management (IPM)-grown and organic foods: insights from three US data sets. *Food Addit. Contam.* 19: 427–446.

BELL, D. D. (2003) Historical and current molting practices in the U.S. table egg industry. *Poult. Sci.* 82: 965–970.

BENBROOK, C., X. ZHAO, J. YÁÑEZ, N. DAVIES and P. ANDREWS (2008) *State of Science Review: New evidence confirms the nutritional superiority of plant-based organic foods*. The Organic Center.

BERKOWITZ, D. B. and I. KRYSPIN-SORENSEN (1994) Transgenic fish: Safe to eat? *Biotechnology* 12: 247–251.

BIONDI, M. and L.-G. ZANNINO (1997) Psychological stress, neuroimmuno-modulation, and susceptibility to infectious diseases in animals and man: a review. *Psychother. Psychosom.* 66: 3–26.

BLOKHUIS, H. J., L. J. KEELING, A. GAVINELLI and J. SERRATOSA (2008) Animal welfare's impact on the food chain. *Trends in Food Science & Technology* 19: S79–S87.

BROOM, D. M. (1993) A usable definition of animal welfare. *J. Agr. Environ. Ethics* 6: 15–25.

BROOM, D. M. (1994) The effects of production efficiency on animal welfare. In E. A. Huisman, J. W. M. Osse, D. van der Heide, S. Tamminga, B. L. Tolkamp, W. G. P. Schouten, C. E. Hollingsworth and G. L. van Winkel (eds), *Biological basis of sustainable animal production. Proc. 4th Zodiac Symp. EAAP Publ. 67*, Wageningen: Wageningen Press, pp. 201–210.

BROOM, D. M. (2006) Behaviour and welfare in relation to pathology. *Appl. Anim. Behav. Sci.* 97: 73–83.

BROOM, D. M. (2009) The roles of industry and science, including genetic selection, in improving animal welfare. *Lucrări stiintifice Zootehnie si Biotehnologii* 42: 532–546.

BROOM, D. M. and R. D. KIRKDEN (2004) Welfare, stress, behaviour and pathophysiology. In R. H. Dunlop and C. H. Malbert (ed.), *Veterinary Pathophysiology*, Blackwell, Ames, IA, pp. 337–369.

BUTLER, W. R. and R. D. SMITH (1989) Interrelationships between energy balance and postpartum reproductive function in dairy cattle. *J. Dairy Sci.* 72: 767–783.

BYRD, J. A., D. E. CORRIER, M. E. HUME, R. H. BAILEY, L. H. STANKER and B. M. HARGIS (1998) Incidence of Campylobacter in crops of preharvest market-age broiler chickens. *Poult. Sci.*77: 1303–1305.

CAGIENARD, A., G. REGULA and J. DANUSER (2005) The impact of different housing systems on health and welfare of grower and finisher pigs in Switzerland. *Prev. Vet. Med.* 68: 49–61.

CALLAWAY, T. R., J. L. MORROW, T. S. EDRINGTON, K. J. GENOVESE, S. DOWD, J. CARROLL, J. W. DAILEY, R. B. HARVEY, T. L. POOLE, R. C. ANDERSON and D. J. NISBET (2006) Social stress increases fecal shedding of *Salmonella* Typhimurium by early weaned piglets. *Current Issues in Intestinal Microbiology* 7: 65–72.

© Woodhead Publishing Limited, 2011

CASEWELL, M., C. FRIIS, E. MARCO, P. McMULLIN and I. PHILLIPS (2003) The European ban on growth-promoting antibiotics and emerging consequences for human and animal health. *J. Antimicrob. Chemoth.* 52: 159–161.

CASTELLINI, C., C. MUGNAI and A. DAL BOSCO (2002) Meat quality of three chicken genotypes reared according to the organic system. *Ital. J. Food Sci.* 14: 401–412.

COONAN, C., C. FREESTONE-SMITH, J. ALLEN and D. WILDE (2002) Determination of the major mineral and trace element balance of dairy cows in organic production systems. In Kyriazakis and Zervas, (ed.), *Proceeding of organic meat and milk from ruminants*, Athens, October 4–6, EAAP Publication 106, pp. 181–183.

DAWKINS, M. S. (2006) A user's guide to animal welfare science. *Trends in Ecology & Evolution* 21: 77–82.

DE PASSILLE A. M. and J. RUSHEN (2005) Food safety and environmental issues in animal welfare. *Rev. Sci. Tech. OIE.* 24: 757–766.

DE VRIES, M., R. P. KWAKKEL and A. KIJLSTRA (2006) Dioxins in organic eggs – a review. *NJAS – Wageningen Journal of Life Sciences* 54: 207–221.

DEVLIN, R. H., D. SAKHRANI, W. E. TYMCHUK, M. L. RISE and B. GOH (2009) Domestication and growth hormone transgenesis cause similar changes in gene expression in coho salmon (*Oncorhynchus kisutch*). *Proc. Natl. Acad. Sci. USA* 106: 3047–3052.

DUBEY, J. P. (2009) Toxoplasmosis in pigs – The last 20 years. *Vet. Parasitol.* 164: 89–103.

DUNCAN, I. J. H. (2001) Animal welfare issues in the poultry industry: Is there a lesson to be learned? *J. App. Anim. Welf. Sci.* 4: 207–221.

DUNHAM R. A. (2003) Status of genetically modified (transgenic) fish: research and application. Working paper topic 2. Food and Agriculture Organization/World Health Organization expert hearings on biotechnology and food safety. Website: ftp://ftp.fao.org/es/esn/food/GMtopic2.pdf (accessed on 13 September 2010).

(EFSA) EUROPEAN FOOD SAFETY AUTHORITY (2005) The welfare aspects of various systems of keeping laying hens. *The EFSA Journal* 197, 1–23.

EKKEL, E. D., E. A. VAN DOORN, M. J. C. HESSING and M. J. M. TIELEN (1995) The specific-stress free housing system has positive effects on productivity, health, and welfare of pigs. *J. Anim. Sci.* 73: 1544–1551.

ELDER, R. O., J. E. KEEN and G. R. SIRAGUSA (2000) Correlation of enterohemorrhagic *Escherichia coli* O157 prevalence in feces, hides, and carcasses of beef cattle during processing. *Proc. Natl. Acad. Sci. USA* 97: 2999–3003.

EMMERSON, D. A. (1997) Commercial approaches to genetic selection for growth and feed conversion in domestic poultry. *Poult. Sci.* 76: 1121–1125.

ENSER, M., K. G. HALLETT, B. HEWETT, G. A. J. FURSEY, J. D. WOOD and G. HARRINGTON (1998) Fatty acid content and composition of UK beef and lamb muscle in relation to production system and implications for human nutrition. *Meat Sci.* 49: 329–341.

FAURE, J. M., W. BESSAI and R. B. JONES (2003) Direct selection for improvement of animal well-being. In: W. Muir and S. Aggrey (ed.), *Poultry Breeding and Biotechnology*. Wallingford, Oxon: CAB International, pp. 221–245.

FISHER, M. W. and D. J. MELLOR (2008) Developing a systematic strategy incorporating ethical, animal welfare and practical principles to guide the genetic improvement of dairy cattle. *N. Z. Vet. J.* 56: 100–106.

FRASER, D. (2001) The 'New Perception' of animal agriculture: Legless cows, featherless chickens, and a need for genuine analysis. *J. Anim. Sci.* 79: 634–641.

FREESTONE, P. and M. LYTE (2010) Stress and microbial endocrinology: prospects for ruminant nutrition. *Anim.* 4: 1248–1257.

© Woodhead Publishing Limited, 2011

FREESTONE, P. P. E., S. M. SANDRINI, D. HAIGH and M. LYTE (2008) Microbial endocrinology: how stress influences susceptibility to infection. *Trends Microbiol.* 16: 55–64.

GALLO, L., P. CARNIER, M. CASSANDRO, R. MANTOVANI, L. BAILONI, B. CONTIERO and G. BITTANTE (1996) Change in body condition score of Holstein cows as affected by parity and mature equivalent milk yield. *J. Dairy Sci.* 79: 1009–1015.

GRANDIN, T. (1997) *Good management practices for animal handling and stunning.* Washington, DC: American Meat Institute.

GRANDIN, T. (2001) *Progress in livestock handling and slaughter techniques in the United States, 1970–2000.* The state of the animals. Humane Society Press. D. Salem, A.N. Rowan (eds), pp. 101–110.

GREGER, M. (2010) Trait selection and welfare of genetically engineered animals in agriculture. *J. Anim. Sci.* 88: 811–814.

GRUMMER, R. R. (1991) Effect of feed on the composition of milk fat. *J. Dairy Sci.* 74: 3244–3257.

HALLERMAN, E. M., E. McLEAN and I. A. FLEMING (2007) Effects of growth hormone transgenes on the behaviour and welfare of aquacultured fishes: a review identifying research needs. *App. Anim. Behav. Sci.* 104: 265–294.

HARPER, G. C. and S. J. HENSON (1998) Consumer Concern about Animal Welfare and the Impact on Food Choice: comparative literature review. EU FAIR CT98-3678.

HARRISON, R. (1964) *Animal Machines: The new factory farming industry.* London: Vincent Stuart Publishers.

HARRISON R. O., S. P. FORD, J. W. YOUNG, A. J. CONLEY and A. E. FREEMAN (1990) Increased milk production versus reproductive and energy status of high producing dairy cows. *J Dairy Sci.* 73: 2749–2758.

HEUER O. E., K. PEDERSEN, J. S. ANDERSEN and M. MADSEN (2001) Prevalence and antimicrobial susceptibility of thermophilic *Campylobacter* in organic and conventional broiler flocks. *Lett. Appl. Microbiol.* 33: 269–274.

HICKS, T. A., J. J. McGLONE, C. S. WHISNANT, H. G. KATTESH and R. L. NORMAN (1998) Behavioral, endocrine, immune, and performance measures for pigs exposed to acute stress. *J. Anim. Sci.* 76: 474–483.

HOLT, P. S., N. P. MACRI and R. E. PORTER JR. (1995) Horizontal transmission of *Salmonella enteritidis* in molted and unmolted chickens. *Avian Diseases* 39: 239–249.

HÖRNING, B. (1998) Tiergerechtheit und Tiergesundheit in ökologisch wirtschaftenden Betrieben. *Dtsche Tierärtzl. Wochenschr.* 105: 313–321.

HUMPHREY, T. (2006) Are happy chickens safer chickens? Poultry welfare and disease susceptibility. *Br. Poult. Sci.* 47: 379–391.

HUNEAU-SALAÜN, A., M. DENIS, L. BALAINE and G. SALVAT (2007) Risk factors for *Campylobacter* spp. colonization in French free-range broiler-chicken flocks at the end of the indoor rearing period *Prev. Vet. Med.* 80: 34–48.

INGVARTSEN K. L., R. J. DEWHURST and N. C. FRIGGENS (2003) On the relationship between lactational performance and health: is it yield or metabolic imbalance that cause production diseases in dairy cattle? A position paper. *Livest. Prod. Sci.* 83: 277–308.

JONES, P. H., J. M. ROE and B. MILLER (2001) Effects of stressors on immune parameters and on the faecal shedding of enterotoxigenic *Escherichia coli* in piglets following experimental inoculation. *Res. Vet. Sci.* 70: 9–19.

JONES, R. B. (1997) Fear and distress. In: M. C. Appleby and B. O. Hughes (eds), *Animal Welfare*, Wallingford, Oxon: CAB International, pp. 75–87.

KAISER, M. (2005) Assessing ethics and animal welfare in animal biotechnology for farm production. *Rev. Sci. Tech. Off. Int. Epiz.* 24: 75–87.

© Woodhead Publishing Limited, 2011

KESSLER, J., I. MOREL, P.-A. DUFEY, A. GUTZWILLER, A. STERN and H. GEYER (2003) Effect of organic zinc sources on performance, zinc status and carcass, meat and claw quality in fattening bulls. *Livest. Prod. Sci.* 81: 161–171.

KHACHATOURIANS, G. G. (1998) Agricultural use of antibiotics and the evolution and transfer of antibiotic-resistant bacteria. *Can. Med. Assoc. J.* 159: 1129–1136.

KIJLSTRA, A. and A. P. BOS (2008) Animal welfare and food safety: danger, risk and the distribution of responsibility. 18th IFOAM Organic World Congress, Modena, Italy, June 16–20, 2008. Website http://orgprints.org/view/projects/conference.html (accessed on September 14th 2010).

KLETER, G. A. and H. A. KUIPER (2002) Considerations for the assessment of the safety of genetically modified animals used for human food or animal feed. *Livest. Prod. Sci.* 74: 275–285.

LEEB, T. and J. BAUMGARTNER (2000) Husbandry and health of sows and piglets on organic farms in Austria. Animal health and welfare aspects of organic pig production. In: T. Alföldi, W. Lockeretz and U. Niggli (eds), Proceedings of the 13th International IFOAM Scientific Conference 28–31 August 2000, Basel, VDF Hochschulverlag an der Eidgenössische Technische Hochschule Zürich, Zürich, p. 361.

LEIFERT, C., U. KOEPKE, M. BONDE, R. STANLEY, U. HAMM, G. WYSS, C. BENBROOK, J. HAJSLOVA, L. LUECK and J. COOPER (2008) QLIF Workshop 2: Safety of foods from organic and low input farming systems. 4th QLIF Congress in the frame of the 2nd ISOFAR Conference, Modena, Italy, June 18–20, 2008 (accessed at http://www.orgprints.org/13379 on 14 September 2010).

LLAMAS MOYA, S., L. A. BOYLE, P. B. LYNCH and S. ARKINS (2006) Part 1 – Stress biology in the pig. *The Pig Journal* 57: 8–29.

LUITING, P. (1990) Genetic-variation of energy partitioning in laying hens – causes of variation in residual reed consumption. *World Poultry Sci. J.* 46: 133–152.

LUTGENDORF, S. K. (2001) Life, liberty and the pursuit of happiness: good welfare in humans. In D.M. Broom (ed.), *Coping with Challenge: Welfare in Animals including Humans*, Dahlem University Press, Berlin, pp. 49–62.

MAES, D., H. DELUYKER, M. VERDONCK, F. CASTRYCK, C. MIRY, B. VRIJENS and A. DE KRUIF (2000) Herd factors associated with the seroprevalences of four major respiratory pathogens in slaughter pigs from farrow-to-finish pig herds. *Vet. Res.* 31: 313–327.

MANTOVANI, A., C. FRAZZOLI and F. CUBADDA (2010) Organic forms of trace elements as feed additives: assessment of risks and benefits for farm animals and consumers. *Pure Appl. Chem.* 82: 393–407.

McDOWELL., S. W. J., F. D. MENZIES, S. H. McBRIDE, A. N. OZA, J. P. McKENNA, A. W. GORDON and S. D. NEILL (2008) *Campylobacter* spp. in conventional broiler flocks in Northern Ireland: epidemiology and risk factors. *Prev. Vet. Med.* 84: 261–276.

McEVOY, J. M., A. M. DOHERTY, M. FINNERTY, J. J. SHERIDAN, L. McGUIRE, I. S. BLAIR, D. A. McDOWELL and D. HARRINGTON (2000) The relationship between hide cleanliness and bacterial numbers on beef carcasses at a commercial abattoir. *Letters in Applied Microbiology* 30: 390–395.

McEWEN, S. and P. J. FEDORKA-CRAY (2002) Antimicrobial use and resistance in animals. *Clin. Infect. Dis.* 34: 93–106.

McPARLAND, S. and D. P. BERRY (2010) Milking the milk. *TResearch* 5: 38–39.

MENCH, J. A. (1999) Ethics, animal welfare and transgenic farm animals. In J. D. Murray, G. B. Anderson, A. M. Oberbauer and M. M. McGloughlin (eds) *Transgenic Animals in Agriculture*. Wallingford, Oxon: CABI Publishing, pp. 251–68.

MIRANDA, J. M., M. GUARDDON, B. I. VÁZQUEZ, C. A. FENTE, J. BARROS-VELÁZQUEZ, A. CEPEDA

© Woodhead Publishing Limited, 2011

and C. M. FRANCO (2008) Antimicrobial resistance in *Enterobacteriaceae* strains isolated from organic chicken, conventional chicken and conventional turkey meat: A comparative survey. *Food Control* 19: 412–416.

MOBERG, G. P. (1987) Problems in defining stress and distress in animals. *J. Am. Vet. Med. Assoc.* 191: 1207–1211.

MOBERG, G. P. (2000) Biological response to stress: Implications for animal welfare. In G. P. Moberg and J. A. Mench (eds), *Biology of Animal Stress: Basic Principles and Implications for Animal Welfare.* Wallingford, Oxon: CABI Publishing.

OLMOS, G., L. A. BOYLE, A. HANLON, J. PATTON, J. J. MURPHY and J. F. MEE (2009a) Hoof disorders, locomotion ability and lying times of cubicle-housed compared to pasture-based dairy cows *Livestock Sci.* 125: 199–207.

OLMOS, G., J. F. MEE, A. HANLON, J. PATTON, J. J. MURPHY and L. A. BOYLE (2009b) Peripartum health and welfare of Holstein-Friesian cows in a confinement TMR system compared to a pasture based system. *Anim. Welf.* 18: 467–476.

(PEW) PEW CHARITABLE TRUSTS AND THE JOHNS HOPKINS BLOOMBERG SCHOOL OF PUBLIC HEALTH (2006) Pew Commission on Industrial Farm Animal Production. http://www.ncifap.org/about/ (accessed 13 September 2010).

PHILLIPS, C. J. C. (1997) Review article: Animal welfare considerations in future breeding programmes for farm livestock. *Anim. Breed. Abstr.* 65: 645–654.

PHILLIPS, I. (2007) Withdrawal of growth-promoting antibiotics in Europe and its effects in relation to human health. *Int. J. Antimicrob. Ag.* 30: 101–107.

PURSEL, V. G., A. D. MITCHELL, G. BEE, T. H. ELSASSER, J. P. McMURTRY, R. J. WALL, M. E. COLEMAN and R. J. SCHWARTZ (2004) Growth and tissue accretion rates of swine expressing an insulin-like growth factor 1 transgene. *Anim. Biotechnol.* 15: 33–45.

RAUW, W. M., E. KANIS, E. N. NOORDHUIZEN-STASSEN and F. J. GROMMERS (1998) Undesirable side effects of selection for high production efficiency in farm animals. A review. *Livest. Prod. Sci.* 56: 15–33.

REGULA, G. J., R. STEPHAN, J. DANUSER, B. BISSIG, U. LEDERGERBER, D. L. WONG and K. D. STARK (2003) Reduced antibiotic resistance to fluoroquinolones and streptomycin in 'animal friendly' pig fattening farms in Switzerland. *Vet. Rec.* 152: 80–81.

REGULA, G. J., B. DANUSER, B. SPYCHER and B. WECHSLER (2004) Health and welfare of dairy cows in different husbandry systems in Switzerland. *Prev. Vet. Med.* 66: 247–264.

ROITNER-SCHOBESBERGER, B., I. DARNHOFER, S. SOMSOOK and C. R. VOGL (2008) Consumer perceptions of organic food in Bangkok, Thailand. *Food Policy* 33: 112–121.

ROSTAGNO, M. (2009) Can stress in farm animals increase food safety risk? *Foodborne Pathog. Dis.* 6: 767–775.

SACHSER, N. (2001) What is important to achieve good welfare in animals? In D.M. Broom (ed.), *Coping with Challenge: Welfare in Animals including Humans*, Berlin: Dahlem University Press, pp. 31–48.

(SAFO) SUSTAINING ANIMAL HEALTH AND FOOD SAFETY IN ORGANIC FARMING (2004) Enhancing animal health security and food safety in organic livestock production. Proceedings of the 3rd SAFO Workshop, 16–18 September 2004, Falenty, Poland.

SANDOE, P. and N. HOLDUG (1993) Transgenic animals – which worries are ethically significant. *Livest. Prod. Sci.* 36: 113–116.

SCHWAIGER, K. E.-M., V. SCHMIED and J. BAUER (2010) Comparative analysis on antibiotic resistance characteristics of *Listeria* spp. and *Enterococcus* spp. isolated from laying hens and eggs in conventional and organic keeping systems in Bavaria, Germany. *Zoonoses Public Hlth.* 57: 171–180.

SIMOPOULOS, A. P. (2002) The importance of the ratio of omega-6/omega-3 essential fatty

© Woodhead Publishing Limited, 2011

acids. *Biomed. and Pharma.* 56: 365–379.

SIMSEK, U. G., I. H. CERCI, B. DALKILIC, O. YILMAZ and M. CIFTCI (2009) Impact of stocking density and feeding regimen on broilers: Chicken meat composition, fatty acids, and serum cholesterol levels. *J. Appl. Poultry Res.* 18: 514–520.

SINSIGALLI, N. A., J. P. McMURTRY, J. A. CHERRY and P. B. SIEGEL (1987) Glucose tolerance, plasma-insulin and immunoreactive glucagon in chickens selected for high and low body weight. *J. Nutr.* 117: 941–947.

SORABJI, R. (1993) *Animal minds and human morals: The origins of the western debate.* Ithaca, NY: Cornell University Press.

SOSSIDOU, E. N. and H. A. ELSON (2009) Hens' welfare to egg quality: A European perspective. *World Poultry Sci. J.* 65: 709–718.

STOOP, W. M., H. BOVENHUIS, J. M. L. HECK and J. A. M. VAN ARENDONK (2009) Effect of lactation stage and energy status on milk fat composition of Holstein-Frieșian cows 92: 1469–1478.

(SVC) SCIENTIFIC VETERINARY COMMITTEE, ANIMAL WELFARE SECTION (1997) The welfare of intensively kept pigs. For the European Commission; Report nr Doc XXIV/B3/ ScVC/0005/1997.

TARR, P. I. and M. A. NEILL (2001) *Escherichia coli* O157: H7 *Gastroenterol Clin. N.* 30: 735.

TERLOUW, E. M. C., C. ARNOULD, B. AUPERIN, C. BERRI, E. LE BIHAN-DUVAL, V. DEISS, F. LEFEVRE, B. J. LENSINK, B. J. and L. MOUNIER (2008) Pre-slaughter conditions, animal stress and welfare: current status and possible future research. *Anim.* 2: 1501–1517.

TOSCANO, M. J., T. J. STABEL, S. M. BEARSON, B. L. BEARSON and D. C. LAY JR. (2007) Cultivation of Salmonella enterica serovar typhimurium in a norepinephrine-containing medium alters *in vivo* tissue prevalence in swine. *J. Exp. Anim. Sci.* 43: 329–338.

VAN REENAN, C. G., T. H. MEUWISSEN, H. HOPSTER, K. OLDENBROEK, T. H. KRUIP and H. J. BLOKHUIS (2001) Transgenesis may affect farm animal welfare: a case for systematic risk assessment. *J. Anim. Sci.* 79: 1763–1779.

VIGNOLA, G., L. LAMBERTINI, G. MAZZONE, M. GIAMMARCO, M. TASSINARI, G. MARTELLI and G. BERTIN (2009) Effects of selenium source and level of supplementation on the performance and meat quality of lambs. *Meat Sci.* 81: 678–685.

VLISIDOU, I., M. LYTE, P. M. VAN DIEMEN, P. HAWES, P. MONAGHAN, T. A. WALLIS and M. P. STEVENS (2004) The neuroendocrine stress hormone norepinephrine augments *Escherichia* coli O157:H7-induced enteritis and adherence in a bovine ligated ileal loop model of infection. *Infect. Immun.* 72: 5446–5451.

WHAY, H. R., D. C. J. MAIN, L. E. GREEN and A. J. F. WEBSTER (2003) Assessment of the welfare of dairy cattle using animal-based measurements: direct observations and investigation of farm records *The Vet. Rec.* 153: 197–202.

WIERUP, M. (2001) The Swedish experience of the 1986 year ban of antimicrobial growth promoters, with special reference to animal health, disease prevention, productivity, and usage of antimicrobials. *Microb. Drug Resist.* 7: 183–190.

© Woodhead Publishing Limited, 2011

Part III

Authenticity and origin of food products

© Woodhead Publishing Limited, 2011

12

Detection and traceability of genetically modified organisms in food supply chains

Y. Bertheau, National Institute for Agricultural Research (INRA), France

Abstract: Traceability in agri-food supply chains is mandatory in all 27 European Union Member States. Novel ingredients or foodstuffs such as genetically modified organisms (GMOs) are subject to specific regulations and traceability, mainly due to their media coverage. This chapter provides an overview of the various regulations, social, economic and particularly technological facts regarding traceability, as defined in the ISO standard, and GMO testing in non-GMO supply chains. It also discusses some of the potential uses of the concepts, methods and strategies used in the field of GMOs in other test areas, a possibly unexpected point of agreement between some third countries and the European Union.

Key words: GMO, genetically modified organism, genetically engineered organism, supply chain coexistence, control, enforcement, detection, regulation.

12.1 History of genetically modified organisms (GMOs)

12.1.1 Background

The 1970s marked the emergence of molecular biology and the periods of intense development that enabled scientists to extend the scope of their work from prokaryotes to eukaryotes, plants in particular. In 1977, MD Chilton demonstrated that crown gall resulted from the transfer of phytopathogenic

© Woodhead Publishing Limited, 2011

bacteria from a fragment of the Ti plasmid to transformed plant cells. In 1983, three groups were able to use this natural plasmid as a vector in plant cell transformation. Other plant transformation methods were subsequently applied, for monocotyledons in particular. Transgenesis in plants became an important instrument in basic research and has since led to a series of commercial applications (Davison and Bertheau, 2008, 2010).

Various techniques are used for genetically modified animals (Gama Sosa *et al.*, 2010). While many applications in the production of pharmaceutical products are foreseeable in the short term, the first commercial application in the food industry is most likely to be the AquAdvantage genetically modified salmon by AquaBounty, currently undergoing deregulation in the United States. A panel of experts commissioned by the US Food and Drug Aadministration has recently issued a mixed opinion recommending further studies. Fourteen years after AquAbounty's initial request, a change to the GM animal inventory remains uncertain.

Genetic engineering is rapidly evolving with the use of more precise transformation systems, plastid transformation, silencing systems, not to mention synthetic biology (Cohan, 2010; Frizzi and Huang, 2010; Glick *et al.*, 2010; Weathers *et al.*, 2010).

Some of these techniques may lead to organisms with 'cleaner' genetic modifications, while others may be used to create completely new organisms for which risk assessment methods must still be defined. Some of the current and new transformation systems could lead to the creation of plants which promote genetically modified (GM) and non-GM crop coexistence by encouraging genetically modified organisms (GMO) bioconfinment. Yet the disastrous 'Terminator' affair has only impeded the availability of crops using technologies such as 'GURT'.

Following the first sales of delayed ripening tomatoes in 1994, the first commercial herbicide-tolerant soybean crops were launched in the United States and the first GMOs arrived in Europe in 1996. This is when the controversy surrounding GMOs really took off and led to the withdrawal in 1998 of the Flavr Savr tomato, which, despite its attractive price, was not popular among all consumers due to its organoleptic properties.

From then on, GMO crop cultivation has significantly developed, covering more than 134 million hectares according to the International Service for the Acquisition of Agri-Biotech Applications (ISAAA; James, 2010). The main producers are, in descending order, the USA, Brazil, Argentina, India, Canada, China, Paraguay and South Africa. In Europe, Spain is the main producer with almost 100 000 hectares of maize used for animal feed. Fourteen million farmers (including thirteen million smallholders, generally in developing countries) allegedly use GMOs.

The main crops in descending order are: soybean, maize, cotton and canola. The most widely adopted traits are, in decreasing order, herbicide tolerance, followed by GMO trait stacking (James, 2010; Que *et al.*, 2010). After naturally crossing transgenic plants containing individual transgenes, these GMOs

© Woodhead Publishing Limited, 2011

combine several traits such as herbicide tolerance and insect resistance. In third place come insect-resistant GMOs.

This data may be impressive, yet it is noteworthy that the ISAAA is the only association to provide worldwide figures. Funded in part by firms which produce GMO seed, various non-governmental organisations (NGOs) have expressed doubts as to the reliability of the data published. This monopoly on information is set to gain momentum as the US Department of Agriculture (USDA) has recently decided to no longer create statistics regarding the GMO adoption rates in the USA.

In 2004, the area used for GMO crops represented 2–4% of total cultivated areas in the world (Semal, 2005), yet the proportion of GMOs seemed to vary, in 2009, from 20% for canola to 75% for soybean (James, 2010). World trade is mostly dependent on the 'ABCD' quartet (i.e. ADM, Bunge, Cargill, Louis Dreyfus) which is present on the GMO and non-GMO product markets (Green and Hervé, 2006).

European consumers have taken up the nutritional, environmental and lifestyle aspects of the global controversy, and have up to now shown reticence as to the benefits of GMOs, unlike Americans (Nielsen, 2001; Nielsen and Anderson, 2001; Arvanitoyannis and Krystallis, 2005; Gaskell *et al.*, 2006; Bonny, 2008; Ervin *et al.*, 2010; Rommens, 2010; Wield *et al.*, 2010).

As a result, a set of directives and regulations has been phased in to protect the freedom of choice of consumers, and producers, through detailed labelling and the recommendation of the implementation of national crop coexistence rules, following the European principle of subsidiarity (European Commission, 2000a,b, 2003a,b, 2010). GMO-specific measures further support a general traceability system in agri-food chains (European Commission, 2002) and labelling of all novel foods and novel food ingredients (European Commission, 1997; Davison and Bertheau, 2007, 2008).

Clearly, there are differences in the opinions expressed and the consumer habits observed, and between Member States (Noussair *et al.*, 2001, 2004, 2007; Rowe, 2004; Costa-Font *et al.*, 2006; Han and Harrison, 2007; Consumerchoice Consortium, 2008; Sheldon *et al.*, 2009).

However, European citizens are generally considered wary of GMO consumption, as are those in other countries such as Japan, Korea and Taiwan, which, alongside Russia, have implemented traceability and labelling systems with thresholds which often reveal dependence on food imports (Yasui, 2004; Park, 2005; Kim and Boyd, 2006; Svec and Vassilieva, 2006; Vassilieva and Svec, 2007; Shineha and Kato, 2009).

Some believe that the anti-GMO feeling among European consumers has been bolstered by supermarkets, which in turn influence their suppliers, including European producers. However, the management of a Europe-based agri-food multinational recently announced publicly that it regretted this hostility among European consumers and claimed to be ready to supply GMO products and derivatives within a short timeframe.

European legislation does not provide for the labelling of animals reared on

© Woodhead Publishing Limited, 2011

GM feed and its position has created much unrest among various NGOs and consumer associations. This position may be justified in that tests are almost impossible to conduct and there is always a chance of falsified documentary traceability (Bertheau *et al.*, 2009). This European position may also be explained by the care taken not to shake up the current animal production model, which in Europe depends on protein imports from third countries. Yet this situation is currently changing in various Member States such as Germany and France.

With regard to its supply of proteins for animal feed, the EU is greatly dependent on soybean and CGF (Corn Gluten Feed) imports which both come predominantly from GMO producing countries. Current world trade in GMO and non-GMO products mostly depends on four traders, ABCD (Green and Hervé, 2006). European Union regions (GMO-free regions: http://www.gmofree-euregions.net:8080/servlet/ae5Ogm) have come together in order to promote the supply of proteins for animal feed that are not genetically modified. The EU's dependence on proteins and asynchronous and isolated authorisations create many problems for compound feed suppliers (Gómez-Barbero and Rodríguez-Cerezo, 2006; European Commission DG-AGRI, 2007; Stein and Rodriguez-Cerezo, 2009). Alongside talks currently under way within the *Codex Alimentarius*, there are plans to define in the near future an EU tolerance level (known as LLP i.e. Low Level Presence) for imports of GM products or derivatives, not authorised in the EU but granted authorisation in some third countries. This LLP would be initially used for products intended for animal feed.

The relatively short history of commercial GMOs has been punctuated by many cases of unauthorised GMO dissemination without checks or with unauthorised mixtures (Starlink and Bt10 maize, LL601, Bt63, KMD1 and KeFeng6 rice, FP967 linseed, Amadea potato; see for instance: http://www.gmcontaminationregister.org/). Recently, the USA has started to share the EU's concerns regarding these unauthorised releases and asynchronous and isolated authorisations (GAO, 2008).

In coming years, there may be a somewhat unexpected convergence between the USA and the EU. This prediction was given more weight recently when the American agriculture secretary reiterated, in the case of sugar beets, that Federal agencies must enable coexistence between conventional, organic and biotechnology production systems.

12.2 The European regulatory framework for genetically modified organisms (GMOs)

In 1990, European regulations began to provide a framework for dissemination in confined areas or into the environment, mainly through risk assessment (European Commission, 1990a,b). Subsequently, a specific labelling procedure was implemented via the general regulation on 'novel foods and novel

© Woodhead Publishing Limited, 2011

ingredients' (European Commission, 1997) and more specific regulations (Davison and Bertheau, 2007, 2008).

The European Food Safety Authority (EFSA), founded in 2002, became the central risk assessor for foodstuffs, including cultivated and imported GMOs. The EFSA 'GMO panel' also assesses environmental risks, which one may think would had to come under the brief of the European Environment Agency.

The creation of a European risk assessor may seem surprisingly slow in coming. As a comparison, the French agency (AFSSA; i.e. Agence Française de Sécurité Sanitaire des Aliments, French Food Safety Agency) was founded in 1998 and the *Commission du Génie Génétique* (genetic engineering, about confined uses) (CGG) and *Commission du Génie Biomoléculaire* (biomolecular engineering about dissemination in the environment) commissions were formed in 1988.

12.2.1 GMO traceability

Following various food crises in the EU (BSE, dioxin, salmonella, etc.), the traceability of foodstuffs has become compulsory in the EU for all operators, one step back – one step forward (European Commission, 2002). This ISO standard on traceability (International Organisation for Standardisation, 2007) was already implemented for GMOs, which were considered as novel foods and novel ingredients (European Commission, 1997), with a 1% threshold label for adventitious or technically unavoidable presence (European Commission, 2000a,b).

Traceability measures evolved again with directive 2001/18/EEC and regulations 1829/2003/EC and 1830/2003/EC, lowering the labelling threshold to 0.9%, and various other more technical regulations (European Commission, 2001, 2003a,b, 2004a,b). Lastly, by signing the Cartagena Protocol, the EU ensures the same quality for products intended for export as those intended for the domestic market through a regulation concerning transboundary movements (European Commission, 2003c).

Following on from the pioneering French and Belgian networks, a GMO detection network, ENGL (European Network of GMO Laboratories, http://mbg.jrc.ec.europa.eu/), chaired by the Joint Research Centre (JRC) was set up in 2002 to share information and draw up guidelines (performance criteria, unauthorised GMO detection strategy, etc.). This network supports the European Union Reference Laboratory (EURL-GMFF), hosted by the JRC's Ispra site, in charge of inter-laboratory validation of methods submitted by applicant companies.

Regulations are set to change following the evaluation of the EU legislative framework on GM food and feed which has just been completed (http://ec.europa.eu/food/food/biotechnology/evaluation/, http://ec.europa.eu/food/food/biotechnology/evaluation/docs/GMO_September_progress_report_en.pdf, http://ec.europa.eu/food/food/biotechnology/evaluation/docs/eval_EU_leg.pdf) and an assessment of the ENGL. It underscores the devolution to the DG Health and Consumers of the entire GMO dossier, as provided for by the second Barroso Commission following the departure of the former European Commis-

© Woodhead Publishing Limited, 2011

sioner for the Environment, Stavros Dimas. Also 'in charge of' the EFSA, the DG Health and Consumers has become, in addition to its role as principal of the JRC IHCP (Joint Research Centre; Institute for Health and Consumer Protection) in charge of validating GMO detection methods, the European spearhead as regards the health and environmental implications of GMOs.

Various national traceability programmes conducted in the 1990s and the European Framework programmes worked on developing detection methods. These programmes developed sampling and detection methods and strategies concerning in particular unauthorised or unknown GMOs and gene stacking, and focused on the costs of such detection methods. They also highlighted operators' practices, such as their contractual threshold being significantly lower than the European labelling threshold (see below). Operators favour supply chain certification by a third party, namely via the main 'traders' or local initiatives, paying specific attention to documentary traceability data. Sound knowledge of the critical points of food chains, generally through the HACCP method, can cut the costs of analytical traceability and controls.

Due to the difficulties encountered in the collection of the necessary GMO seed and the close ties between national and European competent authorities, these research programmes were able to considerably influence European regulations in 2003. These regulations now oblige applicant companies to provide: (i) reference material to the Commission's Joint Research Centre (JRC-IRMM, Joint Research Centre, Institute for Reference Materials and Measurements), (ii) GMO identification and quantification methods (JRC-IHCP) and (iii) a financial contribution to costs incurred.

Many participants in these programmes from public institutions were also members of the ENGL and of the delegations to the CEN (European Committee for Standardisation: www.cen.eu) and the ISO (International Organisation for Standardisation: www.iso.org) which created the current standards on GMO detection. The French standard published in 2000 on the performance criteria of methods was used as a reference for the CEN ISO/DIS 24276 standard, the principle of which (definitions and method performance criteria) was selected in other areas of detection, such as microbiology, before the actual analysis methods were standardised. A consistent set of standards covering sampling and detection methods based on proteins and DNA is now in place. A new working group was recently formed at the ISO to update the standards and consider other aspects of the identification of organisms that are or are not modified by transgenesis.

In order to ensure the interoperability and the quality of results, training courses have been offered for several years by the JRC-IHCP, for third countries in particular. Following on from the ENGL, GMO detection networks have been set up in various regions across the globe, such as around the Black Sea. These networks are in constant contact with each other and the laboratories which conduct tests in emerging countries such as China are invited to ENGL meetings. A database containing all detection methods has been created (Dong *et al.*, 2008), alongside the publication of validated European methods.

© Woodhead Publishing Limited, 2011

A quality assurance system has been set up in control laboratories and authorised national bodies grant accreditations, which may or may not be flexible, to laboratories (Zel *et al.*, 2006, 2008). This contributes to European quality systems (European Commission, 2006).

All analysis methods include various phases (sampling, crushing and reduction of laboratory samples to test portions, analyte extraction, analyses), also known as 'modules' in the field of GMOs. Various taxa or GMO quantification PCR methods each represent a module. A method is validated either by various laboratories (inter-laboratory validation), or internally (in-house validation), the former being the most common and the solution used up to now for GMO detection. Prior to GMO detection developments, analysis methods were validated for the entire analysis process including all modules.

While this 'global' approach could be used to determine measurement uncertainties, it required significant adaptations, or even new and costly validations, when the matrix to be analysed changed. The formalisation by Holst-Jensen and Berdal (2004) of the 'modular approach' for the development and validation of GMO detection methods enabled control laboratories and method developers to focus on the variables of GMO detection, namely the PCR analysis method used for the screening or identification of GMOs and of taxa, whether they represent cultivars or donor organisms of transgene sequences. Since this procedure was made formal, and also adopted in other detection sectors, detection methods developers can concentrate on the PCR modules that are alone validated by the EURL-GMFF. Recent studies have proven the independence of the modules and the ability to determine total measurement uncertainty (Bellocchi *et al.*, 2010b).

GMO labelling is mandatory in the EU. European regulations only state one labelling exemption threshold below a relative value of GMO content, expressed as a percentage, without stating the unit of measurement used. There have been various interpretations regarding the units to be used in the GMO content calculation. Seed manufacturers considered the number of seeds, traders went by mass, and control laboratories used the number of DNA molecules, as this analyte is sufficiently resistant to agri-industrial farm produce processing procedures and can be identified across the entire supply chain. The situation has now been standardised by using the DNA unit expressed as 'haploid genome equivalents' (European Commission, 2004a). The modular approach and the 'haploid genome equivalent' are now used as international standards due to their pragmatism, multiple uses and practicality which suit the needs of analysts and developers.

12.3 Current challenges for traceability in supply chains containing and 'free from' genetically modified organisms (GMOs)

12.3.1 Sampling

In traceability, sampling is the basis of all analytical tests. It aims to provide

© Woodhead Publishing Limited, 2011

control laboratories with a laboratory sample representative of the lot to be analysed.

As world trade is a key factor of many GMO-related supply chains, the first European studies focused on bulk commodities. Studies had demonstrated that, with the exception of certified commercial seed which presented a homogeneous distribution (Remund *et al.*, 2001; Laffont *et al.*, 2005), GMO presence is heterogeneous, particularly in bulk commodity cargoes, as seen for mycotoxins (Paoletti *et al.*, 2003, 2005, 2007). Sampling plans need a number of assumptions with regard to the normal or non-normal distribution of target products. These studies gave rise to a European recommendation concerning sampling, which is not routinely used due to the high level of costs involved. The main advantage of this European recommendation was to state the increment sampling procedures and the possible use of all or part of these increments according to the analysis results on the first increments.

Some countries now use ISO-based sampling plans or mycotoxin sampling plans (Brera *et al.*, 2005; Macarthur *et al.*, 2007). New sampling plans have also been developed that are not based on normal or non-normal distribution assumptions (Bellocchi *et al.*, 2006).

Various experimental and theoretical studies have been conducted with regard to sampling in the field, which is essential for supply chain traceability and coexistence. While random sampling seems to be the best statistical solution, it is too costly for routine applications. Alternative solutions have been put forward, but have proved to be too complex and expensive, or sensitive to different factors such as higher levels of dryness along the edge of fields (DEFRA, 2005; Messeguer *et al.*, 2006; Pla *et al.*, 2006; Sustar-Vrozlic *et al.*, 2010).

Therefore, there is a current lack of consensus sampling plans for bulk commodities, or field sampling, from GAFTA, the EU or ISO, which are often very general. One of the practical repercussions of this lack of consensus, with inter-laboratory measurement uncertainty and the room for manoeuvre that operators wish to protect for the management of later contaminations, is the contractual threshold used by operators, which is considerably lower than the European labelling threshold (0.01%–0.1% depending on the contract). That issue is not specific to the GMO domain.

12.3.2 Cost-benefit analysis for traceability

Operators often complain about the costs of traceability obligations. Despite a number of articles on this issue, some of which are relatively unreliable in the figures proposed, it is currently impossible to have a clear idea, especially in quantitative terms, of the benefit-cost ratio of GMO traceability.

Operators do not consider their obligations under regulation 178/2002 separately from their GMO-specific obligations. Presented analysis cost estimates vary from simple screening, catalogue prices, which do not include price negotiations related to the analysis quantities, to comprehensive estimates

© Woodhead Publishing Limited, 2011

including tests for all possible GMOs. While multinationals claim many analyses across the supply chains, in-depth discussions have revealed that in practice only a few analyses are conducted, most upstream of the supply chain on raw materials that are easier to analyse, while analysis reports alone are used further downstream. Similarly, sampling costs are often overestimated whereas most samples taken are used for various analyses, such as those on mycotoxin or allergen (producing organisms) contents.

Whereas costs are always overestimated, the benefits are generally completely overlooked, such as what can be learned from PCR analysis results also used in the detection of allergen or mycotoxin producing organisms. Operators seldom consider generated value in terms of corporate image and reputation.

The only data that we have been able to collect concerns the premiums of non-GMO products, which include analysis costs but also other aspects such as practical measures for food and feed chain coexistence, which are generally stops at the gates of the farms or silos.

No reliable data is currently available on the actual costs incurred by traceability and tests on GMO products in non-GMO supply chains. Given the practices of operators (analysis of relatively raw materials upstream of the chain followed by documentary traceability), it seems unlikely that the cost of GMO analytical traceability has a real impact on the end price of non-GMO products. The price of raw materials only represents a small share of the end costs of the products on our supermarket shelves, with the exception of direct sales of unprocessed products.

It seems likely that we will never obtain such quantitative data from the interaction of supply chain players as operators consider such data to be a trade secret. The lack of empirical economists with a good understanding of operators' practices favours this situation. Operators can thus justify their higher costs, which are passed on in non-GMO premiums, and are at an advantage in negotiations with buyers and public authorities.

12.3.3 Unambiguous taxonomic identification

The identification of taxa (cultivars or organisms with sequences used in GMO inserts) is necessary for two reasons. Firstly, for relative quantification as requested for in the European regulations (see equation below). Secondly, to prevent false positives in GMO 'screening' methods by detecting 'donor organism'. The determination of the presence of donor organisms is particularly important when the differential quantitative PCR method (dQ-PCR) is used to detect the presence of unauthorised or unknown GMOs.

$$\text{relative GMO content} = \frac{\text{GMO specific sequences number}}{\text{taxon specific sequences number}} \times 100$$

In the case of plant species, or of lower taxonomic levels (such as the potato: *Solanum tuberosum* subsp. *tuberosum*), the genetic and geographical differences between cultivated varieties should be considered when choosing a reference

© Woodhead Publishing Limited, 2011

system that can determine the quantity of the crop. Various criteria in the choice of taxon-specific reference systems are specified in the CEN 24276 and 21570 standards and in the ENGL performance criteria. Therefore, it is required to use at least: (i) twenty cultivated varieties that are representative of geographical and genetic diversity, (ii) related species (to take into account gene introgressions in plant improvement) and (iii) where possible ancestors of the cultivar. This work has been conducted for maize as part of a reliable reference system based on the *adh* gene (Hernandez *et al.*, 2004), yet it has unfortunately not been conducted for another maize reference system, also based on the *adh* gene (Broothaerts *et al.*, 2008). The very fact that two reference systems, one stable and the other not considering mutations between varieties, are based on the same gene leads to regrettable confusion and ultimately analyst errors.

More universal identification systems for the taxa concerned by cultivated GMO species should be made available to keep analysis costs at the lowest possible level (Chaouachi *et al.*, 2007, 2008a,b,d).

This problem is already apparent for a cereal crop such as maize, and is even more complex for 'species complexes' such as canola, sugar beet (Fig. 12.1) or families such as solanaceae. For instance, the GS2 reference system based on the glutathione synthase gene provided by the applicant company for the EH7-1 sugar beet was validated by the EURL-GMFF through an inter-laboratory test despite it is not specific to the sugar beet (Fig. 12.2). The same goes for potato or canola reference systems (for example crossing with the two species from which rapeseed originates).

This highlights two problems concerning the EURL-GMFF:

- The EURL-GMFF's mandate, which should be able to refuse taxon-specific reference systems which do not come under the scope of application requested for by European regulations.
- The pressure put on the EURL-GMFF to validate methods, solely on the basis of inter-laboratory validation statistical criteria, in order to speed up the GMO authorisation process in the EU.

Such conflicts between analyst requirements, European regulatory specifications and the actual possibilities of methods leave room for possible cases of fraud.

The second area requiring unambiguous taxonomic identification is that of GMO screening tests, namely on organisms (generally viruses and bacteria) from which the sequences used in GMO inserts originate. One GMO screening method targets the sequences of these organisms due to their ubiquity in current GMOs. The presence of donor organisms must now be verified to prevent false positives. Such tests were until recently only available for the Cauliflower Mosaic Virus (CaMV) the P35S promoter of which is used in screening and for the detection of unauthorised and unknown GMOs (Feinberg *et al.*, 2005; Fernandez *et al.*, 2005; Cankar *et al.*, 2008; Chaouachi *et al.*, 2008c). Such tests are still lacking for *Agrobacterium* spp. (Pnos, Tnos sequences used in screening), for *Bacillus thuringiensis* (*bt* gene sequences used in screening) or bacteria carrying the *nptII* gene (also used in screening).

© Woodhead Publishing Limited, 2011

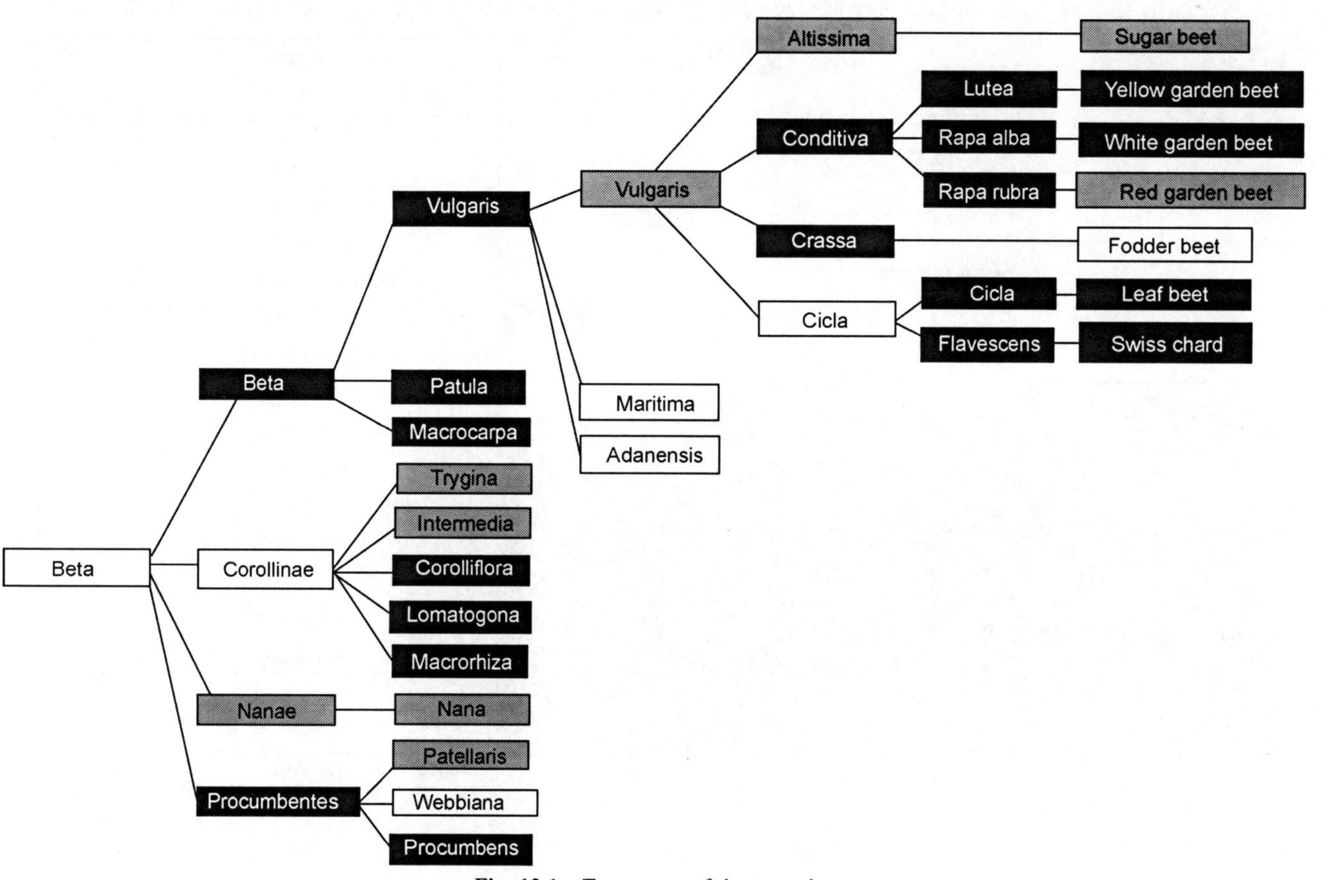

Fig. 12.1 Taxonomy of the sugar beet.

© Woodhead Publishing Limited, 2011

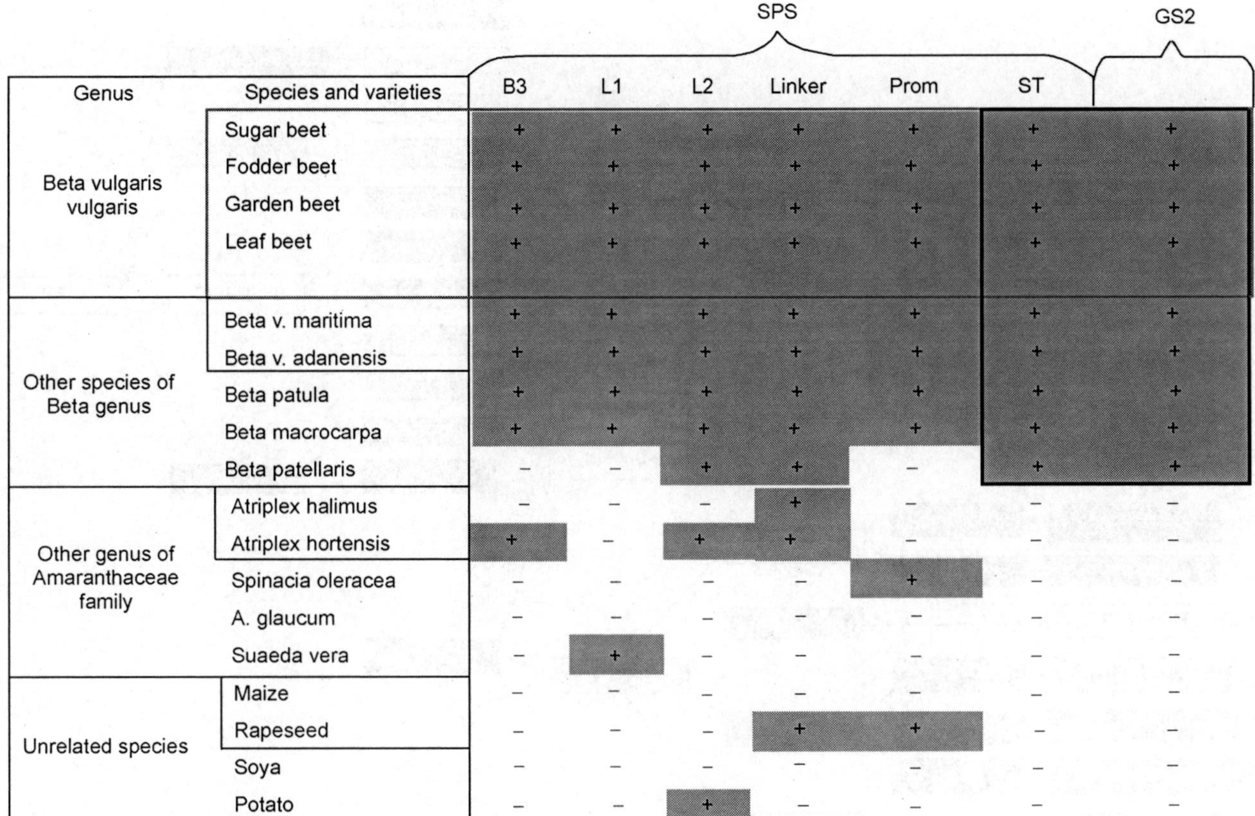

Genus	Species and varieties	B3 (SPS)	L1 (SPS)	L2 (SPS)	Linker (SPS)	Prom (SPS)	ST (SPS)	GS2
Beta vulgaris vulgaris	Sugar beet	+	+	+	+	+	+	+
	Fodder beet	+	+	+	+	+	+	+
	Garden beet	+	+	+	+	+	+	+
	Leaf beet	+	+	+	+	+	+	+
Other species of Beta genus	Beta v. maritima	+	+	+	+	+	+	+
	Beta v. adanensis	+	+	+	+	+	+	+
	Beta patula	+	+	+	+	+	+	+
	Beta macrocarpa	+	+	+	+	+	+	+
	Beta patellaris	–	–	+	+	–	+	+
Other genus of Amaranthaceae family	Atriplex halimus	–	–	–	+	–	–	–
	Atriplex hortensis	+	–	+	+	–	–	–
	Spinacia oleracea	–	–	–	–	+	–	–
	A. glaucum	–	–	–	–	–	–	–
	Suaeda vera	–	+	–	–	–	–	–
Unrelated species	Maize	–	–	–	–	–	–	–
	Rapeseed	–	–	–	+	+	–	–
	Soya	–	–	–	–	–	–	–
	Potato	–	–	+	–	–	–	–

Fig. 12.2 Specific features of two reference systems (SPS, INRA; GS2: applicant company) for sugar beet (INRA).

© Woodhead Publishing Limited, 2011

This absence of qualitative and preferably also quantitative detection methods leads to false positives and compromises analysts' ability to detect the presence of unauthorised or unknown GMOs (Cankar *et al.*, 2008).

The issue of EURL-GMFF's mandate extension once again arises as national research laboratories have not been able to provide the necessary methods.

The problem of identifying and quantifying taxa is already obvious for plant species but it has not yet been addressed for animal species. Ongoing studies of genetically modified fishes demonstrate that the current systems used to identify species and geographical origin use mostly barcoding that is not suited to the relative quantification of genetically modified animals. In an age when genetically modified salmon or pork may be deregulated in North America, it is clear that the animal sector has reached the same stage as plant GMOs in 1996 with regard to detection and therefore in terms of satisfaction of the European regulatory framework.

12.3.4 Reference material

Reference material is of the utmost importance for the application of both qualitative and quantitative detection methods. A certified reference material (CRM) by JRC-IRMM of most GMOs on the market is available in the form of seed powder with known (mass/mass) GMO quantities.

Yet in plant biology a direct relationship cannot be established between masses and DNA content, the unit recommended by the Commission in the form of haploid genome equivalent (Holst-Jensen *et al.*, 2006). Moreover, the availability of these CRMs is dependent on applicant companies and on the cultivation of plants for seed production, for which the varieties used may vary over time. Additional problems of DNA stability (which deteriorates over time), observed by our INRA laboratory as early as 2000, led the JRC-IRMM to modify its CRM preparation by conducting all operations throughout dry steps.

Following on from Japan, Europe has now turned to alternative CRMs for which the commutability between reference materials is set to be demonstrated (Trapmann *et al.*, 2002). After having tested various alternatives, such as seed genomic DNA or random amplified genomic DNA, for instance in the frame of Co-Extra, the JRC-IRMM is now developing an alternative CRM concerning plasmids containing sequences of interest (Charels *et al.*, 2007a,b, 2008; Trapmann *et al.*, 2010).

Such CRMs should ensure the long-term availability of stable reference material, at a low cost and which is easily multipliable even after GM crop withdrawal from the market, and therefore herald the end of the cultivation of some GMOs.

12.3.5 Unauthorised GMOs

The relatively short past of GMOs has been marked by the uncontrolled dissemination of GMOs that are either unauthorised (rice LL601, maize BT10,

© Woodhead Publishing Limited, 2011

linseed FP967, Amadea potato) or limited to certain uses (Starlink^{TM} maize withdrawn from the market in 2001 but still found in cargoes in 2006). The rising production in emerging countries of GMOs cultivated without prior authorisation (rice Bt63, KeFeng6, KMD1) is, alongside 'biohacking' (http://biohack.sourceforge.net/) and possible GMOs for terrorist purposes, a growing concern not only in countries which implement GM traceability and labelling but also in the USA which considers its near monopoly of initial GMO marketing, and therefore testing, threatened (GAO, 2008). Asynchronous and isolated authorisations also have a significant effect on European imports, particularly with regard to animal feed (Lheureux *et al.*, 2003; Gómez-Barbero and Rodríguez-Cerezo, 2006; Stein and Rodriguez-Cerezo, 2009, 2010a,b).

Putting the safety aspects related to the dissemination of unauthorised GMOs aside, such disseminations have had a significant economic effect on supply chains and world trade since the first Starlink^{TM} scandal. The total cost of analyses, various tests and market withdrawals is estimated at over one billion US dollars. In this regard, Greenpeace's alerts on various unauthorised GMOs have been particularly helpful.

GMO detection is therefore a challenge for public authorities and consumers, as it is for the analytical laboratories who have taken up this task at a research level, in particular through the European Co-Extra project, and an ENGL working group.

A wide range of GMO detection methods is now available. This allows analysts to use methods that are directly applicable and at a low cost such as the differential quantitative PCR (dQ-PCR) or more complex methods such as the 'matrix approach' (Cankar *et al.*, 2008; Chaouachi *et al.*, 2008d). At the top end of the scale of complexity and costs, a technique which detects all elements from plant transformation plasmids may be used to detect all GMOs which can currently be engineered (Tengs *et al.*, 2007). As with some other techniques which are also available (anchored PCR, full sequencing, etc.), this method should be limited to cases in which food safety is at risk.

We can currently believe that any GMOs able to go undetected under the current range of unauthorised and unknown GMO detection systems could only be engineered by nations or financially and technically powerful organisations able to develop new types of organism transformation tools.

12.3.6 Gene stacking

Gene stacking in the field of genetic engineering is generally defined as the natural crossing of various GMOs resulting from the initial transformation of an organism. A recent overview demonstrates the complexity of this general terminology (Taverniers *et al.*, 2008). The proportion of areas cultivated with these 'high-technology' GMOs, as some call them, is growing (James, 2010). Their purchase price is also on the rise for farmers. With eight stacked traits, SmartStax^{TM} is currently the commercial GMO with the highest number of traits.

© Woodhead Publishing Limited, 2011

In some third countries, such as the USA, the deregulation of transformation events subsequently allows all combinations in stacks. Regulations in other countries such as Japan and the EU consider GMOs resulting from stacked genes as new GMOs subject to new risk assessments and marketing authorisations. A stacked GMO therefore may not be authorised even though its constituents may be. Conversely, a number of authorisation applications in the EU concern stacked genes while the initial transformation events have not been subject to such an application. Detection methods that are able to distinguish the presence of initial transformation events or stacking in GMO mixtures or stacked GMOs are thus necessary.

The simplest solution, which is, however, the longest and most costly, involves analysing a sample kernel by kernel (Akiyama *et al.*, 2008). A different approach has been developed in Europe as part of the Co-Extra programme, using the sub-sampling technique with qualitative methods to determine the probability of the presence of stacked GMOs. This technique is unfortunately still limited to two stacks but could be improved upon to determine the presence of more traits and therefore provide greater correspondence with ongoing authorisations.

The method developed in Europe uses qualitative detection methods (Ancel *et al.*, 2010). The use of quantitative methods should provide more information, but this approach requires further modelling.

The need to detect several traits within a single organism is not specific to GMOs. It can be used to detect several genes, such as pathogens multiresistant to antibiotics.

12.3.7 'GMO-free' supply chains and detection limits

France is among the Member States which plan to define 'GMO-free' supply chains with a 0.1% threshold, thus lower than the European labelling threshold for adventitious and technically unavoidable presence. This value meets both demands from French organic producers (organic agriculture working group of the INAO; Institut National des Appellations d'Origine Contrôlée; French institute for protected designations of origin) and the limits of the current PCR quantification methods which only aim to increase the reliability of GMO quantification, at a reasonable cost, according to the 0.9% threshold.

If we consider that operators in 'GMO-free' supply chains will work along the same lines as conventional agriculture operators with a contractual threshold one third to one tenth lower than the labelling threshold, operators will consequently work close to the detection limit of most of the current PCR methods.

What are the options for operators working within this range of values? The first commonly used procedure involves requesting a 'negative PCR' from their suppliers. When issued by an accredited laboratory, and supposing that the laboratory sample is representative of the lot to be analysed, this type of result is considered sufficient as, according to CEN/ISO standards, method detection

© Woodhead Publishing Limited, 2011

limits must be close to 0.01%. In a worst case scenario, the laboratory could modify its technique to drop down to 0.01%. This qualitative approach is also advantageous for operators as it only requires positive and negative reference material and low-cost PCR methods. The lack of CRM between 0.1% and 0% should not be an obstacle in this case.

As regards positive qualitative PCR methods set at 0.01%, some operators would like to have an idea of the quantity of GMOs actually present within the 0.1% and 0.01% range, which are usually the detection and quantification limits of standard PCR methods, so that they can draw up a management strategy for positive lots. They have several options, which require at the very least careful preparation of internal standards based on CRMs. The first solution would be to improve the quantification limit of usual quantitative methods, for example by using an optimisation plan by attributes, as the theoretical detection limit of a PCR is 0.00002%. One alternative would be to use a method other than PCR such as the method described by Corbisier *et al.* (2010). Lastly, applying qualitative methods and MPN statistics or a sub-sampling strategy, the GMO content *versus* a predetermined value can be obtained with precision, as it is based on a statistical approach that only depends on the number of (sub-)samples analysed (Kobilinsky and Bertheau, 2005; Berdal *et al.*, 2008). In both cases, only the end cost of the tests, which are individually less costly as they are qualitative, will limit the precision of the result and the detection limit. Both methodologies are widely used in microbiology and seed quality management (Remund *et al.*, 2001; Lee *et al.*, 2009; Schulz *et al.*, 2010).

Cost is therefore the limiting factor for practical detection and quantification limits which can be used in 'GMO-free' or (nationally defined) organic agriculture supply chains.

12.3.8 Multiplexing and qualitative methods

There are approximately one hundred commercially authorised GMOs, not counting stacked genes, despite the first market withdrawals of GMOs with outdated technology. The main component of analysis costs is the time spent by the analyst, from DNA extraction to provision of the results. Multiplex methods have been developed in response to the growing number of GMOs. They may combine screening methods, i.e. screening sequences ubiquitous in GMOs, GMO and taxa identification methods, including testing for donor organism presence in transgenic sequences.

The shortcomings of the multiplexing of real-time quantitative PCR (QRT-PCR) methods have rapidly become apparent due to the overlap in the spectra of the fluorophores used in this technique.

As a result, method multiplexing has turned to a derivative of SNP studies on organisms and DNA chip methods (Chaouachi *et al.*, 2008d; Hamels *et al.*, 2009). These methods, although qualitative, may be validated by inter-laboratory tests and may provide the GMO content *versus* a predetermined value through sub-sampling techniques (Leimanis *et al.*, 2008; Bellocchi *et al.*, 2010a).

© Woodhead Publishing Limited, 2011

Routine use should not depend on the total analysis cost, as when considered in terms of the number of detections made, these technologies prove to be very competitive compared to real-time quantitative PCR methods. Whether such methods will be successful will depend more on the obstacle of analysts' reluctance to change their practices and the initial purchase cost of some apparatus. The Eppendorf DNA chip (DualChip®) is relatively inexpensive, given the amount of data provided, the assistance from the software package and its ability to indicate result discrepancies between screening and identification methods. Based on the 'matrix approach', this software package is the first developed to routinely suspect the presence of unauthorised or unknown GMOs in a sample.

12.4 Conclusions

Traceability, which is documentary in the normative sense of the term, has long been used by European agri-food supply chains, particularly for indicators of quality. There is a developing market segmentation involving even third countries which have up to now shown reticence concerning these labels as they are more attached to commercial brands. Various technologies such as RFID chips have paved the way for its implementation into increasingly complex agri-food chains often operating with just-in-time methods from European or third sources.

Following various health crises, traceability has become mandatory in the EU for all food- and feed-stuffs. This move has gained momentum in some sectors of the food industry, partly to meet health requirements (mycotoxins, allergens, etc.) and to ensure consumers enjoy freedom of choice, as is the case for GMOs.

Analytical tests are an essential corollary to traceability and fraud prevention. An array of test techniques is applied in science. Yet molecular biology, using methods such as PCR, since 1989 – the first time its current description was used, can assuredly claim to be the most changing and plentiful sector in the arsenal of routine test methodologies beside some recent physico-chemical methodologies.

GMOs have undergone a relatively uncommon structuring in the field of testing, mainly due to their media coverage. This is demonstrated by the swift implementation of test systems. GMO testing has also been advanced by the *de novo* creation of a scientific community which rapidly started to work together in research programmes and standardisation bodies in close liaison with national and European competent authorities through information sharing networks, such as ENGL, and inter-laboratory validation.

Some aspects of GMO detection strategies, methods and concepts may seem innovative. They must, however, be considered with regard to the more general detection-related innovations made over the last decade. They have also been developed by using or adapting practices from other sectors, as is the case for sub-sampling plans for example.

© Woodhead Publishing Limited, 2011

A share of the results from the GMO sector should provide useful insights for other test sectors. Examples include sub-sampling to determine target content, inter-laboratory validations of qualitative multiplex methods, decision support systems, differential quantitative PCR and the matrix approach for the detection of unauthorised or unknown GMOs, and the detection of stacked genes at an age when pathogens that are multiresistant to antibiotics are rife.

In conclusion, traceability is essential for supply-chains coexistence that upholds the freedom of choice of consumers and, by extension, that of producers through preserved markets. In addition to providing information directly to consumers on GMOs, traceability is a factor which structures territories by creating a direct relationship between production, production type and location (Conchoy, 2001).

The European Union is far from alone in its attitude to GMOs and could act as a meeting point for the traceability and coexistence of production types (Davison, 2010).

12.5 References

AKIYAMA H., K. SAKATA, K. KONDO, A. TANAKA, M.S. LIU, T. OGUCHI, S. FURUI, K. KITTA, A. HINO and R. TESHIMA (2008) Individual detection of geneticailly modified maize varieties in non-identity-preserved maize samples. *J. Agr. Food Chem.* 56: 1977–1983.

ANCEL V., Y. BERTHEAU, M. FEINBERG, R. LE BOUQUIN, P. PHILIPP and SEILLER M-P (2010) Assessment of differential Quantitative Polymerase Chain Reaction (dQ-PCR) for the detection of unknown GMOs by collaborative studies. Submitted.

ARVANITOYANNIS I.S. and A. KRYSTALLIS (2005) Consumers' beliefs, attitudes and intentions towards genetically modified foods, based on the 'perceived safety vs. benefits' perspective. *Int. J. Food Sci. Technol.* 40: 343–360.

BELLOCCHI G., V. BERTHOLET, S. HAMELS, W. MOENS, J. REMACLE and G. VAN DEN EEDE (2010a) Fuzzy-logic based strategy for validation of multiplex methods: example with qualitative GMO assays. *Transgenic Res.* 19: 57–65.

BELLOCCHI G., M. DE GIACOMO, N. FOTI, M. MAZZARA, E. PALMACCIO, C. SAVINI, C. DI DOMENICANTONIO, R. ONORI and G. VAN DEN EEDE (2010b) Testing the interaction between analytical modules: an example with Roundup Ready soybean line GTS 40-3-2. *BMC Biotechnol.* 10: 55.

BELLOCCHI G., R. CONFALONIERI, M. ACUTIS and G. GENOVESE (2006) SISSI: a resampling-based software for sample size determination. *Proceedings of the 9th European Society for Agronomy Congress. Warsaw Poland, September 4–6*, 741–742.

BERDAL K.G., C. BOYDLER, T. TENGS and A. HOLST-JENSEN (2008) A statistical approach for evaluation of PCR results to improve the practical limit of quantification (LOQ) of GMO analyses (SIMQUANT). *Eur. Food Res. Technol.* 227: 1149–1157.

BERTHEAU Y., J.-C. HELBLING, M.-N. FORTABAT, S. MAKHZAMI, I. SOTINEL, C. AUDÉON, A.-C. NIGNOL, A. KOBILINSKY, L. PETIT, P. FACH, P. BRUNSCHWIG, K. DUHEM and P. MARTIN (2009) Persistence of plant DNA sequences in the blood of dairy cows fed with genetically modified (Bt176) and conventional corn silage. *J. Agric. Food Chem.* 57: 509–516.

© Woodhead Publishing Limited, 2011

BONNY S. (2008) How have opinions about GMOs changed over time? The situation in the European Union and the USA. *CAB Reviews: Perspectives in Agriculture, Veterinary Science, Nutrition and Natural Resources* 3, 1–17.

BRERA C., E. DONNARUMMA, R. ONORI, N. FOTI, B. PAZZAGLINI and M. MIRAGLIA (2005) Evaluation of sampling criteria for the detection of gm soybeans in bulk. *Ital. J. Food Sci.* 17: 177–185.

BROOTHAERTS W., P. CORBISIER, H. SCHIMMEL, S. TRAPMANN, S. VINCENT and H. EMONS (2008) A single nucleotide polymorphism (SNP839) in the adh1 reference gene affects the quantitation of genetically modified maize (*Zea mays* L.). *J. Agric. Food Chem.* 56: 8825–8831.

CANKAR K., V. CHAUVENSY-ANCEL, M-N. FORTABAT, K. GRUDEN, A. KOBILINSKY, J. ZEL and Y. BERTHEAU (2008) Detection of nonauthorized genetically modified organisms using differential quantitative polymerase chain reaction: application to 35S in maize. *Anal. Biochem.* 376: 189–199.

CHAOUACHI M., S. GIANCOLA, M. ROMANIUK, V. LAVAL, Y. BERTHEAU and D. BRUNEL (2007) A strategy for designing multi-taxa specific reference gene systems. Example of application – ppi phosphofructokinase (ppi-PPF) used for the detection and quantification of three taxa: Maize (*Zea mays*), cotton (*Gossypium hirsutum*) and rice (*Oryza sativa*). *J. Agric. Food Chem.* 55: 8003–8010.

CHAOUACHI M., G. CHUPEAU, A. BERAPD, H. McKHANN, M. ROMANIUK, S. GIANCOLA, V. LAVAL, Y. BERTHEAU and D. BRUNEL (2008a) A high-throughput multiplex method adapted for GMO detection. *J. Agric. Food Chem.* 56: 11596–11606.

CHAOUACHI M., R. EL MALKI, A. BERARD, M. ROMANIUK, V. LAVAL, D. BRUNEL and Y. BERTHEAU (2008b) Development of a real-time PCR method for the differential detection and quantification of four Solanaceae in GMO analysis: Potato (*Solanum tuberosum*), tomato (*Solanum lycopersicum*), eggplant (*Solanum melongena*), and pepper (*Capsicum annuum*). *J. Agric. Food Chem.* 56: 1818–1828.

CHAOUACHI M., M-N. FORTABAT, A. GELDREICH, P. YOT, C. KERLAN, N. KEBDANI, C. AUDÉON, M. ROMANIUK and Y. BERTHEAU (2008c) An accurate real-time PCR test for the detection and quantification of cauliflower mosaic virus (CaMV): applicable in GMO screening. *Eur. Food Res. Technol.* 227: 789–798.

CHAOUACHI M., M. ROMANIUK, A. BARTEGI, V. LAVAL, Y. BERTHEAU and D. BRUNEL (2008d) The SNPlex technology: a high throughput method for the detection of authorized and unauthorized GMO with the matrix approach. *Third International Conference on coexistance between Genetically Modified (GM) and non-GM based agricultural supply chains, Seville, Spain, 20–21 November, 2007*, 191–194.

CHARELS D., S. BROEDERS, P. P. CORBISIER, S. TRAPMANN, H. SCHIMMEL and H. EMONS (2007a) Toward metrological traceability for DNA fragment ratios in GM quantification. 3. Suitability of DNA calibrants studied with a MON 810 corn model. *J. Agric. Food Chem.* 55, 3268–3274.

CHARELS D., S. BROEDERS, P. CORBISIER, S. TRAPMANN, H. SCHIMMEL, T. LINSINGER and H. EMONS (2007b) Toward metrological traceability for DNA fragment ratios in GM quantification. 2. Systematic study of parameters influencing the quantitative determination of MON 810 corn by real-time PCR. *J. Agric. Food Chem.* 55, 3258–3267.

CHARELS D., S. BROEDERS, P. CORBISIER, S. TRAPMANN, H. SCHIMMEL and H. EMONS (2008) Towards DNA copy number ratios as GM quantification units. *Third International Conference on coexistance between Genetically Modified (GM) and non-GM based agricultural supply chains, Seville, Spain, 20–21 November, 2007*, 187–190.

© Woodhead Publishing Limited, 2011

COHAN F.M. (2010) Synthetic biology: now that we're creators, what should we create? *Curr. Biol.* 20, R675–R677.

CONCHOY F. (2001) Les effets d'un trop-plein de traçabilité. *La Recherche* numéro spécial 339: 66–68.

CONSUMERCHOICE CONSORTIUM (2008) Do European consumers buy GM foods? European Commission: Framework 6, Project no. 518435. 'Consumerchoice'. Final report.

CORBISIER P., S. BHAT, L. PARTIS, V.R.D XIE and K.R. EMSLIE (2010) Absolute quantification of genetically modified MON810 maize (*Zea mays* L.) by digital polymerase chain reaction. *Anal. Bioanal. Chem.* 396: 2143–2150.

COSTA-FONT L., E. MOSSIALOS and M. COSTA-FONT (2006) Erring on the side of caution? The heterogeneity of public perceptions of biotechnology applications in the European Union. *J. Econ. Issues* 40: 767–777.

DAVISON J. (2010) GM plants: science, politics and EC regulations. *Plant Science* 178: 94–98.

DAVISON J. and Y. BERTHEAU (2007) EU regulations on the traceability and detection of GMOs: difficulties in interpretation, implementation and compliance. *CAB Reviews: Perspectives in Agriculture, Veterinary Science, Nutrition and Natural Resources* 2: 14 pp.

DAVISON J. and Y. BERTHEAU (2008) The theory and practice of European traceability regulations for GM food and feed. *Cereal Food. World* 53: 186–196.

DAVISON J. and Y. BERTHEAU (2010) Genetic engineering: evolution. *Encyclopedia of Biotechnology in Agriculture and Food* 1: 292–296.

DEFRA (2005) Statistical theory and analysis if GMO enforcement (stage). DEFRA project CB0209. DEFRA (ed.), UK.

DONG W., L.T. YANG, K.L. SHEN, B. KIM, G.A. KLETER, H.J.P. MARVIN, R. GUO, W.Q. LIANG and D.B. ZHANG (2008) GMDD: a database of GMO detection methods. *BMC Bioinformatics* 9.

ERVIN D.E., L.L. GLENNA and R.A. JUSSAUME (2010) Are biotechnology and sustainable agriculture compatible? *Renew. Agr. Food Syst.* 25: 143–157.

EUROPEAN COMMISSION (1990a) Council Directive 90/219/EEC of 23 April 1990 on the contained use of genetically modified micro-organisms. *Official Journal of the European Communities* L 117: 1–14.

EUROPEAN COMMISSION (1990b) Council Directive 90/220/EEC of 23 April 1990 on the deliberate release into the environment of genetically modified organisms. *Official Journal of the European Communities* L 117: 15–27.

EUROPEAN COMMISSION (1997) Regulation (EC) No 258/97 of the European Parliament and of the Council of 27 January 1997 concerning novel foods and novel food ingredients. *Official Journal of the European Communities* L 043: 1–6.

EUROPEAN COMMISSION (2000a) Commission regulation (EC) No 49/2000 of 10 January 2000 amending Council Regulation (EC) No 1139/98 concerning the compulsory indication on the labelling of certain foodstuffs produced from genetically modified organisms of particulars other than those provided for in Directive 79/112/EEC. *Official Journal of the European Communities* L 6: 13–14.

EUROPEAN COMMISSION (2000b) Regulation (EC) No 50/2000 of 10 January 2000 on the labelling of foodstuffs and food ingredients containing additives and flavourings that have been genetically modified or have been produced from genetically modified organisms. *Official Journal of the European Communities* L 6: 15–17.

EUROPEAN COMMISSION (2001) Directive 2001/18/EC of the European Parliament and the Council of 12 March 2001 on the deliberate release into the environment of

© Woodhead Publishing Limited, 2011

genetically modified organisms and repealing Council Directive 90/220/EEC. *Official Journal of the European Communities* L 106: 1–38.

EUROPEAN COMMISSION (2002) Regulation (EC) No 178/2002 of the European Parliament and of the Council of 28 January 2002 laying down the general principles and requirements of food law, establishing the European Food Safety Authority and laying down procedures in matters of food safety. *Official Journal of the European Communities* L 31: 1–24.

EUROPEAN COMMISSION (2003a) Regulation (EC) No 1829/2003 of the European Parliament and of the Council of 22 September 2003 on genetically modified food and feed. *Official Journal of the European Union* L 268: 1–23.

EUROPEAN COMMISSION (2003b) Regulation (EC) No 1830/2003 of the European Parliament and of the Council of 22 September 2003 concerning the traceability and labelling of genetically modified organisms and the traceability of food and feed products produced from genetically modified organisms and amending Directive 2001/18/ EC. *Official Journal of the European Communities* L 268: 24–28.

EUROPEAN COMMISSION (2003c) Regulation (EC) No 1946/2003 of the European Parliament and of the Council of 15 July 2003 on transboundary movements of genetically modified organisms. *Official Journal of the European Communities* L 287: 1–10.

EUROPEAN COMMISSION (2004a) Commission recommendation 787/2004 of 4 October 2004 on technical guidance for sampling and detection of genetically modified organisms and material produced from genetically modified organisms as or in products in the context of Regulation (EC) No 1830/2003. *Official Journal of the European Union* L 348: 18–26.

EUROPEAN COMMISSION (2004b) Commission Regulation (EC) No 65/2004 of 14 January 2004 establishing a system for the development and assignment of unique identifiers for genetically modified organisms. *Official Journal of the European Union* L 10: 5–7.

EUROPEAN COMMISSION (2010) Commission recommendation of 13 July 2010 on guidelines for the development of national co-existence measures to avoid the unintended presence of GMOs in conventional and organic crops (2010/C 200/01). *Official Journal of the European Union* C200: 1–5.

EUROPEAN COMMISSION DG-AGRI (2007) Economic impact of unapproved GMOs on EU feed imports and livestock production.

EUROPEAN COMMISSION DJI (2006) Structured inventory of existing food quality assurance schemes within the EU-25. DJI European Commission (ed.). Seville, Spain: DG JRC/IPTS.

FEINBERG M., S. FERNANDEZ, S. CASSARD, C. CHARLES-DELOBEL and Y. BERTHEAU (2005) Quantitation of 35S promoter in maize DNA extracts from genetically modified organisms using real-time polymerase chain reaction, part 2: Interlaboratory study. *J. AOAC Intl.* 88: 558–573.

FERNANDEZ S., C. CHARLES-DELOBEL, A. GELDREICH, G. BERTHIER, F. BOYER, C. COLLONNIER, G. COUÉ-PHILIPPE, A. DIOLEZ, M.-N. DUPLAN, N. KEBDANI, M. ROMANIUK, M. FEINBERG and Y. BERTHEAU (2005) Quantification of the 35S promoter in DNA extracts from genetically modified organisms using real-time polymerase chain reaction and specificity assessment on various genetically modified organisms, part I: Operating procedure. *J. AAOAC Intl.* 88: 547–557.

FRIZZI A. and S.S. HUANG (2010) Tapping RNA silencing pathways for plant biotechnology. *Plant Biotechnol. J.* 8: 655–677.

© Woodhead Publishing Limited, 2011

GAMA SOSA M.A., R. DE GASPERI and G.A. ELDER (2010) Animal transgenesis: an overview. *Brain Struct. Funct.* 214: 91–109.

GAO (2008) Agencies are proposing changes to improve oversight, but could take additional steps to enhance coordination and monitoring. In *Report to the Committee on Agriculture, Nutrition, and Forestry, US Senate*. US Government Accountability Office.

GASKELL G., A. ALLANSDOTTIR, N. ALLUM, C. CORCHERO, C. FISCHLER, J. HAMPEL, J. JACKSON, N. KRONBERGER, N. MEJLGAARD, G. REVUELTA, C. SCHREINER, S. STARES, H. TORGERSEN and W. WAGNER (2006) Europeans and Biotechnology in 2005: Patterns and Trends. Eurobarometer 64.3. A report to the European Commission's Directorate-General for Research. In *Eurobarometer*, European Commission (ed.). Brussels, B.

GLICK B.R., J.J. PASTERNAK and C.L. PATTEN (2010) Molecular biotechnology: principles and applications of recombinant DNA. *Molecular biotechnology: principles and applications of recombinant DNA*. ASM Press.

GÓMEZ-BARBERO M. and E. RODRÍGUEZ-CEREZO (2006) Economic impact of dominant GM crops worldwide: a Review. EUR 22547 EN. In *Technical Report Series*. Seville, Spain: JRC-IPTS.

GREEN R. and S. HERVÉ (2006) IP – Traceability and grains traders: ADM, Bunge, Cargill, Dreyfus. *INRA Les cahiers d'ALISS* 03.

HAMELS S., T. GLOUDEN, K. GILLARD, M. MAZZARA, F. DEBODE, N. FOTI, M. SNEYERS, TE. NUEZ, M. PLA, G. BERBEN, W. MOENS, Y. BERTHEAU, C. AUDÉON, G. VAN DEN EEDE and J. REMACLE (2009) A PCR-microarray method for the screening of genetically modified organisms. *Eur. Food Res. Technol.* 228: 531–541.

HAN J.H. and R.W. HARRISON (2007) Factors influencing urban consumers' acceptance of genetically modified foods. *Rev. Agr. Econ.* 29: 700–719.

HERNANDEZ M., M-N. DUPLAN, G. BERTHIER, M. VAITILINGOM, W. HAUSER, R. FREYER, M. PLA and Y. BERTHEAU (2004) Development and comparison of four real-time polymerase chain reaction systems for specific detection and quantification of *Zea mays* L. *J. Agr. Food Chem.* 52: 4632–4637.

HOLST-JENSEN A. and K.G. BERDAL (2004) The modular analytical procedure and validation approach and the units of measurement for genetically modified materials in foods and feeds. *J. AOAC Intl.* 87: 927–936.

HOLST-JENSEN A., M. DE LOOSE and G. VAN DEN EEDE (2006) Coherence between legal requirements and approaches for detection of genetically modified organisms (GMOs) and their derived products. *J. Agr. Food Chem.* 54: 2799–2809.

INTERNATIONAL ORGANISATION FOR STANDARDIZATION (2007) ISO 22005:2007. Traceability in the feed and food chain – General principles and basic requirements for system design and implementation.

JAMES C. (2010) *Brief 41: Global Status of Commercialized Biotech/GM Crops: 2009*. Ithaca, NY: The International Service for the Acquisition of Agri-biotech Applications (ISAAA).

KIM R. and M. BOYD (2006) Japanese consumers' acceptance of genetically modified (GM) food: An ordered probit analysis. *J. Food Products Marketing* 12: 45–57.

KOBILINSKY A. and Y. BERTHEAU (2005) Minimum cost acceptance sampling plans for grain control, with application to GMO detection. *Chemometr. Intell. Lab.* 75: 189–200.

LAFFONT J.L., K.M. REMUND, D. WRIGHT, R.D. SIMPSON and S. GREGOIRE (2005) Testing for adventitious presence of transgenic material in conventional seed or grain lots

© Woodhead Publishing Limited, 2011

using quantitative laboratory methods: statistical procedures and their implementation. *Seed Sci. Res.* 15: 197–204.

LEE H., L. CHAI, S. TANG, J. SELAMAT, G. FARINAZLEEN MOHAMMAD, Y. NAKAGUCHI, M. NISHIBUCHI and S. RADU (2009) Application of MPN-PCR in biosafety of Bacillus cereus s.l. for ready-to-eat cereals. *Food Control* 20: 1068–1071.

LEIMANIS S., S. HAMELS, F. NAZE, G.M. MBELLA, M. SNEYERS, R. HOCHEGGER, H. BROLL, L. ROTH L, K. DALLMANN, A. MICSINAI, J.L. LA PAZ, M. PLA, C. BRUNEN-NIEWELER, N. PAPAZOVA, I. TAVERNIERS, N. HESS, B. KIRSCHNEIT, Y. BERTHEAU, C. AUDÉON, V. LAVAL, U. BUSCH, S. PECORARO, K. NEUMANN, S. ROSEL, J. VAN DIJK, E. KOK, G. BELLOCCHI, N. FOTI, M. MAZZARA, W. MOENS, J. REMACLE and G. VAN DEN EEDE (2008) Validation of the performance of a GMO multiplex screening assay based on microarray detection. *Eur. Food Res. Technol.* 227: 1621–1632.

LHEUREUX K., M. LIBEAU-DULOS, H. NILSAGÅRD, E. RODRIGUEZ-CEREZO, K. MENRAD, M. MENRAD and D. VORGRIMLER (2003) Review of GMOs under research and development and in the pipeline in Europe. EUR 20680 EN. In *Technical report Series*, pp. 122 [JRCDJ European Commission, Institute for Prospective Technological Studies, editor. Seville, Spain: Joint Research Centre (DG JRC), Institute for Prospective Technological Studies.

MACARTHUR R., A.W.A. MURRAY, T.R. ALLNUTT, C. DEPPE, H.J. HIRD, G.M. KERINS, J. BLACKBURN, J. BROWN, R. STONES and S. HUGO (2007) Model for tuning GMO detection in seed and grain. *Nature Biotechnol.* 25: 169–170.

MESSEGUER J., G. PENAS, J. BALLESTER, M. BAS, J. SERRA, J. SALVIA, M. PALAUDELMAS and E. MELE (2006) Pollen-mediated gene flow in maize in real situations of coexistence. *Plant Biotechnol. J.* 4: 633–645.

NIELSEN C.P. (2001) Global market effects of GMOs: the importance of consumer preferences and policy choices. *DJF Rapport, Markbrug*, 121–140.

NIELSEN C.P. and K. ANDERSON (2001) Global market effects of alternative European responses to genetically modified organisms. *Weltwirtschaftliches Archiv-Rev. World Econ.* 137: 320–346.

NOUSSAIR C., S. ROBIN and B. RUFFIEUX (2001) Consumer behaviour with regard to GMOs in the food supply: an experimental study. *Economie Rurale*, 30–44.

NOUSSAIR C., S. ROBIN and B. RUFFIEUX (2004) Do consumers really refuse to buy genetically modified food? *Econ. J.* 114: 102–120.

NOUSSAIR C., S. ROBIN and B. RUFFIEUX (2007) Measuring preferences for genetically modified food products. *Environ. Econ. Exper. Met.* 8: 344–365.

PAOLETTI C., M. DONATELLI, S. KAY and G. VAN DEN EEDE (2003) Simulating kernel lot sampling: the effect of heterogeneity on the detection of GMO contaminations. *Seed Sci. Technol.* 31: 629–638.

PAOLETTI C., M. DONATELLI, A. HEISSENBERGER, E. GRAZIOLI, S. LARCHER and G. VAN DEN EEDE (2005) European Union perspective – sampling for testing of genetically modified impurities. *Second World Conference on Sampling and Blending 2005* 4: 209–213.

PAOLETTI C., A. HEISSENBERGER, M. MAZZARA, S. LARCHER, E. GRAZIOLI, P. CORBISIER, N. HESS, G. BERBEN, PS. LUBECK, M. DE LOOSE, G. MORAN, C. HENRY, C. BRERA , I. FOLCH, J. OVESNA and G. VAN DEN EEDE (2007) Kernel lot distribution assessment (KeLDA): a study on the distribution of GMO in large soybean shipments. *Eur. Food Res. Technol.* 224: 129–139.

PARK S-H. (2005) Current status of regulation on GM food in Korea. *Shokuhin Eiseigaku Zasshi* 46: J4–J7.

© Woodhead Publishing Limited, 2011

PLA M., J.L. LA PAZ, G. PENAS, N. GARCIA, M. PALAUDELMAS, T. ESTEVE, J. MESSEGUER and E. MELE (2006) Assessment of real-time PCR based methods for quantification of pollen-mediated gene flow from GM to conventional maize in a field study. *Transgenic Res.* 15: 219–228.

QUE Q., M-D.M. CHILTON, C.MD. FONTES, C. HE, M. NUCCIO, T. ZHU, Y. WU, J.S. CHEN and L. SHI (2010) Trait stacking in transgenic crops. Challenges and opportunities. *GM crops* 1: 1–10.

REMUND K.M., D.A. DIXON, D.L. WRIGHT and L.R. HOLDEN (2001) Statistical considerations in seed purity testing for transgenic traits. *Seed Sci. Res.* 11: 101–119.

ROMMENS C.M. (2010) Barriers and paths to market for genetically engineered crops. *Plant Biotechnol. J.* 8: 101–111.

ROWE G. (2004) How can genetically modified foods be made publicly acceptable? *Trends Biotechnol.* 22: 107–109.

SCHULZ S., A. PEREZ-DE-MORA, M. ENGEL, J.C. MUNCH and M. SCHLOTER (2010) A comparative study of most probable number (MPN)-PCR vs. real-time-PCR for the measurement of abundance and assessment of diversity of alkB homologous genes in soil. *J. Microbiol. Meth.* 80: 295–298.

SEMAL J. (2005) Cujus regio ejus Religio-G-M. *Cahiers Agricultures* 14: 493–494.

SHELDON R., N. CLEGHORN, C. PENFOLD, A. BROWN and T. NEWMARK (2009) Exploring attitudes to GM food. Final Report. FSA Social Science Research Unit, UK, editor.

SHINEHA R. and K. KATO (2009) Public engagement in Japanese policy-making: a history of the genetically modified organisms debate. *New Genet. Soc.* 28: 139–152.

STEIN A.J. and E. RODRIGUEZ-CEREZO (2009) The global pipeline of new GM crops. EUR 23486 EN. In *JRC scientific and technical reports*. JRC-IPTS (ed.) Seville, SP.

STEIN A.J. and E. RODRIGUEZ-CEREZO (2010a) International trade and the global pipeline of new GM crops. *Nature Biotechnol.* 28: 23–25.

STEIN A.J. and E. RODRIGUEZ-CEREZO (2010b) Low-level presence of new GM crops: an issue on the rise for countries where they lack approval. *AgBioForum* 13: 173–182.

SUSTAR-VOZLIC J., K. ROSTOHAR, A. BLEJEC, P. KOZJAK, Z. CERGAN and V. MEGLIC (2010) Development of sampling approaches for the determination of the presence of genetically modified organisms at the field level. *Anal. Bioanal. Chem.* 396: 2031–2041.

SVEC K. and Y. VASSILIEVA (2006) Russian Federation. Biotechnology GMO Labeling Requirement. In *GAIN Report*. USDA (ed.), Washington, DC.

TAVERNIERS I., N. PAPAZOVA, Y. BERTHEAU, M. DE LOOSE and A. HOLST-JENSEN (2008) Gene stacking in transgenic plants: towards compliance between definitions, terminology, and detection within the EU regulatory framework. *Environ. Biosafety Res.* 7: 197–218.

TENGS T., A.B. KRISTOFFERSEN, K.G. BERDAL, T. THORSTENSEN, M.A. BUTENKO, H. NESVOLD and A. HOLST-JENSEN (2007) Microarray-based method for detection of unknown genetic modifications. *BMC Biotechnol.* 7.

TRAPMANN S., H. SCHIMMEL, G.N. KRAMER, G. VAN DEN EEDE and J. PAUWELS (2002) Production of certified reference materials for the detection of genetically modified organisms. *J. AOAC Intl.* 85: 775–779.

TRAPMANN S., P. CORBISIER, H. SCHIMMEL and H. EMONS (2010) Towards future reference systems for GM analysis. *Anal. Bioanal. Chem.* 396: 1969–1975.

VASSILIEVA Y. and K. SVEC (2007) Russian Federation. Biotechnology. Russia establishes 0.9% threshold for biotech labeling. In *GAIN Report.* USDA (ed.), Washington DC, USA.

© Woodhead Publishing Limited, 2011

WEATHERS P.J., M.J. TOWLER and J.F. XU (2010) Bench to batch: advances in plant cell culture for producing useful products. *Appl. Microbiol. Biotechnol.* 85: 1339–1351.

WIELD D., J. CHATAWAY and M. BOLO (2010) Issues in the political economy of agricultural biotechnology. *J. Agrar. Change* 10: 342–366.

YASUI A. (2004) New food control system in Japan and food analysis at NFRI. *Accredit. Qual. Assur.* 9: 568–570.

ZEL J., K. CANKAR, M. RAVNIKAR, M. CAMLOH and K. GRUDEN (2006) Accreditation of GMO detection laboratories: improving the reliability of GMO detection. *Accred. Qual. Assur.* 10: 531–536.

ZEL J., M. MAZZARA, C. SAVINI, S. CORDEIL, M. CAMLOH, D. STEBIH, K. CANKAR, K. GRUDEN, D. MORISSET and G. VAN DEN EEDE (2008) Method validation and quality management in the flexible scope of accreditation: an example of laboratories testing for genetically modified organisms. *Food Anal. Meth.* 1: 61–72.

© Woodhead Publishing Limited, 2011

13

The relevance of sampling for the control of genetically modified organisms in the agri-food chain

C. Brera, B. De Santis, E. Prantera, M. De Giacomo and R. Onori, Italian National Institute of Health, Italy

Abstract: Sampling is a critical procedure whenever a decision is taken based on the results of sample measurements. Even though optimizing sampling plans can lead to procedures that are more cost effective, sampling is often not treated with due importance. A feasible sampling procedure that is relevant for all the diverse scenarios in which GMOs (genetically modified organisms) are heterogeneously present in primary production and along feed and food chains has not yet been produced or adopted at the European level. In this chapter existing legislative provisions, norms and guidelines are described and their pros and cons considered. Information on specific provisions for sample preparation and sampling uncertainty is also given.

Key words: genetically modified organism, sampling plans, sampling normative, sampling uncertainty, GMO, heterogeneous distribution.

13.1 Introduction

Sampling is a critical procedure whenever a decision is taken based on the results of sample measurements, and it is recognized worldwide that sampling contributes greatly to overall uncertainty in food control. It is important to consider that in order to correctly conduct discretionary and mandatory food and feed safety and quality control, it is crucial to obtain analytical samples that are representative of an entire batch of material that may weigh up to several

© Woodhead Publishing Limited, 2011

thousand tons. Analytical samples and test portions are very significantly smaller than bulk commodities and to obtain a representative sample, a proper sampling regime must be followed.

Even though optimizing of sampling plans can lead to procedures that are more cost effective, sampling is often not treated with due importance and very often improper 'shortcuts' are taken rendering all the efforts to obtain a representative sample useless. It is therefore common practice for stakeholders in the food industry to use contractual thresholds that are lower than official thresholds in order to manage this issue. This then puts seed companies and farmers under high pressure. Correct sampling procedures are also critical because the factors at stake – food and feed safety and quality – are of primary importance. Fit for purpose sampling procedures are required to implement official control measures and for traceability and labelling. Incorrect sampling leads to conflicts that can cause significant economic losses for producers, lack of trust on the part of consumers, legal suits or annulments of contracts.

These considerations are valid for the wide spectrum of situations in which sampling for food control is carried out, yet are pivotal in the case of GMOs (genetically modified organisms). This is because traceability, labelling systems in official control and coexistence in the agricultural sector are issues strictly linked with sampling. In the wide range of different scenarios where sampling is required for GMO control, fit for purpose sampling procedures are far from available. Effective GMO traceability is imperative to implement GMO labelling regulation, facilitate withdrawal of products where adverse effects on human and animal health or the environment are foreseen, and not least for monitoring of potential effects in the environment. Organizations in the food industry now have to consider the issue of sampling for GMO control since, according to EU Regulations 178/2002/EC and 1830/2003/EC (EC, 2003b) and directive 2001/18/EEC products have to be traceable along all food and feed production and distribution chains and at each point of commercialization.

Farmers now have to deal with coexistence of GM and non-GM crops in agricultural production systems and face the possibility of adventitious or unintended presence of GM crops in non-GM crops and *vice versa*. This induced the EU to set up Commission Recommendation 2003/556/EC (EC, 2003a), amended in 2010 (Commission Recommendation 2010/C 200/01), which give guidelines on the development of national strategies and practices according to the European subsidiarity principle, to ensure that GM production crops can coexist with conventional and organic farming. Farmers should be able to choose to cultivate, after a consultation phase with their neighbours, any kind of agricultural crops, and none of these forms of agriculture should be excluded in the EU territory. However, the EC is currently changing its policy framework as GMO-free regions may now be established (e.g. Madeira). Farmers need to be able to maintain the production system of their choice. For this purpose adequate control and inspection systems have to be implemented and use of proper sampling procedures is essential to ensure successful functioning of co-existence measures.

© Woodhead Publishing Limited, 2011

The point at issue for GMO control is the heterogeneity of all the variables involved (lot, material, distribution). Sampling procedures could be said to be straightforward when the distributions of all the elements in the bulk material are homogeneous. This is the case for commercial seeds, for example, as all the samples taken from the bulk are identical. But homogeneity is neither a characteristic of the real world nor of GM material (Gy, 2004). Furthermore, the outputs of the most relevant study on GMO distribution in soybean grain lots (Paoletti *et al*., 2006), evidenced from analysis of a sound number of lots, that GM material distribution is heterogeneous. At the European level, research on topics associated with GMOs has been of great interest; sampling is one of the aspects frequently covered and there have been significant research outputs on novel sampling approaches and strategies for sampling in food and feed chains, including at the field level.

Sampling processes generate unavoidable sampling errors that must be controlled in order to maintain the representativeness of the sample despite the contribution of the errors. If the approach to the task is not scientific, the risk of uncontrolled amplification of the errors can invalidate the analytical result due to lack of representativeness. The most relevant parameter by which to assess the reliability of GMO control plans (which involve both sampling and analysis of products for GMO measurement), is total uncertainty. Two major components of total uncertainty are commonly distinguished: analytical uncertainty, which is associated with the analytical results and sampling uncertainty, which is associated with the sampling procedure. Analytical uncertainty evaluation is a routine exercise and regulations, different kinds of guidance (such as ISO norms, GAFTA), and sound scientific references relevant to GMOs in this area (https://englnet.jrc.it/sites/engl/Lists/Announcements/Attachments/48/2009-02-10%20MU%20guidance%20doc_v2.pdf; Macarthur *et al*., 2010) have been produced. However the only procedure that can be considered sound and complete for estimating sampling uncertainty in GMO control is the EURACHEM guide (EURACHEM/CITAC, 2007). At present different approaches for sampling GM seeds, grains and processed products exist, but the various circumstances in which testing is necessary are so diverse and not all the approaches are exhaustive, so gaps in knowledge and provision of recommendations still persist. Regardless of the scenario and the stakeholders involved, the unchanging issue at stake is to match appropriate statistical approaches for dealing with uncertainty and representativeness with practical feasibility, costs and personnel requirements.

13.2 Overview of international norms and legislative provisions

When performing official and discretionary control activities associated with heterogeneously distributed analytes such as mycotoxins in bulk lots, it is pivotal to use suitable diagnostic tools and to be able to guarantee the reliability

© Woodhead Publishing Limited, 2011

of the information gained. One of the more common errors is the use of random sampling plans when testing heterogeneous lots (Gy, 1995; Paoletti *et al.*, 2006). This is particularly relevant for GMO control, as presuming randomness is risky, as explained by Lischer (2001) and sampling errors are increased by misinterpretation of the particle distribution. If particles are independent from one another, then the distribution is random. If, however, their occurrence is considered to be characterized by a certain distribution pattern, as is the case for GMOs, then different sampling plans should be used. Furthermore it has to be considered that the judgement on the amount of transgenic material in a lot depends only on the analysis of a test portion of a few milligrams (generally 200 mg), which has to represent tons of the commodity in question in bulk (Kay *et al.*, 2002). With this in mind the need to create accurate and feasible sampling procedures has been taken up at an international level by the most relevant organizations operating in the agri-food chains.

13.2.1 The European approach

To fulfil the requirements set out in Regulation 2003/1830/EC, in 2004 the European Commission put in place Recommendation 787/2004/EC, which provides technical guidance on sampling and detection of GMOs and material produced from GMOs. Furthermore, following the entry of the non-authorized GMO LL RICE 601 into rice products, the European Commission provided sampling and analytical strategies (Commission decision 2006/754/EC) as an emergency measure. Although Recommendation 787/2004/EC did not result in a legally binding act, it provided the first indications of how to collect samples for control purposes based on the results of the KeLDA (Kernel lot distribution assessment) project. This was a European Network of GMO Laboratories' (ENGL) collaborative research project, coordinated by the Molecular Biology and Genomics Unit of the Joint Research Centre's Institute for Health and Consumer Protection (IHCP-JRC). It focussed on assessing the distribution of GM material in fifteen commercial soybeans grain lots (Paoletti *et al.*, 2006). Recommendation 787/2004/EC addressed some of relevant knowledge gaps, such as how to detect the adventitious or technically-unavoidable presence of GMOs in non-GMO lots.

For seeds and other propagating material the Recommendation refers the reader to International Seed Testing Association (ISTA) rules (ISTA, 2007). Specific procedures for sampling bulk agricultural commodities (grains, oilseeds, etc.) are indicated in Recommendation 787/2004/EC based on those described in ISO standards 6644 and 13690, now both replaced by ISO standard 24333. The approach followed in the EC Recommendation is to draw 500 g increment samples and an equal weight and number of file increment samples (to be stored for potential sampling uncertainty measurement). The number of samples is defined in the Recommendation and depends on the lot size. The increment samples have to be combined to form a unique bulk sample which is then analysed. Furthermore, the Recommendation provides an innovative

© Woodhead Publishing Limited, 2011

methodology for decisions as to labelling based on the estimation of measurement uncertainty. Where the analytical result obtained is close to the labelling threshold established by the EU (threshold ± 50% of its value) analysis of the individual file increment samples is recommended to provide a measure of the associated uncertainty. The ± 50% value is consistent with the ISO 21570:2005 definition of consistency of test portions. In addition, the Recommendation refers to ISO standard 2859 for the sampling of pre-packaged food and feed products, even though this ISO standard is not specifically intended for GMO control.

Additionally, Working Group 11 of the European Committee for Standardization Technical Committee 275 (CEN/TC 275 WG11) led the development of Technical Specification CEN/TS 15568:2006 (Foodstuffs – Methods of analysis for the detection of genetically modified organisms and derived products – Sampling strategies). This gives guidance on setting up valid sampling strategies for foodstuffs to be analysed for GMO control and can also be applied to feed and plant samples from the environment. The approach outlined is essentially similar to that of Recommendation 787/2004/EC, but the Technical Specification includes a wider list of terms and definitions, with particular emphasis on sampling division and sampling strategy. The Technical Specification provides a more detailed description of the sampling plans (such as are provided in different standards for other analytes) consisting of three steps:

1. collection of increment samples depending on lot size;
2. reduction of the aggregate sample to a laboratory sample;
3. grinding the laboratory sample to the appropriate particle size and homogenization.

In addition, it clearly states that, whenever technically possible, sampling should be performed in dynamic conditions, for example during loading or unloading of ships, lorries, vans, etc.

The approach to sampling uncertainty is the same as that found in Recommendation 787/2004/EC. The sampling equipment should be accessible to facilitate inspection and cleaning. For dynamic sampling, the increments should be taken from as much of the cross section of the flowing material as possible. For static sampling, the increments should be taken from defined points in the available area throughout the whole height of the lot under investigation. As for pre-packed units, an alternative to ISO 2859 is provided. For pre-packed units up to 50 kg, increments should be taken from specific points, such as the top, middle and bottom of the items using suitable spears. The number of pre-packed items to be sampled depends on the lot size, as shown in Table 13.1.

For bulk commodities, in all the cases, the number of sampling points is defined according to the lot size as shown in Table 13.2.

13.2.2 Gaps and limitations

Recommendation 787/2004/EC has its drawbacks. The difficulties experienced by those following it have been discussed at different national and international

© Woodhead Publishing Limited, 2011

Table 13.1 Number of pre-packed units to be collected following the CEN/TS 15568:2006

No. of pre-packed items	No. of pre-packed items to be sampled
Up to 10	Each pre-packed item
10 to 100	10, gathered at random
More than 100	Square root of total number, collected according to a proper sampling scheme

Table 13.2 Number of sampling points according to lot size following the CEN/TS 15568:2006

Lot size (tons)	Bulk size (kg)	No. of sampling points
≤ 50	5	10
50 to 500	0.01% of lot size	2 times bulk sample size (kg)
≥ 500	50	100

'technical round-tables'. However, to date, six years after the publication of the Recommendation, there has not been agreement on how it should be improved (Davison *et al.*, 2007). Two of the most serious limitations are described below, Firstly, the procedures in ISO standard 2859, recommended for sampling pre-packaged products, indicated in Recommendation 787/2004/EC without any further specifications to make them applicable to GMOs, are completely unfeasible for sampling for GMO control and are not fit for purpose. This ISO standard is a complex standard generally used to guide quality control sampling, which involves inspection of attributes. Until now no studies have verified that it provides reliable results when sampling packaged food and feed products for GMO control. Furthermore competent authorities have not found it to be appropriate for this purpose (Waiblinger, 2008; Onori, 2008: http://gmoglobalconference.jrc.ec.europa.eu/2008/Presentations). Secondly, the procedure to determine the uncertainty of the measurement requires a very large number of analyses. Therefore it is costly and requires significant recruitment of skilled personnel. For the reasons discussed above, Recommendation 787/2004/EC, although statistically sound, fails to be useful from a pragmatic and economic point of view. The Recommendation leaves operators free to choose alternative sampling procedures, though they must keep in mind the representativeness of the aggregate sample to the whole lot. So far, only few EU Member States, if any, use the sampling plan recommended by the EC.

13.2.3 Alternative sampling plans

Currently no other official legislation or norms specifically for GMO sampling are in place. For this reason, in some countries, alternative sampling plans

© Woodhead Publishing Limited, 2011

developed for other analytes heterogeneously distributed in a lot, like EC Regulation 401/2006 for mycotoxins, have been taken into consideration. This regulation is actually implemented by the EU Member States and is based on a similar approach to the one described in Recommendation 787.

The EU Co-Extra project (De Giacomo *et al.*, 2009) aimed to verify whether sampling methodology (EU Regulation 401/2006) in force within EU countries for sampling of mycotoxins could also be used for representative sampling of GMOs and derived products. The results of this study were in agreement with other studies (Johansson *et al.*, 2000; Montesinos *et al.*, 2010) showing that changes in sample size can affect the performance of sampling plans. In fact, sampling according to EU Regulation 401/2006, which results in smaller sample sizes than those generated by the procedures in Recommendation EU 787/2004 (100 g versus 500 g for incremental samples and 17 kg versus 50 kg for bulk samples), was found not fit for purpose for control of GMOs present at very low levels (< 0.01%) in raw bulk grain material. Since the performance of sampling procedures is also influenced by the adopted threshold level (Johansson *et al.*, 2000), though, it is likely that the mycotoxin sampling methodology could also be effective for GMO sampling to verify compliance with the EU labelling threshold (0.9%).

ISO standard 24333:2009 (Cereals and cereal products – Sampling) is applicable to GMO sampling only for the assessment of the quality and condition of GM cereal and cereal product lots. It is considered inappropriate for the determination of the presence of adventitious GM material in non-GM products. No specific studies support the reliability and accuracy of the sampling procedure it describes for this purpose. Codex standard CAC/GL 50-2004 (General Guidelines on Sampling), might also represent another document that could be followed for sampling for GMO control; however, it was not designed specifically for this purpose. The guide states that in case of heterogeneous lots, a stratified sampling procedure should be used, that consists of performing a random sampling procedure in the homogeneous strata that make up the heterogeneous lot. Therefore, it does not seem to be practicable in real situations.

Table 13.3 provides a list of the most prominent guidance on and legislative provisions and recommended procedures for sampling available at an international level for GMO determination in seeds, food and feed products. These could alternatively be adopted. It should be clear that these methods, if adopted, should account for the associated error since in some cases they have not been specifically developed for GMO control.

13.3 Strategies for cost-effective sampling in different scenarios in food and feed chains

Central to the co-existence of GMO supply chains with non-GMO supply chains is the provision of harmonized and validated sampling strategies. These should

© Woodhead Publishing Limited, 2011

Table 13.3 Most relevant sampling norms, guides and legislative provisions

Material	Source	Lot size	Aggregate sample	Laboratory sample	Incremental samples	Incremental size
Seeds	ISTA[a]	Variable depending on species: from 1 t to 4 t	1 kg	3000 seeds	One increment for 300 to 700 kg	Not indicated
Grains	USDA/GIPSA[b]	Up to 10 000 bushel (≅245 000)–10 000 sacks if in bulk	Not lower than 2 kg (≅ 2.5 kg)	See aggregate sample	3 cups or 1 drawing per 500 bushel (about 12 000 kg)	1.25 kg
Grains	USDA/GIPSA StarLink[c]	As above	At least three times laboratory sample	2400 kernels	Not indicated	From 0.2 to 0.5 kg up to 5 kg
All except seeds	ISO norm 24333[d]	500–1500 t	3–10 kg	1–10 kg	3–25	300–1900 g
Cereals	Reg. 401/2006/EC (Mycotoxins)[e]	From $\leq$ 0.01 t to 500 t	1–10 kg	Depending on the aggregate sample from 1 to 3	From 10 to 100	100 g
Feeds	Reg. 152/2009/CE[f]	From $\leq$ 2.5 t to 80 t	4 kg	2 kg (reduced sample)	From 1 to 40	100 g as minimum

[a] http://www.seedtest.org/en/content—1–1083.html
[b] http://www.gipsa.usda.gov/GIPSA/webapp?area=home&subject=lr&topic=hb-hb-1
[c] http://archive.gipsa.usda.gov/rdd/cry9csampling.pdf
[d] ISO/FDIS24333:2009(E) Cereals and cereal products – sampling
[e] Commission Regulation (EC) No. 401/2006 of 23 February 2006 laying down the methods of sampling and analysis for the official control of the levels of mycotoxins in foodstuffs
[f] Commission Regulation (EC) No. 152/2009 of 27 January 2009 laying down the methods of sampling and analysis for the official control of feed

© Woodhead Publishing Limited, 2011

preferably be structured as international standards, which support governmental policies and are suitable for food producers. Sampling plan should be reliable, cost effective (balancing costs and sampling uncertainty well) and feasible to implement in the real world. In this section, sampling at field level and along the food and feed processing chains is described.

13.3.1 Sampling in the field

In principle, farmers should be able to choose the types of agricultural crops they cultivate, be they GM, conventional or organic (unless of course they farm in a dedicated production zone, be it GMO, conventional or organic). However, there is the possibility that GM crops will unintentionally be present in crops thought not to be GM. Flexible co-existence measures (Demont *et al.*, 2009, 2010; Devos *et al.*, 2009) in areas where GMOs are cultivated seek to avoid the unintended presence of GMOs in other products, preventing the potential economic lossses and other impacts of the admixture of GM and non-GM crops (including organic crops). In order to help the Member States to develop national approaches to co-existence at a European level, the Commission adopted in July 2003 Recommendation 2003/556/EC, which was replaced in July 2010 by the new Commission Recommendation 2010/C 200/01.

Both control and monitoring of adventitious GM presence in non-GM crops are needed and can only be achieved through reliable sampling and testing schemes. The results will be used for both validating the gene flow models used in co-existence studies and by farmers and third-party certification companies in the real production environment. Currently no GMO specific in-field sampling standardized methodology is available. In addition the advantages of sampling agricultural commodities in fields have not been investigated as extensively as sampling after harvesting or processing have been.

The adventitious presence of GM crops in non-GM crops is mainly due to cross-pollination between fields. Therefore, the development of reliable sampling methodologies at field level has to be based on data gathered from field experiments and/or computer simulations aimed at determining gene flow in the environment and defining the various parameters that influence the outcrossing rate in different conditions of coexistence. A lot of studies has been performed for different crops: maize (Devos *et al.*, 2005; Messeguer *et al.*, 2006; Weekes *et al.*, 2007; Hüsken *et al.*, 2007; Langhof *et al.*, 2008; Sanvido *et al.*, 2008; Angevin *et al.*, 2008); brassica (Begg *et al.*, 2008; Devaux *et al.*, 2007, 2008; Garnier *et al.*, 2008; Gruber *et al.*, 2007; Ceddia *et al.*, 2007) and sugar beet (Andersen *et al.*, 2005; Stark *et al.*, 2006). So far no consensus on sampling emerged from the several proposals.

GM maize cultures were until 2010 the only GM cultures authorised in the EU. Accordingly, practical sampling work has mostly focused on this species. A field sampling system was developed (Messeguer *et al.*, 2006; Pla *et al.*, 2006) which allows the identification of pollen of foreign origin and the influence of neighbouring fields on the total GM content. The system is accurate and useful

© Woodhead Publishing Limited, 2011

for research purposes but is too expensive to be applied in practical conditions. Moreover, this sampling plan is sensitive to edge row dryness. In France (Bénétrix *et al.*, 2003, 2005), sampling plans were developed in order to accomplish the needs of farmers and cooperatives and to verify the efficiency of the adopted agricultural practices. This practical approach needs further experiments to validate it on the field scale.

Recently, two relevant studies were published describing field sampling approaches for maize, functional for monitoring the adventitious presence of GM crops (Allnut *et al.*, 2008, Šuštar-Vozlič *et al.*, 2010). In the first study a GM dispersal function (based on probabilistic studies deriving from field trials) was used to simulate non-GM crops in the studied zone and the accuracy of five different sampling schemes was tested. In this study, random sampling was found to be the most accurate and least susceptible to the bias by GM spatial structure or gradients. According to this model, the number of samples required increased rapidly as the GMO content approached the GM threshold, implying that accurate labelling when the GMO content is within $\pm$ 17% of the threshold may not be possible with high confidence.

The sampling approach developed by Šuštar-Vozlič *et al.* (2010) can be used for determination of the GMO presence at the field level and for making appropriate labelling decisions. It is a simple decision support system for monitoring the adventitious presence of GMO in the field based on the following steps:

- definition of the prediction model obtained from an estimation of mean out crossing rate in the field, calculated from the geometric data of GM donor and non-GM receptor fields;
- calculation of two distances within the receptor field and sampling scheme set up, if the adventitious presence of GMOs in the field is expected to be near the threshold level;
- collection and analysis of the samples.

If the results of the analyses show the level of GMO being above the threshold (including the standard deviation), then the crop is labelled as GMO; if the results are below the threshold (including the standard deviation), then the crop is not labelled as GMO; and if the results are in between those two values, then additional sampling and analyses is recommended (sampling in the field and/or silo).

The importance of this approach is its applicability to the lower thresholds meeting the demands of food companies and consumers. The efforts in the development of sampling at field level should lead to more feasible and rapid procedures taking into consideration the fact that sampling at this level is costly.

13.3.2 Sampling along food and feed chains

Ascertaining the presence of GMO is mainly relevant to three kinds of materials: raw materials, ingredients and final processed foods. Raw materials, even if they

© Woodhead Publishing Limited, 2011

have been in some way mixed during harvesting, drying, transport and storage, characteristically have a heterogeneous distribution of GMOs, particularly in the case of commodities shipments (Paoletti *et al.*, 2006). Paoletti *et al.* (2006) demonstrated the heterogeneity of the spatial distribution of GM grains in non-GM grains due, for example, to origins of the grain in several different silos. Therefore, sampling at this level is characterized by high sampling uncertainty and costs.

Primary ingredients usually undergo further processing operations (milling, extraction, etc.) and consequently, the distribution of GMOs in them can be more homogeneous (Gilbert, 1999). According to the Theory of Sampling (TOS) (Petersen *et al.*, 2005b) (which presents an explanation of the components of sampling error and also allows them to be minimized or eliminated), the particle size is an important factor for the constitutional heterogeneity of a lot. Thus, particle size reduction by milling represents a dominant factor in reducing heterogeneity. This assumption has been substantiated by a study by Whitaker *et al.* (2004), in which the evaluation of different sampling plans for the analysis of Cry9C protein in maize flour and meal, showed that sampling variability for maize meal was higher than for maize flour due to difference in particle size. Onori *et al.* (2007) evaluated the traceability of Roundup Ready® soybeans (RRS) along a processing chain and showed that flour and expander are the fit for purpose points for GMO control due to the reduction of sampling uncertainty by these points in comparison to grains. Furthermore, De Giacomo *et al.* (2009) established the applicability of a multi-analyte sampling approach for soybean flour at concentration level of RRS 0.1%.

Sampling of final (end-) products is one of the most challenging issues for GMO analysis. It is packaged end food products (i.e. products that will be purchased by the consumer) that are expected to be the main target for control agencies (Kay *et al.*, 2002). Sampling at this level is particularly challenging for many reasons, but principally because of the difficulty of estimating the distribution of GM material in the final products due to the many steps in their production (transport, storage, final processing phase). Another important issue is the fact that DNA and/or proteins can become harder to detect when an ingredient has been highly refined or a food has been processed (Moreano *et al.*, 2005). Finally, without a sampling strategy, the measurement of GM content obtained can only apply to the packets that have been submitted for testing (Kay *et al.*, 2002). For this reason sampling at factory level is more efficient than sampling at retail level. De Giacomo *et al.* (2008) employed a distribution-free statistical procedures supplied by software 'Shortcut in Sample Size Identification' (SISSI) (Confalonieri *et al.*, 2007), for the development of fit for purpose strategies for packed products. According to this study, collection of 20 incremental samples can be considered appropriate to represent a batch of 12 tons of soybean flour packaged products (1 kg each) containing Roundup Ready® Soybean at a level of approximately 5%. Furthermore, statistical modelling could be used as an aid to design retail sampling plans as reported by Macarthur *et al.* (2006) for mycotoxins in food.

© Woodhead Publishing Limited, 2011

13.3.3 Cost of sampling

When choosing a *fit for purpose* sampling strategy, there are two types of considerations to be taken into account: firstly statistical considerations relating to sample representativeness and distribution assumption and, secondly, considerations unrelated to statistics, such as financial, time, equipment and personnel constraints (Paoletti *et al.*, 2005). However, in no case should the second set of considerations be reasons to take unrepresentative samples. Many factors determine the financial cost of a control plan (i.e. a plan that includes both sampling and analysis), such as the total number of increments to be taken, the total weight of material that needs to be sampled, the sample preparation process and the number of required analyses to be undertaken (Gilbert, 1999). The contribution of the cost of sampling to the whole detection process can vary according to the detection plan. For example, the sampling costs might be higher in comparison to the analytical ones if a large sample is taken and homogenized and then just some sub-samples are analysed. Vice versa, the cost of the analysis can be higher if multiple small samples are taken from the lot of origin and analysed. A mathematical model for tuning GMO detection in seed and grain was proposed by Macarthur *et al.* (2007).

It is also to be considered that sampling costs also change depending on the labelling threshold required by the legislative system in force in the country or chosen according to agreements among stakeholders. A more accurate control plan (sampling and analysis) is necessary to detect a lower presence of GM material. A higher number of sub-samples and more sensitive analytical instruments, with a lower limit of detection are needed. Unfortunately it is not possible to have low risk at low cost (Esbensen *et al.*, 2008). The use of a proper, representative sampling strategy planned according to the 'Theory of Sampling' (TOS) that requires greater sampling commitments in terms of time and financial resources than other common inappropriate strategies or grab sampling is recommended.

13.4 Sample preparation strategies

Sampling (bulk sample formation) and sub-sampling (mass reduction) methodologies are used to produce laboratory samples. General guidelines exist on a suitable size for a laboratory sample (ISO 24333:2009; Petersen *et al.*, 2004). In GMO detection, sub-sampling is performed in the laboratory to implement acceptance sampling plans, among other reasons (Schilling *et al.*, 1982). Acceptance sampling plans aim to estimate the GMO content of a lot of seeds (or grains) through the use of qualitative analyses applied to a set of sub-samples. This approach is based on statistical considerations and the GM content is expressed in % of seeds, as units or % of DNA (Remund *et al.*, 2001; Laffont *et al.*, 2005). An example of this approach is the statistical programme ISTA 'Seedcalc' (Remund *et al.*, 2001), which can be used to design sampling and testing plans for purity/impurity estimation of seed lots for adventitious

© Woodhead Publishing Limited, 2011

presence/absence or levels of GM seeds in conventional seed lots. Another example is the program 'Optimal Acceptance Sampling by Attributes' (OPACSA) that allows the method described in Kobilinsky to select the cheapest acceptance sampling plan for control of grain lot purity (Kobilinsky *et al.*, 2005) to be easily put into practice. As when sampling for other purposes, sample size has a significant effect on accuracy (Montesinos *et al.*, 2010). Furthermore, sub-sampling steps such as reduction and/or homogenization, are performed in order to obtain a test portion statistically representative of the laboratory sample.

13.4.1 Test portion preparation from laboratory sample

Standards for GMO detection do not give many indications of the procedure for obtaining the test portion from the laboratory sample using a statistical approach. General recommendations can be found in CEN/TS 15568:2006. This suggests the use of equipment to ensure that the analytical sample is homogeneous and, in the case of material composed of discrete entities, to guarantee the appropriate particle size reduction (a particle number of 10 000 in the test portion as explained in Annex E). Particularly relevant to test portion preparation from a laboratory sample is Pierre Gy's Theory of Sampling (TOS) (Pitard *et al.*, 1993), that provides a description of all errors involved in sampling of heterogeneous materials as well as all the necessary tools for their evaluation, elimination and/or minimization. A set of 'Sampling Unit Operations' (SUOs) that summarize critical aspects of sampling based on the TOS regime is found in Petersen *et al.* (2005a). This paper reviews all the practical aspects of sampling and provides a handy toolbox for sampling operators, laboratory and scientific personnel.

13.5 Estimation of sampling uncertainty

Sampling uncertainty is widely recognized as one of the fundamental parameters for evaluation of the reliability of measurements (EURACHEM, 2007). Uncertainty is usually defined as a 'parameter, associated with the result of a measurement, that characterizes the dispersion of the values that could reasonably be attributed to the measurand' (ISO, 2008). The concept of uncertainty is strictly connected with others concepts like accuracy, error, trueness, bias and precision. The two major components of uncertainty are analytical uncertainty and sampling uncertainty. Analytical uncertainty refers to the contribution of the analysis process to the overall uncertainty. Sampling uncertainty encompasses uncertainty related to all aspects of the sampling process; in fact it is defined as 'The part of the total measurement uncertainty attributable to sampling' (EURACHEM, 2007).

Analytical uncertainty has been widely investigated and reviewed in the literature (ISO norms, GAFTA, CAC, codex guidelines, etc.). A method for

© Woodhead Publishing Limited, 2011

estimating the size of uncertainty associated with the measurement of materials derived from GMOs based on the 'uncertainty profile' has recently been presented (Macarthur *et al*., 2010). The 'uncertainty profile' is a development of a graphical statistical tool called the accuracy profile, which was developed for the validation of quantitative analytical methods. 'Uncertainty profiles' can be useful as aids in decision making and assessment of fitness for purpose.

Sampling uncertainty has not received the same attention. However, in 2007 a very exhaustive and sound reference document was produced, the EURACHEM/CITAC Guide Measurement uncertainty arising from sampling (EURACHEM, 2007). The relevance of the contribution of sampling uncertainty to the whole uncertainty budget is universally recognized. It is especially significant when the analyte distribution is heterogeneous, as is the case for GMOs, as demonstrated by the KeLDA project. The project aimed to improve understanding of the characteristics of GM distribution in oilseed (soybean) lots, since an incorrect assumption about the distribution is misleading for sampling uncertainty assessment. Indeed some sampling protocols assume that GM materials are distributed randomly. Therefore the project was established to verify whether this assumption, which was suspected to be far from reality, was actually incorrect and to assess the actual distribution of GM material in soybean lot. Randomness may probably be considered as only a limit case for this kind of matrix.

The outputs of the KeLDA project (Paoletti *et al*., 2006) evidenced with a sound number of analyses of GM seed lots that the distribution of GM material in lots is often heterogeneous and that the lots display significant spatial structuring. According to these findings, sampling protocols were developed based on statistical models free of distribution assumption requirements. To date, the only published sampling protocols specifically developed for GM material that are free of distribution assumption are the Recommendation 2004/787/EC and the CEN/TS 15568:2006.

Sampling uncertainty is one of the main issues (together with inter-laboratory measurement uncertainty) justifying the practical contractual threshold (generally at 0.1% level) used by stakeholders to ensure compliance with the GMO labelling threshold. It also partly explains the use of AQL (Acceptable Quality Level) and LQL (Low Quality Level) in the ISTA seeds purity testing schemes (Remund *et al*., 2001).

Two approaches for uncertainty estimation have been identified by the EURACHEM guide (EURACHEM, 2007): top-down approaches and bottom-up approaches. In general top-down approaches are easier and cheaper to implement than bottom-up ones. Indeed, they do not require information on the nature of the material to be sampled.

13.5.1 Bottom-up approaches

Bottom-up approaches, commonly defined as modelling, theoretical or predictive approaches, try to identify all sources of uncertainty and combine and

© Woodhead Publishing Limited, 2011

quantify them using a proper model. The identification and quantification of every possible source of uncertainty is of course necessary to build up the mathematical model. It must also be kept in mind that measurements may be affected by some bias, derived, for example, from an improper analytical practice or an incorrect application of the sampling procedure (EURACHEM, 2007).

Pierre Gy's Theory of Sampling (TOS)

One of the more widely used bottom-up approaches is the TOS, which was conceived for mineral matrixes and subsequently proved to be valid in other fields by Pierre Gy in the 1950s. The TOS allows the identification of the specific components of sampling errors, delineating these as seven error components, and provides practical suggestions for reducing them. The geometry of the lot is crucial for TOS. This theory is applicable to one-dimensional lots, but. can feasibly be employed for lots with more than one dimension, provided that the lot is reduced to one dimension. In moving streams (fluids and slurries), the theory allows the sampling error to be reduced and in some cases the sampling bias eliminated (Petersen, 2005a). TOS provides a scientific definition of sampling representativeness and a description of the components of the sampling errors in order to minimize or eliminate them (Petersen, 2005b).

13.5.2 Top-down approaches

Top-down approaches (also called 'empirical approaches') use some levels of replication of the whole measurement process to deduce an estimate of the uncertainty of the final result of the measurement.

Re-sampling techniques

There have been interesting recent attempts to evaluate the sampling uncertainty and determine a sample size for the estimation of the mean with a certain level of precision without distribution requirements using (virtual) re-sampling techniques already in use in other sectors, such as phylogenetics. These techniques are particularly useful in cases where the sample size has to be small, hence is difficult to make an estimation of the uncertainty associated with sampling due to the failure of the distribution assumptions of normality on the sample mean. Re-sampling techniques repeatedly estimate statistical parameters drawing samples randomly from the available data thereby also obtaining the sampling uncertainty. The sole assumption is that the dataset used is representative of the population under study. The principal techniques of re-sampling are bootstrap, jack-knifing and Monte Carlo methods (Robert *et al.*, 2004).

Two approaches are possible to implement re-sampling methods: to program in a computer language such as C++, R, Re-sampling Stats, or SAS (business analytics software), or to use a menu-driven program such as Excel, SISSI (Shortcut In Sample Size Identification), S-Plus, Stata (Data Analysis and

© Woodhead Publishing Limited, 2011

Statistical Software) or StatXact. SISSI is a program based on a re-sampling procedure named jack-knife for sample size determination created by Confalonieri (Confalonieri *et al.*, 2007). The algorithm systematically takes randomly chosen virtual sub-samples of the available data and calculates the mean and the standard deviation for different sample sizes (i.e. number of increment samples) while providing a visual evaluation system for optimal sample size determination. The program may also be used to estimate the sampling uncertainty associated with different sample sizes. The sole, but essential, assumption, as noted above, is that the dataset used is representative of the population under study.

A Monte Carlo algorithm has recently been used to determine the sample sizes necessary to obtain estimations of the mean of contamination of aflatoxin B1 (AFB1) at different percentages of precision for contaminated corn fields. AFB1 is an analyte characterized by a highly heterogeneous distribution very similar to GMO analytes (Brera *et al.*, 2010). The Monte Carlo method is an iterative, non-parametrical technique that has been used to: (i) theoretically sample the field, (ii) evaluate the effect of the sample size on the normality assumption, and (iii) determine the variability (precision) among sample means of different sizes (Robert *et al.*, 2004; Brera *et al.*, 2010).

These kind of virtual re-sampling programs are useful, in particular, because they do not require any particular assumptions about the distribution of the statistical parameters as required by classical methods, except that the data set is representative of the population distribution. In particular a virtual re-sampling program may be used when the available resources only allow reduced sample sizes, for instance when working on high-value seeds in the first steps of a breeding process. In those cases is not possible to make assumption about the distribution of the statistical parameters. A limitation of these techniques is that they require data sufficiently representative of the target population and this is not always easy to achieve. Moreover, often the obtained results refer specifically to situations where the mean and the variance are comparable to those of the dataset used; to extend the results to other situations it is necessary to investigate the nature of the relationship between means and relative variances in the matrix of interest.

13.5.3 Geostatistic models

Geostatistic models are also useful when the assumption of a random distribution is not realistic, as they allow eventual spatial patterns to be highlighted in the available data. Specifically, geostatistic models, which take into account autocorrelation (the correlation between proximal locations – for instance close units are more related than distant ones in function of the distance in time or space), allow situations where a non-random distribution exists to be investigated. Those approaches allow the pattern of the variability of the analyte, the number of incremental samples to take and the frequency of sampling to be defined (Petersen *et al.*, 2005a).

© Woodhead Publishing Limited, 2011

13.6 Statistical programs/software

The wide spectrum of problems associated with sampling necessitate the creation and dissemination of different pieces of software for determination of uncertainty and sample size, to study possible spatial correlation or to use re-sampling techniques, among others. Tables 13.4–13.8 illustrate some of the most widely used pieces of software for sampling. Each table focuses on a specific field of application. These programs created for specific purposes are very useful tools for estimation of uncertainty components, studying such aspects

Table 13.4 Software for estimating measurement uncertainty

Name	Link	Description
ROBAN	http://www.rsc.org/Membership/Networking/InterestGroups/Analytical/AMC/Software/ROBAN.asp	Powerful modelling tool for estimating total measurement uncertainty through classic analysis of variance (ANOVA) and robust analysis of variance (RANOVA).

Table 13.5 Software to calculate the sample size/power effect

Name	Link	Description
POWER and PRECISION	http://www.power-analysis.com/home.htm	Calculates the sample size required for any level of power.
Wessa	http://www.wessa.net/stat.wasp	The Free Statistics Calculator offers sample-size calculations, descriptive and explorative analysis.
PC-Size	ftp://ftp.simtel.net/pub/simtelnet/msdos/statstcs/size102.zip	Uses exact and approximate methods for sample size/power calculations for t-tests, 1-way and 2-way ANOVA, simple regression, correlation, and comparison of proportions.
DSTPLAN	http://biostatistics.mdanderson.org/SoftwareDownload/SingleSoftware.aspx?Software_Id=41	Calculates sample size/power for t-tests, correlation, a difference in proportions, 2xN contingency tables, and various survival analysis designs.
Splus	http://www.insightful.com/products/splus/default.asp	Provides functions for sample size/power for t-tests and comparing proportions.
Stata	http://www.stata.com/	It has some simple built-in power and sample size functions.

© Woodhead Publishing Limited, 2011

Table 13.6 Software for re-sampling techniques

Name	Link	Description
C++	http://www.cplusplus.com/doc/tutorial/	One of the most popular programming languages ever created. It is considered as a *middle-level* language, C++ is widely used in the software industry.
R	http://cran.r-project.org/doc/manuals/R-intro.html	Is a one of the most used programming languages for statistical computing, is widely used for statistical software development and data analysis.
Resampling Stats	http://www.resample.com/content/about.shtml	Implements re-sampling methods in statistics (including simulations, as well as bootstrap and permutation procedures). It is available as an independent product, as well as for Excel and Matlab.
SAS	http://www.sas.com/	The leader software in business analytics software and services. It includes macros for re-sampling methods.
SISSI	http://hdl.handle.net/2434/26500	A re-sampling-based software for sample size determination.
S-Plus	http://stat.ethz.ch/~www/SandR.html	Is a high-level language and an environment for statistical analysis and graphics development, including application for re-sampling methods.
StatXact	http://www.cytel.com/Software/StatXact.aspx	Is a package for exact nonparametric statistical inference on continuous or categorical data, data analysis, modelling, visualization, data mining, fitting.

Table 13.7 Software to study sampling strategies and distribution properties

Name	Link	Description
KeSTE	http://www.sipeaa.it/tools/KeSTE/What_use_for_KeSTE.htm	An exploratory tool developed to investigate the possible consequences of different sampling strategies on the structure and properties of bulk samples when contaminants are not randomly distributed. KeSTE allows the evaluation of sampling strategies on both simulated lots and on real lots, through the analyses of data collected from a maximum of 100 increments.

© Woodhead Publishing Limited, 2011

Table 13.8 Software for determining the minimum cost for acceptance sampling plans for grain control in GMO detection (Kobilinsky and Berthau, 2005)

Name	Link	Description
OPACSA	http://www.coextra.eu/researchlive/reportage851.html	Helps to find the cheapest and most reliable mode of analysis by sub-sampling, taking into account economic factors and enables the use of inexpensive qualitative methods.

involved in the sampling process as sample size and distribution properties and assessing cost issues related to acceptance sampling plans, etc.

13.7 Conclusions

A number of factors have to be taken into consideration when developing or approving sampling plans for GMO control in the food chain. In particular, it is pivotal to keep in mind that unavoidable sampling errors must be controlled to maintain sample representativeness at as high a level as possible. Several sampling methods have been standardized by competent authorities such as the European Commission and international organizations such as ISO, but all of these are products of different viewpoints and were created with different criteria in mind. A harmonized official sampling procedure exists for seeds, but a suitable sampling procedure for testing GMO presence in food and feed commodities as well as in the field that fulfils both accuracy and feasibility requirements is still lacking. Adopting 'multi-analyte' sampling plans could reduce the costs of sampling, improve versatility, lead to improvements in the quality of control activities and increase transparency in commercial transactions. It is also worth mentioning that sampling issues are not viewed as particularly relevant by those approving and financing research projects at national, European and international level. Further funding could lead to improvements in practice in this area.

13.8 References

ALLNUTT T.R., M. DWYER, J. McMILLAN, C. HENRY and S. LANGRELL (2008) Sampling and modelling for the quantification of adventitious genetically modified presence in maize. *J. Agr. Food Chem.* 56 (9): 3232–3237.

ANDERSEN N.S., H.R. SIEGISMUND and R.B. JØRGENSEN (2005) Low level of gene flow from cultivated beets (*Beta vulgaris L. ssp. vulgaris*) into Danish populations of sea beet (*Beta vulgaris L. ssp. maritima (L.) Arcangeli*). *Molecular Ecology* 14: 1391–1405.

ANGEVIN F., E.K. KLEIN, C. CHOIMET, A. GAUFFRETEAU, C. LAVIGNE, A. MESSÉAN and J.M. MEYNARD (2008) Modelling impacts of cropping systems and climate on maize

© Woodhead Publishing Limited, 2011

cross-pollination in agricultural landscapes: the MAPOD model. *Eur. J. Agron.* 28: 471–484.

BEGG G.S., M.J. ELLIOTT, D.W. CULLEN, P.P.M. IANNETTA and G.R. SQUIRE (2008) Heterogeneity in the distribution of genetically modified and conventional oilseed rape within fields and seed lots. *Transgenic Research*, doi:10.1007/s11248-008-9166-7.

BÉNÉTRIX F. (2005) Managing the co-existence of conventional and genetically modified maize from field to silo – A French initiative. Proceedings of the 2nd International Conference on Co-existence between GM and non GM based agricultural supply chains, GMCC-05, 14–15 November 2005, Montpellier, France.

BÉNÉTRIX, F., D. BLOC, X. FOUEILLASSAR and B. NAÏBO (2003) A study to evaluate co-existence of GM and conventional maize on the same farm. Proceedings of the 1st European Conference GMCC-03 on the Co-existence of Genetically Modified Crops with Conventional and Organic Crops, 13–14 November 2003, Helsingør, Denmark.

BRERA C., B. DE SANTIS, E. PRANTERA, F. DEBEGNACH, E. PANNUNZI, F. FASANO, C. BERDINI, A. B. SLATE, M. MIRAGLIA and T.B. WHITAKER (2010) Effect of sample size in the evaluation of 'in-field' sampling plans for aflatoxin B1 determination in corn. *J. Agric. Food Chem.* 58: 8481–8489.

CAC/GL 50-2004 Codex General Guideline on Sampling.

CEDDIA M.G., M. BARTLETT and C. PERRINGS (2007) Landscape gene flow, coexistence and threshold effect: The case of genetically modified herbicide tolerant oilseed rape (*Brassica napus*). *Ecological Modelling*, 205: 169–180.

CEN/TS 15568 (2006) Foodstuffs – Methods of analysis for the detection of genetically modified organisms and derived products – Sampling strategies.

CONFALONIERI R, M. ACUTIS, G. BELLOCCHI and G. GENOVESE (2007) Resampling-based software for estimating optimal sample size. *Environmental Modelling & Software* 22: 1796–1800.

DAVISON J. and Y. BERTHEAU (2007) EU regulations on the traceability and detection of GMOs: difficulties in interpretation, implementation and compliance. CAB *Perspectives in Agriculture, Veterinary Science, Nutrition and Natural Resources* 2: 77: 1–14.

DE GIACOMO M., R. ONORI, C. DI DOMENICANTONIO, G. BELLOCCHI, E. GRAZIOLI, M. MIRAGLIA and G. VAN DEN EEDE (2008) Novel methodology for sampling of GM soybean packed products. Proceedings of the 1st Global Conference on GMO Analysis, 27 June, Como, Italy.

DE GIACOMO M., R. ONORI, E. PALMACCIO, M. DE VIVO, C. DI DOMENICANTONIO, C. BRERA, M. MIRAGLIA, A. MALCEVSKY and N. MARMIROLI (2009) GMO versus mycotoxins sampling plan: a pragmatic approach. Poster presentation, Co-Extra international conference, Paris, 2–5 June 2009.

DEMONT M., K. DILLEN, W. DAEMS, C. SAUSSE, E. TOLLENS and E. MATHIJS (2009) On the proportionality of EU spatial *ex ante* coexistence regulations. *Food Policy* 34: 508–518.

DEMONT M., Y. DEVOS and O. SANVIDO (2010) Towards flexible coexistence regulations for GM crops in the EU. *Eurochoices* 9: 18–24.

DEVAUX C., C. LAVIGNE, F. AUSTERLITZ and E.K. KLEIN (2007) Modelling and estimating pollen movement in oilseed rape (*Brassica napus*) at the landscape scale using genetic markers. *Molecular Ecology* 16: 487–499.

DEVAUX C., E. K. KLEIN, C. LAVIGNE, C. SAUSSE and A. MESSÉAN (2008) Environmental and landscape effects on cross-pollination rates observed at long-distance among

© Woodhead Publishing Limited, 2011

French oilseed rape (*Brassica napus*) commercial fields. *Journal of Applied Ecology* 45: 803–812.

DEVOS Y., D. REHEUL and A. DE SCHRIJVER (2005) Review. The co-existence between transgenic and non-transgenic maize in the European Union: a focus on pollen flow and cross-fertilization. *Environ. Biosafety Res.* 4: 71–87.

DEVOS Y., M. DEMONT, K. DILLEN, D. REHEUL, M. KAISER and O. SANVIDO (2009) Coexistence of genetically modified (GM) and non-GM crops in the European Union. A review. *Agronomy for Sustainable Development* 29: 11–30.

EC (EUROPEAN COMMISSION) (2001) Directive 2001/18 on the deliberate release into the environment of genetically modified organisms and repealing Council Directive 90/220/EEC. *Official Journal of the European Communities*, L 106: 1–38.

EC (EUROPEAN COMMISSION) (2002) Regulation 2002/178 laying down the general principles and requirements of food law, establishing the European Food Safety Authority and laying down procedures in matters of food safety. *Official Journal of the European Communities*, L 31: 1–23.

EC (EUROPEAN COMMISSION) (2003a) Recommendation 2003/556 on guidelines for the development of national strategies and best practices to ensure the co-existence of genetically modified crops with conventional and organic farming. *Official Journal of the European Communities*, L 189: 36–47.

EC (EUROPEAN COMMISSION) (2003b) Regulation 1830/2003 of the European Parliament and of the Council of 22 September 2003 concerning the traceability and labelling of genetically modified organisms and the traceability of food and feed products produced from genetically modified organisms and amending Directive 2001/18/ EC. *Official Journal of the European Union*, L 268: 24–28.

EC (EUROPEAN COMMISSION) (2004) Recommendation 787/2004 on technical guidance for sampling and detection of genetically modified organisms as or in products in the context of Regulation (EC) 1830/2003. *Official Journal of the European Union* L 348: 18–26.

EC (EUROPEAN COMMISSION) (2006) Regulation 401/2006 of 23 February 2006 laying down the methods of sampling and analysis for the official control of the levels of mycotoxins in foodstuffs. *Official Journal of the European Union*, L 70: 12–34.

EC (EUROPEAN COMMISSION) (2006) Commission Decision 2006/754 amending Decision 2006/601/EC on emergency measures regarding the non-authorised genetically modified organism 'LL RICE 601' in rice products. *Official Journal of the European Union*, L 306: 17–20.

EC (EUROPEAN COMMISSION) (2009) Regulation 152/2009 laying down the methods of sampling and analysis for the official control of feed. *Official Journal of the European Union* L 54: 1–130.

EC (EUROPEAN COMMISSION) (2010) Recommendation (EC) 2010/C200/01/EC Guidelines for the development of national coexistence measures to avoid the unintended presence of GMOs in conventional and organic crops *Official Journal of the European Union*, C 200: 1–5.

ESBENSEN K.H., C. PAOLETTI and P. MINKKINEN (2008) Reliable assessment of total errors in sampling large kernel lots: a new variographic approach (theory of sampling). Presented at the 1st Global Conference on GMO Analysis, Como, Italy.

EURACHEM/CITAC (2007) *Guide Measurement uncertainty arising from sampling: A guide to methods and approaches*, M. H. Ramsey and S. L. R. Ellison.

GARNIER A., S. PIVARD and J. LECOMTE (2008) Measuring and modelling anthopogenic secondary seed dispersal along road verges for feral oilseed rape. *Basic and*

© Woodhead Publishing Limited, 2011

Applied Ecology, 9: 533–541.

GILBERT J. (1999) Sampling of raw materials and processed foods for the presence of GMOs. *Food Control* 10: 363–365.

GRUBER S. and W. CLAUPEIN (2007) Fecundity of volunteer oilseed rape and estimation of potential gene dispersal by a practice-related model. *Agriculture, Ecosystems & Environment* 119: 401–408.

GY P.M. (1995) Introduction to the theory of sampling I. Heterogeneity of a population of uncorrelated units. *Trends in Analytical Chemistry* 14 (2): 67–76.

GY P. (2004) Sampling of discrete materials II. Quantitative approach-sampling of zero-dimensional objects. *Chemometrics and Intelligent Laboratory Systems* 74: 25–38.

HÜSKEN A., K. AMMANN , J. MESSEGUER, R. PAPA, P. ROBSON, J. SCHIEMANN, G. SQUIRE, P. STAMP, J. SWEET and R. WILHELM (2007) A major European synthesis of data on pollen and seed mediated gene flow in maize in the SIGMEA project. In: Stein A.J., E. Rodriguez Cerezo (Eds) Third International Conference on Coexistence between Genetically Modified (GM) and non-GM based Agricultural Supply Chains, Seville (Spain), 20–21 November 2007, Book of abstracts: 53–56.

ISO 98-3:2008 Uncertainty of measurement – Part 3: Guide to the expression of uncertainty in measurement (GUM:1995)

ISO 2859:1995 Sampling procedures for inspection by attributes.

ISO 13690:1999 Cereals, pulses and milled products – Sampling of static batches. International organization of standardization, Geneva, Switzerland.

ISO 6644:2002 Flowing cereals and milled products. Automatic sampling by mechanical means.

ISO 24333:2009 Cereals and cereal products – Sampling.

ISO 21570 (2005) Foodstuffs: methods of analysis for the detection of genetically modified organisms and derived products: quantitative nucleic acid based methods, 1st edition. International Standard ISO 21570, International Organisation for Standardisation, Geneva, Switzerland.

ISTA (2007) International rules for seed testing. The International Seed Testing Association (ISTA), Bassersdorf, Switzerland.

JOHANSSON A.S., T.B. WHITAKER, F.G. GIESBRECHT, W.M. JR. HAGLER and J.H.YOUNG (2000) Testing shelled corn for aflatoxin, Part III: Evaluating the performance of aflatoxin sampling plans. *J. AOAC International* 83 (5): 1279–1284.

KAY S. and C. PAOLETTI (2002) Sampling strategies for GMO detection and/or quantification. EUR 20239 EN. 8.

KOBILINSKY A. and Y. BERTHEAU (2005) Minimum cost acceptance sampling plans for grain control, with application to GMO detection. *Chemometrics and Intelligent Systems* 75: 189–200.

LAFFONT, J.L., K.M. REMUND, D. WRIGHT, R.D. SIMPSON and S. GREGOIRE (2005) Testing for adventitious presence of transgenic material in conventional seed or grain lots using quantitative laboratory methods: statistical procedures and their implementation. *Seed Science Research* 15: 197–204.

LANGHOF M., B. HOMMEL, A. HÜSKEN, J. SCHIEMANN, P. WEHLING, R. WILHELM and G. RUHL (2008) Coexistence in maize: Do non-maize buffer zones reduce gene flow between maize fields. *Crop Sci.* 48: 305–316.

LISCHER P. (2001) Sampling procedures to determine the proportion of genetically modified organisms in raw materials. Part I: Correct sampling, good sampling practices. *Mitt. Lebensm. Hyg.* 92: 290–304.

MACARTHUR R., S. MACDONALD, P. BRERETON and A. MURRAY (2006) Statistical modelling

© Woodhead Publishing Limited, 2011

as an aid to the design of retail sampling plans for mycotoxins in food. *Food Additives and Contaminants* 23 (1): 84–92.

MACARTHUR R., A.W.A. MURRAY, T.R. ALLNUTT, C. DEPPE, H.J. HIRD, G.M. KERINS, J. BLACKBURN, J. BROWN, R. STONES and S. HUGO (2007) Model for tuning GMO detection in seed and grain. *Nat. Biotechnol.* 25: 169–170.

MACARTHUR R., M. FEINBERG and Y. BERTHEAU (2010) Construction of measurement uncertainty profiles for quantitative analysis of genetically modified organisms based on interlaboratory validation data. *J. AOAC Int* 93(3): 1046–1056.

MESSEGUER J., G. PENAS, J. BALLESTER, M. BAS, J. SERRA, J. SALVIA, M. PALAUDELMAS and E. MELE (2006) Pollen-mediated gene flow in maize in real situations of coexistence. *Plant Biotechnol. J.* 4(6): 633–645.

MONTESINOS L, O. A., A. MONTESINOS L, J. CROSSA, K. ESKRIDGE and C.M. HERNANDEZ SUAREZ (2010) Sample size for detecting and estimating the proportion of transgenic plants with narrow confidence intervals. *Seed Science Research* 20: 123–136.

MOREANO F., U. BUSCH and K.H. ENGEL (2005) Distortion of genetically modified organism quantification in processed foods: influence of particle size compositions and heat-induced DNA degradation. *J. Agric. Food Chem.* 53: 9971–9979.

ONORI R. (2008) Novel methodology for sampling of GM soybean packed products. Presented at the 1st Global Conference on GMO Analysis, Como, Italy.

ONORI R., M. DE GIACOMO, R. LOPARDO, B. DE SANTIS, L. TANCREDI and M. MIRAGLIA (2007) Traceability of Round-up Ready Soybean along processing chain. Third International Conference on Coexistence between Genetically Modified (GM) and non-GM based Agricultural Supply Chains. Seville (Spain), November 2007. JRC European Commission.

PAOLETTI C., M. DONATELLI, A. HEISSENBERGER, E. GRAZIOLI, S. LARCHER and G. VAN DEN EEDE (2005) European Union Perspective – Sampling for Testing of Genetically Modified Impurities. Australasian Institute of Mining and Metallurgy, Australia. Second World Conference on Sampling and Blending Book, 209–213.

PAOLETTI C., A. HEISSENBERGER, M. MAZZARA, S. LARCHER, E. GRAZIOLI, P. CORBISIER, N. HESS, G. BERBEN, P.S. LUBECK, M. DE LOOSE, G. MORAN, C. HENRY, C. BRERA, I. FOLCH, J. OVESNA and G. VAN DEN EEDE (2006) Kernel lot distribution assessment (KeLDA): A study on the distribution of GMO in large soybean shipments. *Eur. Food Res. Technol.* 224: 129–139.

PETERSEN L., K.H. ESBENSEN and C.K. FDAHL (2004) Representative mass reduction in sampling: A critical survey of techniques and hardware. *Chemometrics and Intelligent Systems* 74: 95–114.

PETERSEN L. and K. ESBENSEN (2005a) Representative process sampling for reliable data analysis – a tutorial. *Journal of Chemometrics* 19: 625–647.

PETERSEN L., P. MINKKINEN and K. ESBENSEN (2005b) Representative sampling for reliable data analysis: theory of sampling. *Chemometrics and Intelligent Laboratory Systems* 77: 261–277.

PITARD, F.F. (1993) *Pierre Gy's Sampling Theory and Sampling Practice: Heterogeneity, Sampling Correctness and Statistical Process Control*, 2nd edn. CRC Press, Boca Raton, FL.

PLA, M., J.L. LA PAZ, G. PENAS, N. GARCIA, M. PALAUDELMAS, T. ESTEVE, J. MESSEGUER and E. MELE (2006) Assesment of real-time PCR based methods for quantification of pollen-mediated gene flow from GM to conventional maize in a field study. *Transgenic Research* 15: 219–228.

REMUND K., D.A. DIXON, D.L. WRIGHT and L.R. HOLDEN (2001) Statistical considerations in

© Woodhead Publishing Limited, 2011

seed purity testing for transgenic traits. *Seed Science Research* 11: 101–119.

ROBERT C.P. and G. CASELLA (2004) *Monte Carlo Statistical Methods*, 2nd edn. Springer-Verlag, New York.

SANVIDO O., F. WIDMER, M. WINZELER, B. STREIT, E. SZERENCSITS and F. BIGLER (2008) Definition and feasibility of isolation distances for transgenic maize cultivation. *Transgenic Res.* 17: 317–335.

SCHILLING E.G. (1982) *Acceptance Sampling in Quality Control, Statistics: Textbooks and Monographs*, vol. 42, Marcel Dekker, New York.

STARK C., S. LIEPELT, M. DIECKVOSS, D. BARTSCH, B. ZIEGENHAGEN and A. ULRICH (2006) Fast and simple monitoring of introgressive gene flow from wild beet into sugarbeet. *Journal of Sugar Beet Research* 43: 145–154.

ŠUŠTAR-VOZLIČ J., K. ROSTOHAR, A. BLEJEC, P. KOZJAK, Z. ČERGAN and V. MEGLIČ (2010) Development of sampling approaches for the determination of the presence of genetically modified organisms at the field level. *Anal. Bioanal. Chem.* 396: 2031–2041.

WAIBLINGER H.U. (2008) A sampling procedure for the detection of non-authorized GMO at low levels. Presented at the 1st Global Conference on GMO Analysis, Como, Italy.

WEEKES R., T. ALLNUTT, C. BOFFEY, S. MORGAN, M. BILTON, R. DANIELS and C. HENRY (2007) A study of crop-to-crop gene flow using farm scale sites of fodder maize (*Zea mays* L.) in the UK. *Transgenic Res.* 16: 203–211.

WHITAKER T.B., M.W. TRUCKSESS, F.G. GIESBRECHT, A.B. SLATE and F.S. THOMAS (2004) Evaluation of sampling plans to detect Cry9C protein in corn flour and meal. *J. Assoc. Off. Analytical Chem. Int.* 87: 950–960.

© Woodhead Publishing Limited, 2011

14

New approaches to determining the origin of food

S. Kelly and P. Brereton, Food and Environment Research Agency, UK, C. Guillou, Joint Research Centre of the European Commission, Italy, H. Broll and I. Laube, Federal Institute for Risk Assessment, Germany, G. Downey, Teagasc, Ireland, A. Rossman, Isolab GmbH, Germany, S. Hözl, Bavarian State Collection for Palaeontology and Geology, Germany and G. van der Veer, Wageningen University, Netherlands

Abstract: Recent food scares and the globalisation of food and feed markets combined with the relative ease with which food commodities are transported through and between countries and continents, means that consumers are increasingly concerned about the origin of the foods they eat. Over the last five years an increasing number of genetic, spectroscopic, molecular and isotopic methods that permit independent verification of the origin of the foods we consume have been described in the literature. Paper trails of food products can be falsified by experienced criminals, or errors could be introduced unintentionally, leading to incorrect information on product labels. With increasing legislative demands on food producers to provide clear and concise origin information, the new approaches to independent verification techniques described here have the potential to contribute significantly to the food origin record process.

Key words: food authenticity, geographical origin, provenance, isotope ratio, spectroscopy, molecular biology, isoscape, protected designation of origin, PDO, protected geographical indication, PGI.

© Woodhead Publishing Limited, 2011

14.1 Introduction

In Europe, an ever more discerning consumer is prepared to pay extra for 'added value' products (FSA, 2010). In many cases the product is essentially similar but with particular 'quality' attributes which are used to promote sales through specific labelling. These attributes such as local/regional/country of origin, free range, organic, species, blended or single variety, protected designation of origin (PDO), Fair Trade to name but a few, provide food scientists with an enormous challenge in terms of confirming labelling compliance that will satisfy consumers, food industry and control bodies alike. This challenge also has varying degrees of complexity depending upon the question asked of the food or feed scientist. For example, the question 'is this product adulterated?' is often followed by the more difficult questions 'what is the level of adulterant in the authentic product?' and 'where does this counterfeit product originate?'. Consequently, it is easy to see why food scientists need to resort to sophisticated methods and techniques to authenticate food origin. Moreover, food and feed scientists are requested to provide scientific methods and perform measurements that establish the compliance of food and feed products with the regulatory requirements. Food safety aspects are generally easily controlled with official and standardised methods determining physico-chemical or biological parameters against well-established specifications and tolerances, e.g. maximum residue limits, using measurement uncertainties determined by, generally, inter-laboratory trials. The analytical testing for authentication purposes such as recognition of geographic origin, or detection of adulteration is generally significantly more complex.

The specifications for foods with geographic indications are generally based on consolidated traditional practices and often on subjective characteristics such as organoleptic properties. Although physico-chemical parameters are sometimes included in the specifications, they are generally not specific to the geographic origin and do not allow differentiation from other similar products. Consequently, it is likely that imitations or adulterated products are present on the market and compete in an unfair manner with the genuine products. It is in the interest of the consumers, the producers and also control bodies to utilise analytical methods and objective parameters that enable a better, quantitative in numerous instances, control of 'origin' of food products (see other chapters of this book). Isotopic methods have been shown to be an efficient tool to fight against adulteration of food products since the beginning of the 1990s after the European Council decided to establish an official European isotopic wine databank for controls in the wine sector (EC Regulation 2048/1989/EC). Several European funded research projects further developed these isotopic techniques and also spectroscopic fingerprinting methods on a broader range of food products, investigating applications for the control of the botanical, animal and geographic origins of food products. The techniques developed in molecular biology have also led to the possibility of characterisation of species and origins. This chapter illustrates the value of these researches relating to the origin of a broad range of food and feed products.

© Woodhead Publishing Limited, 2011

Many of these works are based on isotopic techniques eventually in combination with trace elements (Kelly *et al.*, 2005). The fingerprinting techniques are also important tools for confirmation of the origin of many food products provided that a reference databank of genuine products can be established. The research works presented in this chapter briefly highlight some of the analytical possibilities available today to the food and feed chemists to provide objective physico-chemical parameters helping the recognition of origin provenance of food and feed products and also to detect adulteration or mis-labelling. Much progress has been achieved in the past years, through the European Integrated Project TRACE (www.trace.eu.org). It is against this background that this chapter will present some of the latest research in this area of determining food provenance using molecular biology, spectroscopic fingerprinting, bio-element and heavy element stable isotope ratio analysis and isotope mapping.

14.2 Molecular biological methods

Molecular biological methods target the inheritance present in any living cell. The so-called DNA (deoxyribonucleic acid) is identical in every cell of an individual regardless of the origin of the tissue but differs between individuals. The greater the phylogenetic distance between individuals, the greater the differences in the sequence of the DNA that can be targeted to identify individuals and thus phylogeny is also often related to geographic separation. When a distinction among animal or plant species is needed, the DNA sequence may differ sufficiently to develop a suitable molecular biological method for detection.

DNA has the potential to provide more information than protein due to the degeneracy of the genetic code and the presence of many non-coding regions. In addition DNA is much more resistant and thermostable in comparison to proteins and, despite being altered through processing of food products in general, it is still possible to amplify small DNA fragments with sufficient information to allow identification. DNA is a long-lived biological molecule. For forensic purposes DNA has even been successfully extracted from ancient mummies (Pääbo *et al.*, 1989; Pääbo, 1989). Typical amplification conditions are well documented and include the target DNA, a thermostable DNA polymerase, two oligonucleotide molecules called primers, desoxynucleotide triphosphates (dNTPs), reaction buffer and magnesium. In theory a single target molecule is sufficient to result in a positive amplification. However, along with the potential of PCR (polymerase chain reaction), the risk of cross-contamination resulting in false-positive samples is high and needs to be monitored for each individual analysis. In order to identify such contaminations, appropriate positive (analyte present) and negative (analyte absent) controls are included at each step of the analysis.

Another problem specifically related to food analysis is the co-extraction of inhibitory substances such as starch or glucose from composed food products along with the DNA. It is necessary to identify such inhibition by applying

© Woodhead Publishing Limited, 2011

appropriate amplification controls. A suitable amplification control might be a PCR system targeting a declared ingredient (e.g. soya in case of tofu product). In case the respective PCR is negative it is necessary to subsequently remove them in order to avoid false-negative results. Specific isolation procedures are developed for this problem and provided by various suppliers. Independent from the starting copy number of target molecules or number of cycles performed in PCR, the amplified products need to be subsequently analysed by using gel electrophoresis in order to identify the predicted amplified product and to assess the sample under investigation.

Real time PCR offers a more advanced application in which an oligonucleotide probe is additionally added and hybridises between both primers (Logan *et al.*, 2009). This probe is labelled with a fluorescent dye. Another possibility (less expensive, but also less specific) is a DNA-binding dye which binds to all double-stranded (ds) DNA in PCR, causing fluorescence of the dye. The accumulation of fluorescence during PCR cycles is directly proportional to the amount of DNA amplified by the polymerase. The essential advantages over the classical PCR are the ability to omit the subsequent electrophoresis step and that real time PCR drastically reduces the potential for contaminations. The assay is rapid, and a large number of food samples can be analysed in a single run. Moreover, quantification is possible using the well-established standard curve approach as used, e.g., in GMO detection (Querci *et al.*, 2009). The bottleneck of PCR analysis is still the DNA extraction. While it is not a problem to extract DNA from any unprocessed product, it could be a very difficult task to extract DNA from highly processed products such as refined oils or acidic samples. These products usually no longer contain any traces of DNA and consequently no PCR can be applied. The degree of degradation of DNA is also an important aspect when calculating the amount of individual animal or plant species in foods (Laube *et al.*, 2007).

For species identification in food products, the mitochondrial DNA genome (*mt*DNA) is preferentially targeted compared to nuclear DNA. Indeed, because there are several copies of *mt*DNA inside a cell (approximately 500–1000 times the copies of nuclear DNA) it is more likely to amplify a fragment within the mitochondrial genome rather than within the genome of the nucleus. In addition, *mt*DNA generally evolves much faster than nuclear DNA and thus enables even closely related species to be differentiated and identified, which is important for species differentiation. Although it is an advantage that it is present in high copy numbers per cell, it is almost not suitable for quantitative purposes, since the exact copy number is not known. For quantification it is necessary to choose a target present only once per haploid genome. Several detection methods have been published fulfilling this requirement (Laube *et al.*, 2007, 2010). PCR has the potential for standardisation. Today PCR is established in many laboratories around the world for control purposes. Primer and probes as well as all other reagents needed are available from different suppliers. There is no dependence on a single provider, which is usually the case if immunological methods are applied using specific antibodies.

© Woodhead Publishing Limited, 2011

Alternative techniques to PCR are include RAPD (random amplified polymorphic DNA), SCAR (sequence characterised amplified region), AFLP (amplified fragment length polymorphism), barcoding systems such as SNP (single-nucleotide polymorphism) and microsatellites, SSCP (single-strand conformation polymorphism), RFLP (restriction fragment length polymorphism), which are used along with PCR and real time PCR, e.g. for the detection of the breed in food labelled PDO (protected designation of origin) or PGI (protected geographical indication). The same markers can be used to differentiate wild and domestic animals (Woolfe and Primrose, 2004). A significant amount of molecular biological research has been dedicated to species and breed identification. Case studies have focused on the development and validation of molecular markers able to trace PGI beef products, PGI and PDO honeys by pollen and bee identification, identification of cereal species (Prins *et al.*, 2010) as well as the differentiation between Organic and non-Organic (or conventional) farming (Schneider *et al.*, 2010). One of several major achievements in this area of research has been the establishment of a database containing the whole set of information about DNA-based detection methods developed within the framework of the TRACE project. Moreover, all information on DNA-based methods from previous European projects (e.g. MOLSPEC-ID) Development of quantitative and qualitative molecular biological methods to identify plant and animal species in foods; http://ec.europa.eu/research/quality-of-life/ka1/volume1/qlk1-2001-02373.htm) has been incorporated, resulting in a comprehensive database providing methods for control purposes as well as for quality assurance as part of an identity preservation system established in food processing companies. This database is publicly available (http://www.trace.eu.org/mbdb/). In total 51 DNA-based detection methods are included in the database. For each individual method the whole internationally agreed set of validation characteristics is described. Parameters such as the limit of detection (LOD), specificity and sensitivity have been investigated intensively and determined for each individual method. This information is the basis for the analysts and stakeholders to evaluate if the method selected is 'fit-for-their-purpose' and could be validated through a collaborative trial and thus standardised. In addition, three of the methods ('Detection of plant species in honey using real time PCR') included in the database have been validated thoroughly through international collaborative studies providing the most robust validation characteristics.

The prerequisite for running PCR is a specifically equipped laboratory with trained staff under quality assurance according to ISO 17025 or an equivalent accreditation scheme based on that standard. Today all EC Member States are equipped with enforcement laboratories fulfilling these quality assurance standards and level of accreditation (Zel *et al.*, 2006, 2008). ISO and CEN standards exist describing the establishment of laboratories. Several standards have also been established on specific methods for the purpose of the identification of pathogens, plant species and genetically modified foods (GMO) including specific methods or performance criteria (see e.g. CEN ISO 2476).

© Woodhead Publishing Limited, 2011

The determination of the measurement uncertainty (MU) associated with entire procedures is also described in a document prepared by the Joint Research Centre of the European Commission (JRC-EC) and can be applied to the methods developed for origin determination (Trapmann *et al.*, 2009; Macarthur *et al.*, 2010; Gustavo González and Ángeles Herrador, 2007; European Commission DG Health and Consumer Protection, 2004).

By combining the identification of different targets in a single reaction ('multi-plexing'), the costs may be reduced as personnel time is the most important factor of analysis costs. This also reduces the time and increases the efficiency of sample analysis for both qualitative and quantitative analysis. The individual steps can be automated to a very large extent. Instruments are available for the DNA extraction as well as for the subsequent PCR step. This automation will also enhance the reliability of results but initially requires capital investment. The methods developed described here are an important milestone in order to support the application of such PCR methods for the purpose of traceability of foods of plant and animal origin. Many useful methods are fully described with regard to their performance and freely available on the TRACE web-site (www.trace.eu.org) for interested stakeholders equipped with the necessary laboratories and trained staff.

14.3 Spectroscopic and fingerprinting techniques

Rapid screening techniques for the confirmation of food authenticity or the detection of adulteration are of critical importance in any food traceability system to justify claims of food provenance, processing practice or brand identity. Suitable food screening methods may be divided into two categories – spectroscopic or separation techniques. They include near infrared, mid-infrared, Raman and nuclear magnetic resonance (NMR) spectroscopies, UPLC-ToF-mass spectrometry, LC/MS/mass spectrometry and ambient mass spectrometry (DART).

Infrared and Raman spectroscopy are characterised by rapid spectral acquisition, little or no sample preparation and relatively simple instrument operating procedures. Consequently, they lend themselves to the rapid screening of foods at the process level. Spectra, however, vary in the level of fine detail they contain and are commonly composed of multiple overlapping peaks. Therefore, these spectra are usually treated as molecular 'fingerprints' and require manipulation by multivariate mathematical techniques for information extraction, e.g. principal component analysis. Nuclear magnetic resonance (NMR) spectroscopy is based on the observation that most atomic nuclei, oriented by a strong magnetic field, absorb radiation at a characteristic frequency. Different NMR experiments and equipment give different information but a common technique in food authenticity analysis is proton (^{1}H) NMR. NMR analysis allows many different compounds to be detected in a single experimental acquisition, thereby obtaining a complete metabolic profile of a food stuff or

© Woodhead Publishing Limited, 2011

ingredient (Wishart, 2008). Ultra-performance liquid chromatography coupled to a quadrupole time-of-flight (UPLC-QTof)-mass spectrometry is a high-resolution, high-throughput mass spectrometry system which employs time-of-flight mass spectrometry technology for the accurate mass determination of chemical compounds in a wide range of sample matrices (Ni *et al.*, 2010). The main advantages of Direct Analysis in Real Time (DART) ambient mass spectrometry compared to conventional mass spectrometry techniques include direct sample examination in the open atmosphere, minimal or no sample preparation and remarkably high sample throughput (Vaclavik *et al.*, 2009). The results of the application of several of these techniques to a number of food provenance issues are summarised below.

Fingerprinting and profiling techniques characteristically do not provide obvious information on specific analytes in a mixture. In order to use these fast but non-specific methods, it is necessary to interrogate the datasets using complex mathematical procedures – multivariate data analysis (MVA) (Leardi, 2003). MVA can ensure the integrity of a dataset and also has the ability to detect patterns among samples, for example samples of the same or similar type. While very many MVA procedures exist, a limited number has found wide-spread use in food authenticity applications: PLS-DA (partial least squares – discriminant analysis) and NOPLS (non-orthogonalised partial least squares), SIMCA (soft independent modelling of class analogy), artificial neural networks (ANN) and support vector machines (SVM).

14.3.1 Case study on honey

Honey has a special status among consumers not only as a food of natural origin, but also for its purported health benefits (Traynor, 2008). A special part of consumer expectation is associated with the geographical origin of honey and this is due in part to previous food scandals involving honey. For example, in 1984 Chinese honey was discovered within the European Community (EC), adulterated with high fructose corn syrup (HFCS) and there have been other incidents where excessive antibiotic residues have been found in honey derived from third-world countries (Ochi, 2005). Therefore consumers may prefer to spend more money for honey with a declared origin from their regional producer or from a European Community Member State rather than purchasing cheaper honey with no declared origin.

In this case study honey samples (a total of 381) were obtained from the Corsican Island (France) and a number of countries in the circum-Mediterranean region from two consecutive years of harvest. Statistical models to discriminate between honey from Corsica and all other geographical origins were developed using the same sample sets. Models were developed on a set of samples comprising 2/3 of the Corsican and an approximately equal number of non-Corsican honeys; these models were then applied to the remaining honeys in the sample set and prediction results recorded. A number of data sets were used, i.e. raw data and data pre-treated by both first and second derivative transforms

© Woodhead Publishing Limited, 2011

involving a variety of gap sizes as well as a standard normal variate (SNV) pre-treatment. Using NIR spectra, the PLS-DA model recommended for confirming the provenance of honey used a 1st derivative (21 data points) data pre-treatment. A correct classification rate of 86.3% for both Corsican and non-Corsican honeys was achieved. Applying mid-infrared spectroscopy to the same sample sets, values of 72.6 and 89.8% correct classification for Corsican and non-Corsican honey were obtained. In the case of Raman spectra, correct classification rates of 85.5 and 94.6% for Corsican and non-Corsican honeys respectively were obtained following the application of SVM. Using NMR spectroscopy on these samples, the best approach to confirm honey provenance used a two-stage genetic programming model and produced an excellent overall correct classification rate of 92.4% (Donarski *et al.*, 2010).

14.3.2 Case study on Trappist beer

Trappist beer is a beer brewed by or under control of Trappist monks. Of the world's 171 Trappist monasteries (as of April 2005), seven produce beer, six in Belgium (Orval, Chimay, Westvleteren, Rochefort, Westmalle and Achel) and one in the Netherlands (La Trappe) (Weeranantanaphan and Downey, 2010). Only these seven breweries are authorised to label their beers with the authentic Trappist product logo that indicates compliance with various rules edicted by the International Trappist Association. The aim of this association is to prevent non-Trappist commercial companies from abusing the Trappist name.

The issue of Trappist beer brand was examined to confirm the identity of Rochefort 8° beers produced in Belgium. Datasets obtained by analysis of 265 beer samples were examined using linear discriminant analysis (LDA) and artificial neural networks (ANNs). Both LDA and ANNs allowed discrimination between Trappist and non-Trappist beers based on either mass spectrometric profiles generated under the conditions of ambient mass spectrometry (DART–HR-TOF-MS). The best model (ANN) produced correct classification rates of 96.4 and 87.9% for Rochefort 8° and non-Rochefort 8° beers respectively. The encouraging data obtained using these techniques show that it is possible to answer the question 'is this beer a Rochefort?' with a high percentage of correct classification.

14.3.3 Case study on chicken origin

In the case of chicken meat, public awareness has been increased by recent food scares such as Avian Flu and the malpractice of some food producers (e.g. the use of banned nitrofurans in chicken production). The validity of the declared quality, safety and origin of the chicken meat is in question.

The development of methods that can provide a characteristic metabolomic profile and that can be used for the determination of the geographical origin, therefore, can provide a valuable tool for the authentication and traceability of foods such as chicken. In this case study 383 samples of dried, defatted chicken

© Woodhead Publishing Limited, 2011

breast meat were obtained from Europe, South America, Thailand and China and analysed by ^{1}H NMR. Applying NOPLS to the results obtained revealed that all of the groups could be discriminated from all of the others. Thus, 99% of Chinese samples could be accurately distinguished from those from all other countries and 98% of the relevant samples could be classified correctly as either European or Chinese. Interestingly, two separate groups of samples could be discerned within the Chinese set. The compounds responsible for the discriminations were identified using the original NMR data. It was also found that the European samples had relatively higher levels of alanine, glycine, branched chain amino acids (valine, leucine, isoleucine), betaine and dimethylglycine and lower levels of carnosine and anserine. Comparing China with 'the rest', Chinese samples had higher levels of lactate, carnosine and anserine and lower levels of amino acids including those listed above plus phenylanine, tyrosine and tryptophan.

In summary, fingerprint and profiling methods are being applied to an increasing range of food authentication problems including non-targeted analyses. Varying levels of success have been reported in the literature depending on the authentication issue and the analytical technique. The major strength of fingerprinting methods lies in their generally rapid data acquisition and result delivery times. The major difficulty in their acceptance by the non-scientific community, which may be their greatest user, lies in the delivery of food and feed origin prediction results and associated uncertainty values in a manner which is easily comprehensible. Pending developments in the acceptance of multivariate mathematical models by legal authorities, these methods seem likely to be restricted to a screening role by enforcement labs in supporting labelling claims on food concerning provenance (Isphording, 2004). A potential way through the presentation of multivariate data in a court of law would be to standardise the methodology under CEN or ISO standards.

14.4 Bio-element and heavy element stable isotope ratio analysis

14.4.1 Bio-element (HCNOS) stable isotopes

The measurement of the stable isotope ratios of hydrogen and oxygen in water are applicable to the characterisation of geographical origin because they are strongly latitude dependent (Merlivat and Jouzel, 1979). Meteoric water that has passed through the meteorological cycle of evaporation, condensation and precipitation finally constitutes a groundwater that exhibits a systematic geographical isotope variation. Decreasing temperatures cause a progressive heavy-isotope depletion of the precipitation when the water vapour from oceans in equatorial regions moves to higher latitudes and altitudes. Evaporation of water from the oceans is a fractionating process that decreases the concentration of the heavy isotopomers of water ($^{1}H^{2}H^{16}O$, $^{1}H^{1}H^{18}O$) in the clouds compared to the sea. As the clouds move inland and gain altitude, further evaporation,

© Woodhead Publishing Limited, 2011

condensation and precipitation events occur decreasing the concentration of deuterium and ^{18}O. Consequently, the ground water reflects this isotopic gradient from the coast to inland areas. The variation of ^{18}O in the hydrosphere follows an analogous pattern to that of ^{2}H. These facts have been applied in hydrological and ecological studies for nearly 50 years (Araguás-Araguás *et al.*, 2000).

More recent, previous studies have clearly shown that the oxygen-18 and deuterium content of water, which is taken in by plants or consumed by animals, exhibits a strong correlation with the oxygen-18 and deuterium content of organic compounds present in plant and animal products such as sugars, lipids, proteins and milk, butter, and meat, respectively (Kelly et al., 2005). This correlation, which is found in biological samples, has been exploited by ecologists to determine, e.g., the migratory patterns of birds and insects (Hobson, 1999).

Climatic conditions, species specific biology, local agricultural practices and animal diet affect $^{15}N/^{14}N$ and $^{13}C/^{12}C$ (and sometimes $^{34}S/^{32}S$) ratios respectively. C_3 plants use the Calvin photosynthetic pathway to assimilate CO_2. During this process the plants discriminate against ^{13}C and therefore possess relatively lower $^{13}C/^{12}C$ ratios than C_4 plants that utilise the more energy consuming Hatch-Slack pathway (Hobbie and Werner, 2003). Since C_3 plants predominate at higher latitudes and C_4 plants are more common in warmer climates at lower latitudes (such as the tropics or the Mediterranean area), and the carbon isotope fractionation of C_3 plants increases with the humidity, there is a gradient of decreasing $^{13}C/^{12}C$ in plant material from the equator to the poles, which can also be used as a proxy for geographical origin determination. In addition to ^{15}N signatures providing information about regional agricultural practices (Bateman and Kelly, 2007), supplementary information may be obtained by measuring stable sulfur isotope ratios. Sulfate reduction in plants does not usually produce significant fractionation, such that organic sulfur is clearly related to its source. Therefore, the soil or sulfate fertiliser from which it is derived can provide useful geographical origin information. The influence of sea-salt sulfate precipitated from aerosols in coastal environments can significantly affect both plant and animal sulfur isotope composition due to the enriched ^{34}S marine signal (Camin *et al.*, 2007).

Taking the above-mentioned general aspects of stable isotope distribution in nature as a basis, the use of stable isotope analysis of H and O in water and C and N in food components for food authenticity control has been in common practice since the 1970s (Bricout *et al.*, 1974). Stable isotope data from one or two elements have been used to detect addition of components from external origin to food products, e.g. addition of water to wine or of beet or cane sugars to fruit juices. Such methods have been adopted in the EU and the US as official procedures to detect adulteration of fruit juices, honeys, wines, maple syrup, vinegars, and flavours since 1990 (Rossmann, 2001). However, the analysis of H and O isotopes in organic compounds and S isotopes in sulfates and organics, has evolved significantly since the new millennium due to the availability of

© Woodhead Publishing Limited, 2011

advanced instrumentation, which enabled the fully automated, routine application of stable isotope determinations for H, O and S in organics.

A large fundamental study was undertaken between 2004 and 2009 of the relevant bio-element stable isotopes in mineral and surface waters, soil extracts and food commodities (honey, cereals, olive oils, lamb, beef, chicken) from 21 regions across Europe. The results have been evaluated using statistical methods such as Linear Discriminant Analysis and Principal Component Analysis and published in international peer reviewed journals (e.g. *Food Chemistry*, Volume 118, issue 4, 15 February 2010). Unlike what is being done for authenticity checks of food products using one or two stable isotope parameters, as mentioned above, for geographical origin assignment the use of multi-element stable isotope pattern is essential. Some publications even combine results for light elements' isotopes with those from heavy elements' isotopes, minerals and trace element composition (Rummel *et al.*, 2010). The methods developed were applied in several case studies to food commodities from overseas regions (China (chicken) and Argentina (wine, olive oil, cereals, beef)) and results from these international studies are now being applied in national control laboratories and commercial laboratories for improved control of food geographical origin and authenticity, which is important for the protection of premium products such as PDO olive oils, cheeses, meat, and honeys. On the basis of the data supplied, prediction models that are now coming on-line such as Google Earth applications, are discussed in Section 14.5.

14.4.2 Heavy element stable isotopes

In search of reliable parameters for tracing the origin of foodstuff, isotopes came early in the field of vision. Compared to chemical compositions with respect to the absolute quantity of main or trace elements or certain compounds, isotopes often have the advantage of being more consistent and finally more reliable intrinsic markers. Isotope analysis of the so called 'bio-elements' H, C, N, O and S has been applied successfully in this field for more than 20 years (Kelly *et al.*, 2005). Besides those 'stable isotopes' which use low mass elements up to around 40 also 'geoisotopes' like Sr, Pb or Nd came into use in this field of application during the last 10 years (Hölzl *et al.*, 2004). These 'heavy' isotope systems with nuclide masses from 80 up to more than 200 do not compete with stable isotopes, they have to be regarded mainly as a completion to the 'less expensive' stable isotopes.

The basic precondition for tracing geographical origins with isotopes is the (natural or not) local variation of isotope ratios. For heavy elements, isotope effects and related mass fractionations are, according to the much smaller relative mass differences, insignificant. The reason for their variation in isotope abundances is spontaneous nuclear transmutation, which is a natural or artificial conversion of so-called mother nuclides into daughter nuclides. Natural slow radioactive decay (half-lives of millions or billions of years) is the most dominant process which leads to variable isotope abundance ratios of some

© Woodhead Publishing Limited, 2011

heavy (daughter)-elements in nature (Faure, 1986). The local isotopic variation is an outcome from the fact that the enrichment of 'radiogenic' daughter nuclides is subject to: 1) the abundance ratio mother- to daughter-element and 2) the time the decay was active. Both factors are the result of individual geological history so that we can speak rightly of 'geo isotopes'. Millions and billions of years of isotopic evolution are enclosed in rocks and reach through weathering – with certain shifts – soil and water and finally the food chain (Hölzl *et al.*, 2004).

Only a very restricted number of 'heavy' decay systems are able to leave usable traces in nature: the abundance of the mother-nuclide and its half-life have to be sufficiently high and mother and daughter element have to exhibit some differences in (geo-) chemical properties to experience changes in elemental abundance ratios during geological processes. So in practice only three elements are left which are used typically. In order of decreasing importance: Sr, Pb and Nd. For TRACE we restricted ourselves to Sr as its geochemistry is relatively well understood and this element is sufficiently abundant for analysis in most commodities (ppb or ppm range).

Sr-isotope signals are noted as isotope ratios: the radiogenic isotope ^{87}Sr is is noted relative to the non-radiogenic isotope ^{86}Sr as $^{87}Sr/^{86}Sr$. $^{87}Sr/^{86}Sr$ in soils and plants depends on the Rb/Sr and geological ages of the underlying rocks. For example, young basaltic rocks with low Rb/Sr have low $^{87}Sr/^{86}Sr$-values, whereas old crystalline rocks with high Rb/Sr, like gneisses or granites show high $^{87}Sr/^{86}Sr$-values. Seawater shows a very uniform isotope ratio today and marine carbonates which were deposited in ancient oceans display large and well-known variation in $^{87}Sr/^{86}Sr$. The values reflect an average of the strontium isotopic signatures of the eroded continental crust at the time of deposition. Bioavailable Sr, i.e. the water soluble part, passes from rock to soil into the atmosphere, aerosols and water and from there follows the food chain without any significant fractionation.

Isotope ratio determinations of Sr are performed with thermal ionisation mass spectrometers (TIMS) or multi collector inductively coupled plasma mass spectrometers (MC-ICPMS). Before measurement, decomposition of the sample and preferably complete separation of Sr from matrix elements by chemical methods is inevitable, mainly because of the isobaric interference of ^{87}Sr and ^{87}Rb. As organic materials often contain much more Rb compared to Sr and numerical correction for ^{87}Rb via ^{85}Rb results in large uncertainties, this is a demanding issue. Here TIMS has the advantage over MC-ICPMS that remains of Rb can be heated off before measurement of Sr in most cases due to the lower evaporation temperature of this element. Rb-rich commodities like beef or chicken are often limited to analysis by TIMS. Also commodities extremely poor in Sr like olive oils can only be measured by TIMS due to its higher sensitivity. To ensure the comparability of results provided by TIMS and MC-ICPMS inter laboratory tests, which included also external labs, have been performed. It turned out that in the project both methods and different labs were in agreement only within some hundred ppm of the values, depending on the

© Woodhead Publishing Limited, 2011

commodity. This is dissatisfying from the analytical point of view, for most practical questions, however, it is sufficient (unpublished data).

In a new case study around 4000 samples have been analysed for Sr isotopes. As expected, most of the data for mineral water, cereals and honey showed a relatively clear relation to the underlying geology. For animal products like meat this relationship was not so evident. We found significant variations in Sr isotope ratios in most of the TRACE regions. Also the ranges between the regions were overlapping. This was not surprising as most of the animals were fed at least partly with industrial fodder. This leads to a convergence of values towards intermediate values near that of recent seawater and thus represents a more or less average 'worldwide' mean value. In some regions this value could also indicate the influence of seaspray (e. g. Ireland, Sicily, Cornwall). In some cases, however, Sr ratios were in good agreement with the geological settings of the region. And as expected, samples from limestone areas gave the lowest Sr isotope ratios and exhibited relatively small variations.

In conclusion, as expected, Sr isotope ratios prove to be valuable parameters for tracing the origin of food and feed with strengths and failings. Commodities with a very strong genetic relationship to geology (e.g. mineral water, plants, products from naturally feeding animals) Sr isotopes proved to be very strong parameters. For other commodities, exclusion of certain regions was often possible only by means of Sr isotopes alone or they at least proved or refuted possibilities. A sufficient assignment of unknown samples to unknown regions, however, will mostly demand the knowledge of additional parameters.

14.5 Food isotope maps

A common approach to verify the geographical origin of a suspect food sample in an objective way is by statistical comparison of its chemical composition to a database of samples of which the geographical origin is known (Kelly *et al.*, 2005). Because this approach requires a representative database of samples from all the relevant production areas, it is especially useful to discern between food commodities that are produced in a limited number of confined areas or by a limited number of producers, which is, for example, the case for food products that have a protected geographical indication (PGI) or protected designation of origin (PDO) status. This approach, however, may quickly become too expensive for a food commodity produced in many different areas. For such cases, the food isotope map approach may provide a cost-effective alternative to the database approach.

The food isotope map approach is a geographical information system (GIS)-based approach in which the isotopic composition of food is predicted for the relevant production areas by using additional information in the form of one or more ancillary variables. Ancillary variables are variables that are available at a higher spatial density than the original data that are used to improve the final interpolation results. Although isotope maps can in principle also be derived by

© Woodhead Publishing Limited, 2011

direct spatial interpolation of the data, interpolation using ancillary variables reduces the need for an exhaustive sample database and furthermore puts constraints on the predicted isotope values in sparsely sampled areas (Bowen and Revenaugh, 2003). The food isotope map concept is based on the observation that local climatic and geological characteristics of a production area are often reflected in the isotopic composition of food produced in that area. Provided that there is a clear relation between both, the isotopic composition of food can be predicted for un-sampled areas using ancillary spatial information about the local climate and geology. This information is usually available in the form of high resolution digital maps. In general, food isotope maps can be used to answer two basic questions about a suspected food sample:

1. Testing for compliance: Can the food sample possibly come from the claimed production area at a certain level of confidence?
2. Determination of the area of possible origin: From what area(s) can the food sample possibly come?

The latter possibility allows for determining how specific these findings are in a spatial sense, and provides an additional advantage to the database approach. As for all statistical techniques the method does not allow us to *prove* whether a suspect food sample comes from a certain production area, but can only confirm its possible origin at a specified level of confidence and spatial specificity of the method. The approach should therefore be regarded as a screening method.

14.5.1 Developing food isotope maps

The first step in the development of a food isotope map is to establish an empirical relationship between the isotopic composition of a food commodity and some geo-climatic factor(s) that are available at a much higher spatial density. Typical isotope systems used for this purpose are deuterium, carbon-13, oxygen-18, which often have a relation with local climate and $^{87}Sr/^{86}Sr$ ratios, which often have a relation with the local geological setting (see Section 14.3). In principle also other variables (e.g. trace elements) can be used provided they have a relation with some known ancillary variable.

An example of such an empirical relationship is shown in Fig. 14.1, where the oxygen-18 composition of bottled mineral water from Europe is plotted against the annual minimum temperature of the coldest month at the production sites. A food isotope map of the oxygen-18 composition in bottled mineral waters in Europe can now be derived using a digital map of the annual mean temperature as an ancillary variable. A similar approach has been applied to predict the deuterium and oxygen-18 composition of precipitation on a global scale (see Bowen and Revenaugh, 2003; Van der Veer *et al.*, 2009).

To be able to test whether a suspect food sample complies with the predicted isotopic composition for its presumed production area, a measure of uncertainty around the predicted isotope values is required. It is important, therefore, to use

© Woodhead Publishing Limited, 2011

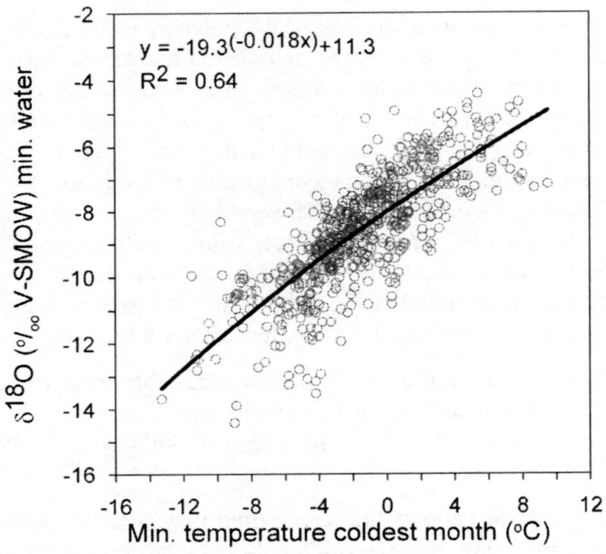

Fig. 14.1 The minimum temperature during the coldest month at the production site plotted against the oxygen-18 composition of bottled European mineral waters ($n \approx 600$). An exponential function is fitted through the data.

a spatial interpolation technique that allows determination of the confidence intervals around the predicted values, for example co-kriging or kriging with an ancillary variable (Oliver and Webster, 1990). Alternatively, a bootstrapping, jack-knifing or similar approach can be used to determine the model uncertainty.

Often, a food isotope map based on one single isotope system is spatially not very discriminative, because in areas where similar climatic or geological conditions prevail, similar isotopic values will be found (Voerkelius *et al.*, 2010). Furthermore, there will be an overlap of the predicted confidence intervals in areas with only slightly different conditions. As a result, the area in which a measured isotope value of a suspect food sample will fall within the confidence intervals – here referred to as area of possible origin (APO) – can be quite extensive. The size of this area depends both on the width of the confidences intervals and also on the climatic or geological variation encountered within the area considered.

To increase the spatial discrimination power of the isotope map approach, it is thus advantageous to use a combination of several food isotope maps in order to reduce the combined area of possible origin (multiple isotope approach; see Fig. 14.2). This approach is similar to fingerprinting techniques in which multiple markers are considered simultaneously. For optimal performance this approach requires that the isotope systems used in the different map layers are independent or at best only weakly dependent, which should be tested in advance. To further confine the area of possible origin, an additional

© Woodhead Publishing Limited, 2011

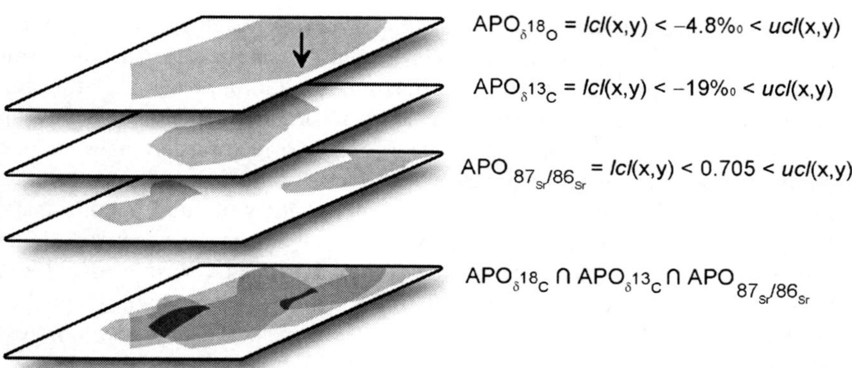

$APO_{\delta^{18}O}$ = $lcl(x,y) < -4.8‰ < ucl(x,y)$
$APO_{\delta^{13}C}$ = $lcl(x,y) < -19‰ < ucl(x,y)$
$APO_{^{87}Sr/^{86}Sr}$ = $lcl(x,y) < 0.705 < ucl(x,y)$
$APO_{\delta^{18}C} \cap APO_{\delta^{13}C} \cap APO_{^{87}Sr/^{86}Sr}$

Fig. 14.2 Schematic representation showing the individual areas of origin for a hypothetical food sample with measured isotope ratios: $\delta^{18}O = -4.8$, $\delta^{13}C = -19$ and $^{87}Sr/^{86}Sr = 0.705$. *lcl* and *ucl* refer to the lower and upper confidence limits of the predicted isotope values. The claimed origin is indicated with an arrow, and the intersection of the three areas is shown in dark grey. Obviously, it is unlikely that the sample does come from the claimed area because only its $\delta^{18}O$ composition complies with the observed areas of possible origin.

overlay of the production areas can be added, provided that this information is available.

To use the food isotope maps for compliance testing of a suspect food sample, its measured isotopic composition should be compared with the lower and upper confidence interval predicted for the acclaimed production area. This will indicate whether the sample can possibly come from that area at a given level of confidence. However, when combining multiple map layers for testing it is important to notice that the collective α-error of the test increases while at the same time the overall confidence level decreases (see Table 14.1). To retain an acceptable overall confidence level when working with two or more maps, one could define high confidence levels for the individual map layers (e.g. 98% or 99%). This will in turn increase area of possible origin, and therefore a fair trade-off should be sought between the desired level of confidence and spatial specificity when using this method.

Table 14.1 Overall confidence levels as a function of the number of map layers used for compliance testing at individual α-errors of 10%, 5%, 2% and 1%. In this example the map layers are assumed to be strictly independent

No. of map layers	10%	5%	2%	1%
1	90%	95%	98%	99%
2	81%	90.2%	96%	98%
3	72.9%	85.7%	94.1%	97%
4	65.6%	81.5%	92.2%	96.1%

© Woodhead Publishing Limited, 2011

Successful application of the food isotope map approach requires the presence of a clear relationship between the isotopic composition of a food commodity and some geo-climatic or other factors. In addition, such a relationship should exist for two or more isotope systems in order to develop several map layers to confine the possible area of origin. The relationship between the isotopic composition of groundwater and the local climatic conditions such as temperature clearly illustrates the potential application of food isotope maps (see Fig. 14.1). Mineral water as well as beverages that contain local groundwater often contain a pronounced climatic signal (West *et al.*, 2007; Chesson *et al.*, 2010; Rummel *et al.*, 2010). In addition, the geological signal as reflected by the $^{87}Sr/^{86}Sr$ ratio is generally well expressed in mineral waters (Montgomery *et al.*, 2006; Voerkelius *et al.*, 2010). For this reason, the isotope map approach is well suited for mineral waters as well as locally produced beverages. At present, the food isotope map approach has been successfully applied and tested for mineral waters and olive oil from Europe (TRACE Tool, 2009). In vegetable and animal products, the reflection of the local climate and geological signal is, in general, somewhat more complex compared to mineral water and locally produced beverages. Modelling these more complex relationships presents a challenge and requires a thorough understanding of the mechanisms behind the geographical variation of the isotope patterns in food. The ongoing development of compound specific isotope methods will improve our understanding in this respect, resulting in a wider application of the food isotope map approach.

14.6 Conclusions

The strong global interest in authentication and traceability of food and feed products will continue to stimulate research in this field. In principle, the most powerful methods are generally those that combine as many of the genetic, spectroscopic, elemental and isotopic signatures as possible, into a multi-dimensional authenticity matrix where natural variation is more readily constrained and detection of mis-description is almost impossible to subvert. In practice a single or sub-set of these methods are likely to be used based on their likelihood of success. If multiple techniques and multivariate analysis are used to interpret origin then it is critical that the robustness of multivariate models is tested with techniques such as cross-validation and 'blind-testing'. Furthermore, it is extremely important to quantify the repeatability and reproducibility of genetic, spectroscopic, multi-element and multi-isotopic measurements. In order to make meaningful origin assignments it is necessary to know the measurement uncertainty so that differences in elemental concentrations or isotope ratios, for example, in commodities from different geographical origins can be deemed as statistically significant or not. Control of long-term measurement repeatability with matrix matched in-house reference materials is essential if authentic databases are to be established over a number

© Woodhead Publishing Limited, 2011

years. Furthermore, if databases are to be developed on a European-scale for enforcement purposes, in the same way as the wine databank, then long-term reproducibility is also of paramount importance. In a comparable way, food markets, the globalisation of feed markets and the desire to obtain cheap sources of feed may mean that certain molecular and isotopic signals in animal products will not reflect the region of production. It is therefore important to recognise these constraints on geographical interpretation. Similarly, it is important not to use geographical models that are heavily dependent on trace element markers from manufacturers' processes as this can lead to 'unstable' models for predicting origin. In the longer term it is hoped that isotope mapping and a deeper understanding of how meteorological and geochemical signatures are transferred into food systems may provide an alternative strategy or enhancement to comparative databases of authentic foods. For the moment though, there is a need for comparative databases to be established for premium foods and this is highly desirable for brand protection of PDO and PGI products.

14.7 References

ARAGUÁS-ARAGUÁS, L., FROEHLICH, K. and ROZANSKI, K. (2000) Deuterium and oxygen-18 isotope composition of precipitation and atmospheric moisture. *Hydrological Processes*, 14, 8, 1341–1355.

BATEMAN, A.S. and KELLY, S.D. (2007) Fertilizer nitrogen isotope signatures. *Isotopes in Environmental and Health Studies*, 43, 3, 237–247.

BOWEN, G.J. and REVENAUGH, J. (2003) Interpolating the isotopic composition of modern meteoric precipitation. *Water Resources Research*, 39, 10, 1299–1311.

BRICOUT, J., FONTES, J.C. and MERLIVAT, L (1974) Detection of synthetic vanillin in vanilla extracts by isotopic analysis. *Journal of the Association of Official Analytical Chemists*, 57, 3, 713–715.

CAMIN, F., BONTEMPO, L., HEINRICH, K., HORACEK, M., KELLY, S.D, SCHLICHT, C., THOMAS, F., MONAHAN, F.J., HOOGEWERFF, J. and ROSSMANN, A. (2007) Multi-element (H,C,N,S) stable isotope characteristics of lamb meat from different European regions. *Analytical and Bioanalytical Chemistry*, 389, 1, 309–320.

CHESSON, L.A., VALENZUELA, L.O., O'GRADY, S.P., CERLING, T.E. and EHLERINGER, J.R. (2010) Links between purchase location and stable isotope ratios of bottled water, soda, and beer in the United States. *Journal of Agricultural and Food Chemistry*, 58, 7311–7316.

DONARSKI, J.A., JONES, S.A., HARRISON, M., DRIFFIELD, M. and CHARLTON, A.J. (2010) Identification of botanical biomarkers found in Corsican honey. *Food Chemistry*, 118, 4, 989–994.

EUROPEAN COMMISSION. DG HEALTH AND CONSUMER PROTECTION (2004) Report to the Standing Committee of the Food Chain and Animal Health on the relationship between analytical results, the measurement uncertainty, recovery factors and the provisions in the EU food and feed legislation.

FAURE, G. (1986) *Principles of Isotope Geology*, 2nd edn, John Wiley & Sons, New York, pp. 117–140.

© Woodhead Publishing Limited, 2011

FOOD STANDARDS AGENCY (2010) Country of Origin Labelling: A Synthesis of Research. Available at http://www.food.gov.uk/multimedia/pdfs/coolsyn.pdf (accessed September 2010).

GUSTAVO GONZÁLEZ, A. and ÁNGELES HERRADOR, M. (2007) A practical guide to analytical method validation, including measurement uncertainty and accuracy profiles. *TrAC Trends in Analytical Chemistry*, 26, 227–238.

HOBBIE, E.A. and WERNER, R.A. (2003) Intramolecular, compound-specific, and bulk carbon isotope patterns in C3 and C4 plants: a review and synthesis. *New Phytologist*, 161, 2, 371–385.

HOBSON, K.A. (1999) Tracing origins and migration of wildlife using stable isotopes: a review. *Oecologia*, 120, 3, 314–326.

HÖLZL, S., HORN, P., ROSSMANN, A. and RUMMEL, S. (2004) Isotope-abundance ratios of light (bio) and heavy (geo) elements in biogenic tissues: methods and applications. *Analytical and Bioanalytical Chemistry*, 378, 2, 270–272.

ISPHORDING, W.C. (2004) *The right way and the wrong way of presenting statistical and geological evidence in a court of law (a little knowledge is a dangerous thing!)*. Geological Society, London, Special Publications; 232, 281–288.

KELLY, S.D., HEATON, K. and HOOGEWERFF, J. (2005) Tracing the geographical origin of food: the application of multi-element and multi-isotope analysis. *Trends in Food Science and Technology*, 16, 555–567.

LAUBE, I., ZAGON, J. and BROLL, H. (2007) Quantitative determination of commercially relevant species in foods by real-time PCR. *International Journal of Food Science and Technology*, 42, 336–341.

LAUBE, I., BRODMANN, P., HIRD, H., ULLMANN, S., SCHÖNE-MICHLING, M., CHISHOLM, J. and BROLL, H. (2010) Development of primer and probe sets for the detection of plant species in honey. *Food Chemistry*, 118, 4, 979–986.

LEARDI, R. (2003) Chemometrics in data analysis. In Lees, M. (ed.), *Food Authenticity and Traceability*, Woodhead Publishing, Cambridge, pp. 156–183.

LOGAN, J., EDWARDS, K. and SAUNDERS, N. (EDS) (2009) *Real-Time PCR: Current Technology and Applications*. Caister Academic Press.

MACARTHUR, R., FEINBERG, M. and BERTHEAU, Y. (2010) Construction of measurement uncertainty profiles for quantitative analysis of genetically modified organisms based on interlaboratory validation studies. *Journal of AOAC International*, 93, 1–11.

MERLIVAT, L. and JOUZEL, J. (1979) Global climatic interpretation of the deuterium-oxygen 18 relationship for precipitation. *Journal of Geophysical Research*, 84, C8, 5029–5033.

MONTGOMERY, J., EVANS, J.A. and WILDMAN, G. (2006) $^{87}Sr/^{86}Sr$ isotope composition of bottled British mineral waters for environmental and forensic purposes. *Applied Geochemistry*, 21, 1626–1634.

NI, S., QIAN, D., DUAN, J., GUO, J., SHANG, E., SHUA, Y. and XUE, C. (2010) UPLC–QTOF/MS-based screening and identification of the constituents and their metabolites in rat plasma and urine after oral administration of Glechoma longituba extract. *Journal of Chromatography B*, 878, 28, 2741–2750.

OCHI, T. (2005) Former Japanese beekeeper speaks out about antibiotics in Chinese honey. *American Bee Journal*, 145, 12, 973–938.

OLIVER, M.A. and WEBSTER, R. (1990) Kriging: a method of interpolation for geographical information systems. *International Journal of Geographical Information Science*, 4, 3, 313–332.

© Woodhead Publishing Limited, 2011

PÄÄBO, S. (1989) Ancient DNA: extraction, characterization, molecular cloning and enzymatic amplification. *Proceedings of the National Academy of the Science of the USA*, 86, 1939–1943.

PÄÄBO, S., HIGUCHI, R.G. and WILSON, A.C. (1989) Ancient DNA and the polymerase chain reaction. *Journal of Biological Chemistry*, 264, 9709–9712.

PRINS, T.W., VAN DIJK, J.P., VAN HOEF, A.M.A., VOORHUIJZEN, M.M., BROEDERS, S., TRAPMANN, S., SEYFARTH, R., PARDIGOL, A., SCHOEN, C.D., AARTS, H.J.M. and KOK, E.J. (2010) Towards a multiplex cereal traceability tool using padlock probe ligation on genomic DNA. *Food Chemistry*, 118, 4, 966–973.

QUERCI, M., VAN DEN BULCKE, M., ZEL, J., VAN DEN EEDE, G. and BROLL H. (2009) New approaches in GMO detection. *Analytical and Bio-analytical Chemistry*. DOI 10.1007/s00216-009-3237-3 (available online).

ROSSMANN, A. (2001) Determination of stable isotope ratios in food analysis. *Food Reviews International*, 17, 3, 347–381.

RUMMEL, S., HOELZL, S., HORN, P., ROSSMANN, A. and SCHLICHT, C. (2010) The combination of stable isotope abundance ratios of H, C, N and S with $^{87}Sr/^{86}Sr$ for geographical origin assignment of orange juices. *Food Chemistry*, 118, 4, 890–900.

SCHNEIDER, S., HARTMANN, M., ENKERLI, J. and WIDMER, F. (2010) Fungal community structure in soils of conventional and organic farming systems. *Fungal Ecology*, 3, 3, 215–224.

TRACE TOOL (2009) web application available at http://update.maritech.is/tracetool/publish.htm (accessed September 2010).

TRAPMANN, S., BURNS, M., BROLL, H., MACARTHUR, R., WOOD, R. and ZEL, J. (2009) *Guidance Document on Measurement Uncertainty for GMO Testing Laboratories*. European Commission Joint Research Centre Institute for Reference Materials and Measurements.

TRAYNOR, K.S. (2008) Sweet solutions for good health. *American Bee Journal*, 148, 3, 205–208

VACLAVIK, L., CAJKA, T., HRBEK, V. and HAJSLOVA, J. (2009) Ambient mass spectrometry employing direct analysis in real time (DART) ion source for olive oil quality and authenticity assessment. *Analytica Chimica Acta*, 645, 56–63.

VAN DER VEER, G., VOERKELIUS, S., LORENTZ, G., HEISS, G. and HOOGEWERFF, J.A. (2009) Spatial interpolation of the deuterium and oxygen-18 composition of global precipitation using temperature as ancillary variable. *Journal of Geochemical Exploration*, 101, 2, pp. 175-184.

VOERKELIUS, S., LORENZ, G.D., RUMMEL, S., QUéTEL, C.R., HEISS, G., BAXTER, M., BRACH-PAPA, C., DETERS-ITZELSBERGER, P., HOELZL, S., HOOGEWERFF, J., PONZEVERA, E., VAN BOCXSTAELE, M. and UECKERMANN, H. (2010) Strontium isotopic signatures of natural mineral waters, the reference to a simple geological map and its potential for authentication of food. *Food Chemistry*, 118, 933–940.

WEERANANTANAPHAN, J. and DOWNEY, G. (2010) Identity confirmation of a branded, fermented cereal product by UV spectroscopy: a feasibility study involving a Trappist beer. *Journal of the Institute of Brewing*, 116, 1, 56–61.

WEST, J.B., EHLERINGER, J.H. and CERLING, T.E. (2007) Geography and vintage predicted by a novel GIS model of wine $\delta^{18}O$. *Journal of Agricultural and Food Chemistry*, 55, 7075–7083.

WISHART, D.S. (2008) Metabolomics: applications to food science and nutrition research. *Trends in Food Science & Technology*, 19, 9, 482–493.

WOOLFE, M. and PRIMROSE, S. (2004) Food forensics: using DNA technology to combat

© Woodhead Publishing Limited, 2011

misdescription and fraud. *Trends in Biotechnology*, 22, 222–226.

ZEL, J., CANKAR, K., RAVNIKAR, M., CAMLOH, M. and GRUDEN, K. (2006) Accreditation of GMO detection laboratories: improving the reliability of GMO detection. *Accreditation and Quality Assurance*, 10, 531–536.

ZEL, J., MAZZARA, M., SAVINI, C., CORDEIL, S., CAMLOH, M., STEBIH, D., CANKAR, K., GRUDEN, K., MORISSET, D. and VAN DEN EEDE, G. (2008) Method validation and quality management in the flexible scope of accreditation: an example of laboratories testing for genetically modified organisms. *Food Analytical Methods*, 1, 61–72.

© Woodhead Publishing Limited, 2011

15

Tracing fish and fish products from ocean to fork using advanced molecular technologies

J. T. Martinsohn, EU Commission DG Joint Research Centre, Italy, A. J. Geffen, University of Bergen, Norway, G. E. Maes, Katholieke Universiteit of Leuven, Belgium, E. E. Nielsen, Technical University of Denmark, Denmark, R. Ogden, TRACE Wildlife Forensics Network, UK, R. S. Waples, Northwest Fisheries Science Center, USA and G. R. Carvalho, Bangor University, UK

Abstract: The ability to determine authenticity and provenance of fish and fish products throughout the international fish trade distribution chain is of paramount importance, and in many countries traceability in the fisheries sector is based on labelling rules. As shown by numerous fraud cases worldwide, however, and the relentless global problem of illegal, unreported and unregulated (IUU) fishing, independent control technologies are urgently needed to ensure appropriate implementation of traceability schemes. Here, we discuss opportunities and challenges arising from the rapid progress in research and technology pertinent to traceability. In support of an integrative approach, several technologies will be considered, though emphasis is placed on DNA technology as an approach witnessing major recent development.

Key words: mislabelling, DNA technology, trace elements, IUU, forensics, species identification, origin assignment, technology transfer.

15.1 Introduction

Marine fish are a common natural resource and the scenario described in Garrett Hardin's famous essay on 'The tragedy of the commons' (Hardin, 1968), which depicts how individuals driven by self-interest tend to destroy the very resource

© Woodhead Publishing Limited, 2011

upon which everyone depends, fits unfortunately all too well into global fisheries.

Massive fishing activity has had a dramatic impact on capture fisheries: according to the United Nations Food and Agriculture Organisation (FAO), virtually 70% of the world's fisheries are fully exploited, overexploited or even in a state of collapse and therefore being in need for rebuilding (FAO, 2009). Our dependence on fish and fish products is well illustrated by trade data. World fish imports reached a new peak of more than US$75 billion in 2004, after a steep rise of 25% in only four years (FAO, 2006). Also the European Union is heavily dependent on import of fishery products, accounting for more than 60% of its fish consumption in 2008 (European Commission, 2009).

These data illustrate the importance of the international fish trade. Efforts to regulate fishing activities so as to reach sustainable levels are greatly compromised by illegal, unreported and unregulated (IUU) fishing (Fig. 15.1). The estimated annual global value for IUU fishing amounts to as much as US$23.5 billion, representing up to 26 million tons of fish, which is equivalent to one-fifth of the global reported catch (Agnew *et al.*, 2009; Fig. 15.2). Illegal fisheries imports into the EU have been conservatively estimated at €1.1 billion annually (European Commission, 2007). IUU fishing causes unsustainable harvest of fish stocks and other marine wildlife, but its drastic impact goes beyond mere fishing activity. In addition to depleting fish stocks to the point where they become commercially unviable or even on the brink of extinction, IUU fishing can perturb entire ecosystems (Scheffer *et al.*, 2005). It also leads to loss of nutrition, and hampers socio-economic development and stability due to loss of income and employment for legitimate fishermen. Furthermore, IUU fishing weakens labour standards, and distorts markets of legally harvested fish (OECD, 2005). Criminal activities in the fisheries sector extend into the supply chain, where fish is marketed under the wrong species name, or with falsified origin specifications, thereby obstructing consumer information and protection.

Traceability, the ability to trace the history, application or location of an entity by means of recorded identification (ISO 9000:2005), is an important component in the legal framework underlying control and enforcement in the fisheries sector. It is also indispensable for consumer protection and information, as it requires tracking and surveillance throughout the food supply chain, from onboard vessels to the sale of processed products ('ocean to fork'). This aspect is relevant also for eco-labelling schemes attesting that fish products derive from sustainably exploited stocks, and assuring consumers that a product has been produced according to defined environmental standards (Brécard *et al.*, 2009). However, there remains a risk of failure of such eco-labelling schemes, caused by falsified labels and certificates, which would undermine credibility and consumer confidence.

Currently, traceability in the fisheries sector is mainly assured by documentation and labelling of products. For example, for the EU, the 'Common Fisheries Policy (CFP) Control regulation' (EC) No 1224/2009, lays down that fisheries and aquaculture products on the market in the Community shall be

© Woodhead Publishing Limited, 2011

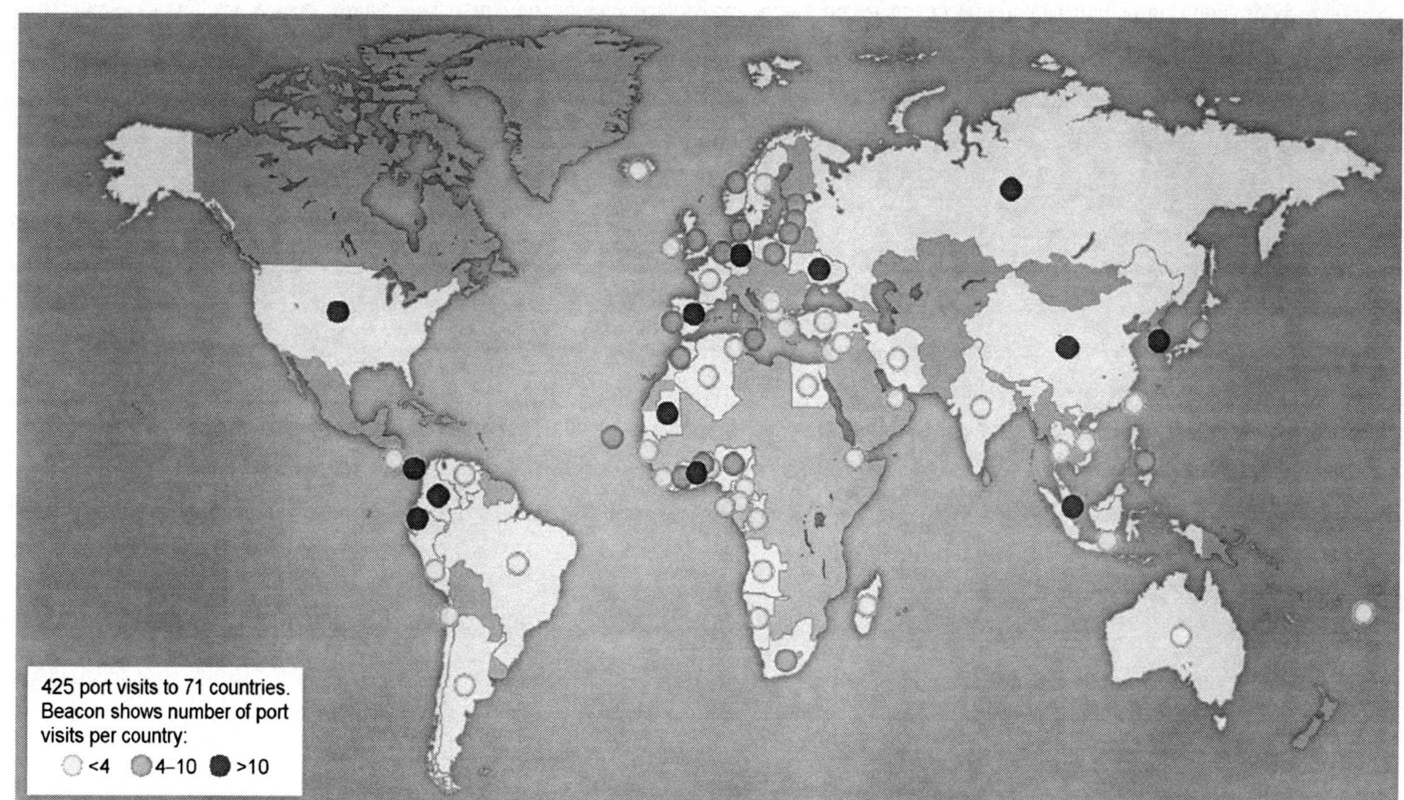
425 port visits to 71 countries.
Beacon shows number of port
visits per country:
<4 4–10 >10

Fig. 15.1 The global dimension of IUU fishing. Shown are the port visits by IUU-listed vessels worldwide. Countries where port visits were recorded are marked in light grey. (with kind permission of Flothmann *et al.* 2010).

© Woodhead Publishing Limited, 2011

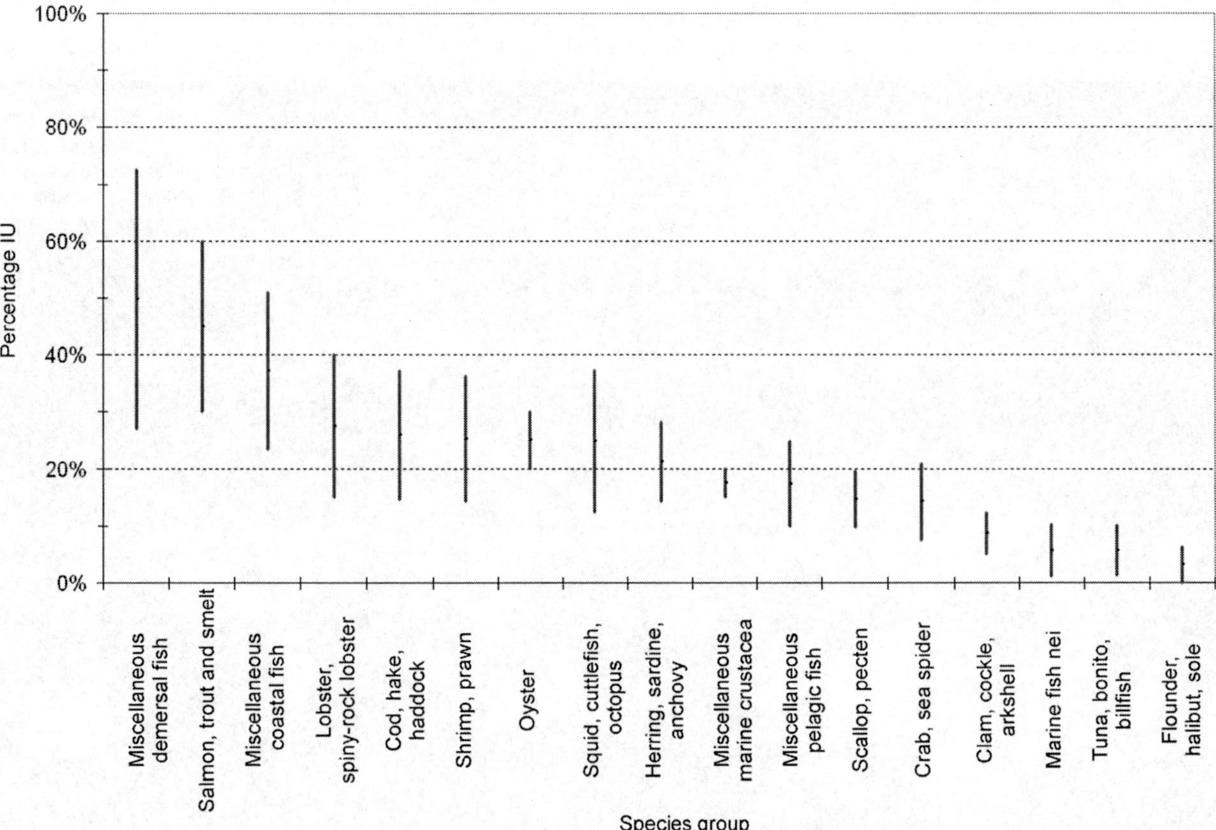

Fig. 15.2 The impact of IUU fishing. Illegal and unreported catch, expressed as a percentage of reported catch, and including upper and lower bounds, by species group in the period from 2000 to 2003. Reproduced with kind permission of Agnew *et al.* (2009).

© Woodhead Publishing Limited, 2011

adequately labelled to ensure traceability (European Council, 2009) in line with regulation (EC) 178/2002 (European Parliament and Council, 2002). Minimum labelling requirements include the FAO alpha-3 code of each species and information for consumers provided for in regulation (EC) No 2065/2001 such as commercial designation, scientific name, relevant geographical area and production method (European Commission, 2001). Additionally, according to the recently implemented 'IUU Regulation' (European Council, 2008), fishery products can only be imported into the European Union when accompanied by a catch certificate. The certification scheme also aims to ensure product traceability at all stages of production, from catch – to processing, and transport – to marketing, and to be a tool to enhance compliance with conservation and management rules.

To protect such a system from fraud or error, an independent and robust procedure for species and geographic origin assignment is required. It is the application of tools derived from advanced molecular and chemical methods that can provide an objective and unequivocal test for traceability of diverse fish products, and it is these methods that we focus on here.

This chapter introduces briefly the available modern molecular technologies with high potential for practical traceability in the fisheries sector. We demonstrate how such technologies can effectively contribute to tackle IUU and support sustainable fisheries management as well as consumer protection. To do so, such technologies have to address the following key questions (Fig. 15.3):

- What species is it?
- Where was it caught?
- And in light of increasing aquaculture activity, as well as breeding for marine stock enhancement, is it derived from wild sources or aquaculture/fish farms?

While the first question about the species identity is nowadays vastly routine, due to DNA barcoding and related methods (see below), the identification of origin remains a challenging task. To provide a general background, available techniques will first be introduced, followed by a description of how species identification and origin assignment are enabled by advanced technologies. Finally, we discuss to what extent these technologies are already applied in the fisheries sector by authorities and the fishing industry, and how to facilitate a coherent global transfer of such tools for easy uptake by stakeholders and end-users worldwide.

15.2 The traceability tool box: an overview of available molecular technologies for species identification and origin assignment of fish products

Traceability in the 'ocean to fork' sense relies on efficient, reliable, cost-effective technologies, enabling the independent control of compliance with rules. In the fisheries sector, this encompasses the ability to determine whether

© Woodhead Publishing Limited, 2011

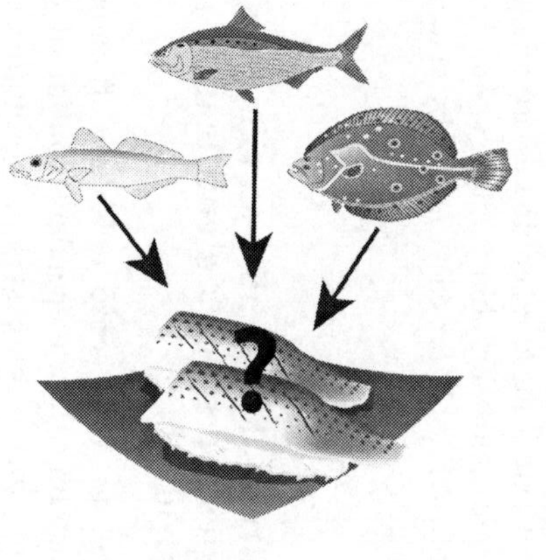

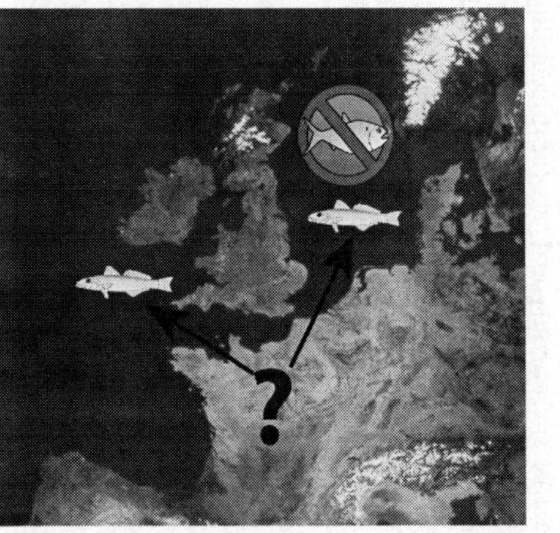

Fig. 15.3 The three principal questions of a traceability framework for fish and fish products: a) What? b) Where from? c) Wild or cultured? (Fish symbols courtesy of the Integration and Application Network, University of Maryland Centre for Environmental Science. Map: European Union, 2010.)

© Woodhead Publishing Limited, 2011

labels on fish and fish products identify the correct species, correct origin, and whether fish are derived from aquaculture or the wild. Ideally, such methods should be adapted to end-users such as staff of control authorities, be applicable on whole fish as well as processed products, and swiftly lead to results. Moreover, if they are to be utilised for enforcement, these methods should be validated applying forensic standards and generate levels of confidence based on statistical analysis certainty, which are considerably higher than that required for purely scientific inference (Murphy and Morrison, 2007).

Historically, a panoply of diverse methods have been used in identification of fish species or in analyses supporting origin assignment. These include the description of morphology used to identify a species. Markers, determined by the environment such as parasite load (MacKenzie and Abaunza, 1998), can be used to determine where a fish comes from (origin assignment). However, most morphological markers are also influenced by environmental factors (e.g. temperature), which complicates their interpretation, in contrast to molecular genetic markers (see below) which are transmitted across generations independently of environmental factors.

With the advent of molecular biology, molecular and genetic markers are increasingly employed for both species identification and origin assignment. It is important to understand that traceability tools for species identification and origin assignment require comprehensive reference data sets ('baselines'): When control authorities wish to test whether a fish fillet derives from the species indicated on the product label, their analytical data must be comparable to a set of validated data for species identification. This has been achieved to a great extent by the DNA barcoding approach (see below). Likewise, if the origin of a fish (product) is under investigation, fish deriving from different geographical regions must have been formerly analysed and distinct features, robust 'population-level signatures', characterising these groups must be recorded. We now consider some fundamental principles on the use of advanced analytical technologies based on genetics, genomics and chemistry for traceability of fish species identification and origin assignment.

15.2.1 Fatty acid analysis

In marine animals, about 20 different fatty acids are present at relative amounts of greater than 1%, serving as sources of energy and structural elements at the cellular level. The fatty acid profile (the amount of each particular fatty acid in a tissue) may vary significantly at both species and population levels, and can be analysed by spectrometric methods. The fatty acid composition of fish tissues appears to be largely genetically determined (Kwetegyeka *et al.*, 2008), but is also influenced by biological (e.g. age, maturity) and environmental features (e.g. diet, water temperature; reviewed in Grahl-Nielsen, 2005).

Populations or stocks of a fish species inhabiting different geographical areas are typically characterised by distinct environmental conditions and nutriments, which may be distinguishable by qualitative and quantitative analysis of the fatty

© Woodhead Publishing Limited, 2011

acid composition. Such an approach has been followed for stock identification in a presumed fisheries fraud case investigating alleged false reporting of herring landed by a Norwegian purse seiner. Charges against the fisherman were dropped after comparison of fatty acid profiles of fish under investigation with fish of known provenance (Grahl-Nielsen, 2005). Also, nuclear magnetic resonance (NMR) spectroscopy of lipids has been used successfully to distinguish farmed from wild Atlantic salmon, to discriminate between different geographical origins, and to verify the origin of market samples (Aursand *et al.*, 2009).

Fatty acid profiling can be performed on fresh or frozen fish tissues, but its application on processed fish products, such as smoked or canned fish, requires further evaluation. FishPopTrace, a project funded through the EU 7th Framework programme, is currently exploring the potential of fatty acid profile analysis as a geographical origin identification tool for traceability purposes.

15.2.2 Microchemistry and stable isotope analysis

Chemical elements from the surrounding water are incorporated into mineralised tissues such as otoliths (the fish ear stones) and scales (Fig. 15.4). The relative concentrations of various elements, especially Sr, Ba, Mn, and heavy metals, as well as the isotopic ratios of O, Sr, and Ba, can be used as a 'chemical fingerprint' which is characteristic of fish living in different geographical locations. Otoliths grow throughout the life of a fish, and the composition of each layer is discreet and permanent so that it is possible to determine origin and movements of fish at different points in time (Campana, 2005; Elsdon *et al.*, 2008).

Techniques for elemental analysis include bulk analysis, in which the whole otolith is dissolved and assayed to give an average composition, and surface analysis, where specific zones are targeted for analysis of specific life stages or time periods. Both approaches have been employed to examine population structure and traceability of various marine fish species (Geffen *et al.*, 2003; Elsdon *et al.*, 2008; Cuveliers *et al.*, 2010). Otolith analyses can be used to

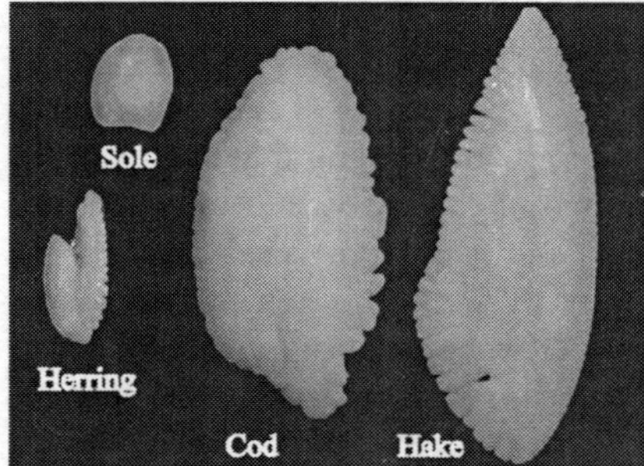

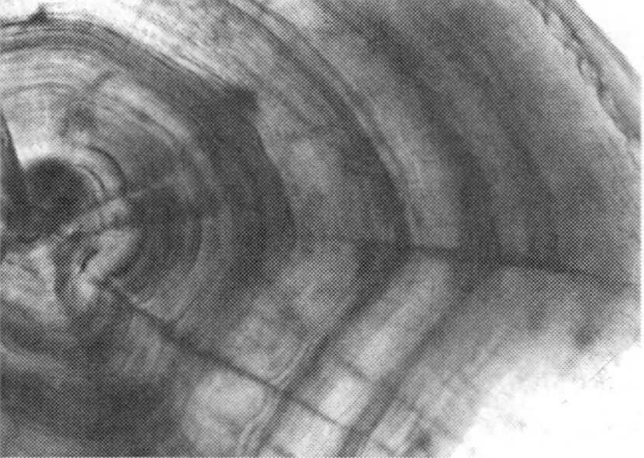

Fig. 15.4 Otoliths (fish ear stones). On the left-hand side otoliths from common sole, herring, hake and cod are displayed. On the right-hand side a sole (*Solea solea*) otolith of which a section is magnified clearly showing the growth increments.

© Woodhead Publishing Limited, 2011

characterise both the population of origin and the location of harvest, and results in high discriminatory powers even over relatively short geographical distances. For example, the composition of plaice (*Pleuronectes platessa*) otoliths differed significantly over locations less than 100 km apart within the eastern Irish Sea (Geffen *et al.*, 2003). Microchemistry data, in combination with otolith shape, DNA markers, and other parameters, allowed separation of wild North East Atlantic cod from farmed cod, together with the harvest origin of individuals (Higgins *et al.*, 2010). For traceability purposes, otolith element analysis is suitable for all stages along the supply chain provided the heads remain with the fish. An advantage of otoliths is that the elemental composition is resilient, and does not degrade or change over time after death (Thresher, 1999).

Other hard tissues, such as scales or spines, have also been used to establish elemental fingerprints (Gillanders, 2001), including the discrimination of wild and farmed Atlantic salmon (*Salmo salar*) (Adey *et al.*, 2009). It was also shown that trace element analysis on soft (muscle) tissue, can be used to distinguish groupings of pink ling (*Genypterus blacodes*) from different areas in New Zealand (Bremner, 2009).

15.2.3 Genetic markers

Genetic markers are heritable, discrete and stable. Their use in fisheries science dates back to the 1950s, and encompasses a wide scope of applications. Many of those uses, however, are rather academic and their relevance for a modern and efficient traceability scheme, are discussed in detail elsewhere (Kochzius, 2008).

Broadly speaking, the analysis of genetic markers can be separated into DNA-analysis, revealing DNA sequence variation, and protein analysis, that reveals differences in amino acids. For traceability purposes, samples of fish (or products) have to be collected, followed by DNA or protein extraction from tissue, nowadays a routine process facilitated by commercial kits. Protein analysis, such as isoelectric focusing (IEF) and related techniques (Rehbein, 2003), or assays based on antibodies binding to proteins with high specificity (Taylor *et al.*, 1994), can provide high species resolution and are still applied for fish food authenticity control. However their application is progressively being replaced by DNA-based technologies (see below).

Recently, the use of DNA markers has increasingly become the standard approach in fisheries genetics and fish food research (Hauser and Carvalho, 2008; Rasmussen and Morrissey, 2008). Both for fish species identification and population structure studies, which ultimately underlie origin assignment, the genome of the mitochondria – sub-cellular organelles – and the nuclear genome have been useful (Kochzius, 2008). Particularly, the use of the mitochondrial cytochrome oxidase I (COI) gene for species identification by 'DNA barcoding' has a high potential for routine fish species identification purposes (Ward *et al.*, 2009). For the analysis of species identifying DNA markers, a panoply of different technologies has been employed, which is reviewed in detail elsewhere (Teletchea, 2009).

© Woodhead Publishing Limited, 2011

Here we will focus on those DNA markers most amenable to traceability purposes. Microsatellites, also called short tandem repeats (STRs) in forensics, are tandem sequence repeats of one to six nucleotides (e.g. 'cgtacgtacgtacgtacgta') in the genome. Their high polymorphism is characterised by variable repeat numbers (between 5 and 100) even between individuals. Microsatellites are the standard marker for human identity testing by DNA profiling and for forensic genetic crime scene investigation (Butler, 2005). They have also been extensively used in fish population studies, and their potential value as traceability markers for origin assignment is very high. However, despite the widespread application of microsatellites, there are drawbacks, particularly scoring error and lack of comparability among laboratories (Dewoody *et al.*, 2006). Nevertheless, numerous examples exist where microsatellites are used for fish population/stock analysis, management, and also origin assignment (Manel *et al.*, 2005; Hauser and Carvalho, 2008), including Atlantic salmon (Primmer *et al.*, 2000), Pacific salmon (Fisheries and Oceans Canada, DFO[1]) and cod (Nielsen *et al.*, 2001).

Meanwhile, single nucleotide polymorphisms (SNPs) entered the realm of fisheries genetics, offering great potential for origin assignment (Hudson, 2008). SNPs are genome sites where more than one nucleotide (A, C, G or T) is present in a species. They are the most abundant polymorphism in the genome (Brumfield *et al.*, 2003), but per locus normally only two alleles exists (biallelic markers), thus they are less variable than microsatellites, where often many alleles exist. The lack of potential information per SNP marker is outweighed by their high abundance. Compared to other genetic markers, where routine genotyping and transfer of protocols between laboratories proves difficult, the information retrieved from SNPs is categorical, and data can be standardised across laboratories for forensic applications (Sobrino *et al.*, 2005). However, as discussed below, a substantial research effort targeting all commercial marine fish species will be necessary before SNPs can be employed routinely for origin assignment.

Despite this, available studies on marine fish using SNPs are encouraging. SNPs as markers to distinguish stocks of Atlantic cod (*Gadus morhua*) provided a high resolution power for stock identification, comparable to that of microsatellite loci (Wirgin *et al.*, 2007). Another example for the increasing use of SNPs is the North Pacific Anadromous Fish Commission (NPAFC) that is developing SNP arrays for Pacific salmon (http://www.npafc.org).

The rapid progress of DNA analysis technologies will have significant effects on the development of traceability tools. High-throughput sequencing has declined dramatically in cost, while speed and quality of analysis has increased by orders of magnitude. This is well illustrated by human genome sequencing, which can currently be accomplished in a few weeks at costs below US$5000, compared to more than a year and more than US$100 million in 2007 (Metzker,

1. DFO website: http://www.pac.dfo-mpo.gc.ca/science/facilities-installations/pbs-sbp/mgl-lgm/proj/index-eng.htm

© Woodhead Publishing Limited, 2011

2010). New sequencing technologies provide great opportunities for genetic fish population analysis and consequently also for traceability in the fisheries sector as shown by FishPopTrace (https://fishpoptrace.jrc.ec.europa.eu) where in a few weeks over 100 million bases of sequence data have been generated for European hake (*Merluccius merluccius*), Atlantic herring (*Clupea harengus*) and common sole (*Solea solea*).

For species identification, DNA microarrays ('DNA-Chips') can here be of great value. Microarrays consist of a surface with thousands of covalently attached DNA oligonucleotides. This allows monitoring of thousands of different (i.e. species identifying) DNA sequences simultaneously with one array (size about 1 × 1 cm). Theoretically, just one chip would enable the screening for all major economic fish species simultaneously (Kochzius *et al.*, 2008). While the development of DNA-microarrays is laborious, the running costs are moderate. Other high-throughput and parallel processing methodologies for fish species identification have also been developed (Dooley *et al.*, 2005). Such technologies might ultimately lead to the development of handheld analytical devices, enabling field use, which is critical with respect to the response time (the period between starting an investigation and the receipt of analytical results). For example, inspectors in the fisheries sector carry huge responsibility: if they decide to put landings 'on hold' because of suspect content, there can be severe consequences for fishers and stakeholders. Engineering of such machines is carried out in support of forensic genetic analysis at crime scenes (Liu *et al.*, 2008). However, while recent publications show that progress has been made in this area (Arnaud, 2008), currently no cost-effective handheld analytical device supporting fisheries control and enforcement or traceability is available.

15.2.4 Forensics in support of traceability

Forensics, the application of scientific methods to answer questions in relation to a crime or to a civil action, have become increasingly relevant in wildlife law enforcement (Ogden *et al.*, 2009). That forensic evidence can support investigations and deter illegal fishing or fraud along the food supply chain has been demonstrated: forensic genetic and chemical analyses contributed to the body of evidence during various recent crime cases, such as the illegal import and sale of over 1000 tons of falsely labelled catfish, worth US$15.5 million, into the USA. The investigation of this large-scale conspiracy was undertaken in cooperation between diverse US federal agencies. Species identification was undertaken with forensic DNA testing carried out by scientific laboratories. The test results provided evidence during the court trials and led to convictions (Personal communication: Paul Raymond, NOAA Office of Law Enforcement and Fiske, 2008).

Before criminal courts will routinely accept DNA testing, or chemical analysis, in cases involving animals, rigorous validation of the techniques, together with appropriate sampling, evidence handling, and statistical evaluation of analytical results are required. A major challenge is to present evidence which

© Woodhead Publishing Limited, 2011

is robust, lucid and unequivocal, since all elements such as sample collection and transfer, data compilation and analysis, as well as reference data sets are liable to be scrutinised for flaws during court trials and must be validated (Ogden, 2008). The handling of physical evidence is one of the most important factors in an investigation. Samples must be collected, inventoried, preserved, transported, and submitted for testing without compromising the evidential chain of custody. In this respect, guidelines established for processing human crime scenes, such as approved by the Technical Working Group on Crime Scene Investigation (Reno *et al.*, 2000), may be transferred into fisheries enforcement, to avoid errors that result in inadmissibility of evidence. Quality assurance is also obtained through Standard Operating Procedures (SOPs); documents containing instructions to perform procedures that are routine, standardised, and for which no *ad-hoc* modification is acceptable. SOPs help to ensure quality and integrity of data and provide for uniformity and accountability (Bowker, 2006).

Apart from guaranteeing acceptance of evidence in courts, applying forensic standards to traceability technologies will greatly facilitate international collaboration and technology transfer, particularly to developing countries, which urgently need support in this area. The transfer of molecular technologies into a forensic framework for traceability, fisheries control and enforcement is specifically addressed by FishPopTrace, and was also a central topic during a recent interdisciplinary FAO workshop on forensic technologies for fisheries control and enforcement.[2]

15.3 Species identification and origin assignment of fish: key components of an efficient traceability framework

Mislabelling of seafood undermines fisheries management and conservation, impacts the economy and deceives the consumer (Jacquet and Pauly, 2008). Worldwide traceability schemes are in place to counteract such effects. However, as mentioned at the outset, traceability is presently predominantly built on labels and certificates.

Customs and fisheries authorities control for compliance with traceability requirements throughout the supply chain, by establishing correctness and consistency of information accompanying the products. However, control technologies, which are entirely independent from information provided, would be an invaluable asset to traceability schemes. We now consider how the technologies described above can provide efficient independent tools to control for compliance, and support enforcement of laws underlying traceability in the fisheries sector.

2. Informal Workshop on the Use of Forensic Technologies in Fisheries Monitoring, Control and Surveillance; FAO, Rome, Italy, 9–10 December 2009.

© Woodhead Publishing Limited, 2011

15.3.1 Fish species identification

A study of food fish in the United States revealed that three-quarters of fish sold as 'red snapper' were mislabelled (Marko *et al.*, 2004). DNA-barcoding of fish samples from markets and restaurants in North America revealed that about 25% of analysed specimens were mislabelled (Wong and Hanner, 2008). In South Africa, genetic tests on fish fillets revealed that half of all fillets examined were mislabelled (Von Der Heyden *et al.*, 2010). These and other recent examples (e.g. Pepe *et al.*, 2007; Schwartz, 2008; Bertoja *et al.*, 2009; Miller and Mariani, 2010) highlight how common false labelling of fish is, and stress the urgency of providing authorities and the industry with cost-effective efficient species authentication tools.

For whole fish, the species can generally be determined by visual inspection. However, fish species identification on a routine basis as part of a traceability scheme, must be possible on processed fish (fillets, canned and cured, etc.) and not require expert taxonomic knowledge. Molecular methods provide such assets.

Formerly, authorities often used protein analysis for authentication purposes (reviewed by Rehbein, 2003). Meanwhile, DNA-based methods are increasingly used to reveal species substitution in fish and seafood product (Rasmussen and Morrissey, 2008, 2009). DNA-based methods have advantages over protein-based analysis: DNA is less sensitive to degradation, e.g. during processing, it can be extracted at all stages from egg to adult, from processed products and even historical samples (Nielsen and Hansen, 2008). Genetic variation at the DNA level can be revealed by sequencing, and PCR amplification makes it possible to analyse minute amounts of tissue. Also – important in a forensic context – DNA sequence data are considered as easier to replicate and standardise across laboratories (Ward *et al.*, 2009).

Genetic markers for species identification must exhibit low diversity between individuals of the same species (low intraspecies polymorphism), but high diversity between individuals of different species (high interspecies polymorphism: so-called 'barcoding gap'). These conditions are generally met by the mitochondrial *cytochrome c oxidase subunit 1* (COI) gene, of which a segment is employed as a marker in the DNA-barcoding approach steered by the Barcode of Life initiative, which is devoted to developing DNA-barcoding as a global standard for species identification (Costa and Carvahlo, 2007).

As part of CBOL, the Fish Barcode of Life Initiative (FISH-BOL) co-ordinates the assembly of a standardised reference sequence library for fish species, derived from taxonomically identified voucher specimens, with more than 7000 fish species currently barcoded (June 2010). DNA barcoding has repeatedly been applied successfully to reveal mislabelling of seafood and the US Food and Drug Administration (FDA) considers replacing protein isoelectric focusing on fish products by barcoding as an authentication tool (Yancy *et al.*, 2008). Also the mitochondrial *cytochrome-b* gene has successfully been used for species determination (Logan *et al.*, 2008): a data base exists within the FishTrace database (www.fishtrace.org), where more than 200 commercial marine fish species are recorded in a genetic catalogue, and moreover all

© Woodhead Publishing Limited, 2011

compiled DNA sequences are linked to a voucher stored in a Natural Museum. Importantly, both the FISH-BOL and FishTrace databases are public, and can therefore be used by control and enforcement authorities to access reference DNA sequences.

As discussed above, microarray platforms provide the possibility of screening many fish species simultaneously and a DNA microarray has been developed to identify 11 fish species from European seas based on mitochondrial 16S rDNA sequences (Kochzius *et al.*, 2008). Also, SNPs can be used as species identification markers. This is currently being tested in an international collaboration[3] for the identification of caviar, a product notorious for being subject to illegal trade (European Commission, 2006).

15.3.2 Origin assignment of fish

Determining the geographical origin of fish and fish products throughout the supply chain also addresses the question of whether a fish originates from the wild or from aquaculture. Assignment depends on matching individual fish to groups of fish with similar characteristics, driven by common environmental factors, common genetic material or both. Often rather loosely, such groups are referred to as populations or, in a fisheries management context, as stocks. The terms 'stock' and 'population' are not necessarily interchangeable, which in the context of traceability is important to consider: 'stock' is a technical term describing a group of individuals under management considerations for exploitation (Carvalho and Hauser, 1994). Stocks might not be genetically distinct groups of fish, but reflect differences in phenotypic life history parameters in response to environmental variation and fishing pressure. Normally, stocks occupy well-defined spatial areas (fishing areas) independent of other stocks of the same species, but random dispersal or migration due to seasonal or reproductive activity can occur. The term 'population' takes into account the reproductive relatedness of organisms. Simplified, it describes individuals of a particular species, which can be defined as a local interbreeding group (a more elaborate account of the term 'population' has been provided by Waples and Gaggiotti, 2006). Such a grouping has reduced genetic exchange (gene flow) with other groups of the same species, meaning that mating between individuals of different groups occurs only rarely. This interbreeding restriction leads to isolation from other groups from the same species and genetic differentiation occurs (Waples *et al.*, 2008). Thus, groupings based on demographic relationships (e.g. microchemistry) might not necessarily reveal identical patterns to those dependent on interbreeding and resultant population structure (genetic markers). Such potentially discordant patterns revealed by different tools necessarily demands sufficient data on salient features of the population and demographic structure of fish for interpretation.

3. SturSNiP – Sturgeon Product Traceability based on SNPs; http://stursnip.jrc.ec.europa.eu

© Woodhead Publishing Limited, 2011

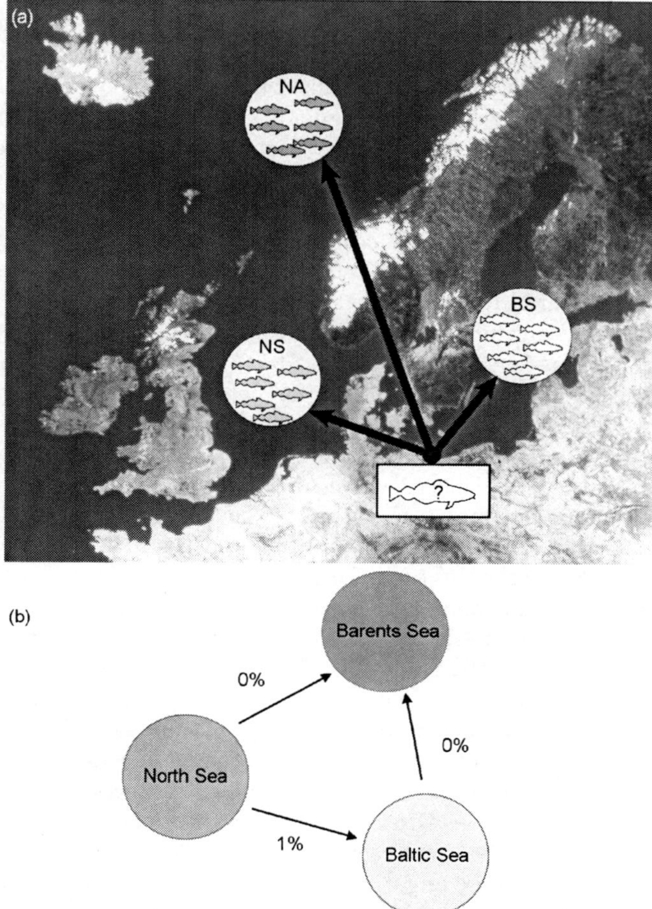

Fig. 15.5 The principle of individual assignment (IA): (a) Individual fish of unknown origin are assigned to a set of baseline stocks or populations. In this scenario, using the three baseline populations North Atlantic (NA), North Sea (NS) and Baltic Sea (BS) the fish in question will be assigned to BS. (Map: European Union, 2010). (b) Percentage misidentification of cod from each of three geographic areas based on individual-based assignment. Each individual of unknown origin is assigned probabilistically to a number of known baseline populations based on its multi-locus genotype (here for cod and using 9 microsatellites). The arrows indicate percentage of misassignment (Nielsen, 2001; courtesy of E.E. Nielsen).

Assigning individuals to their origin is based on the probability estimation of encountering a combination of characteristic features (e.g. genotype or chemical composition), established for these individuals, in a number of potential source populations (Fig. 15.5). As pointed out, this can only be accomplished if a baseline exists, that is, the composition of the potential source populations has formerly been determined. Genetic markers especially combine various assets

© Woodhead Publishing Limited, 2011

with regards to creating such baselines. They are heritable and the genetic structure of populations typically remains stable over generations. Also, genetic loci are particularly accessible by statistical analysis and a large body of underlying population genetic theory supporting the statistical analysis is readily available (Waples, 1990). Moreover, a variety of software applications using multilocus genotype data to infer population structure and assign individuals to populations, have been developed (Cornuet *et al.*, 1999; Banks and Eichert, 2000; Pritchard *et al.*, 2000; Piry *et al.*, 2004), and recently approaches have been devised to analyse population structure in the context of the seascape and environmental variables (Hansen and Hemmer-Hansen, 2007). These developments will greatly support traceability approaches.

Using genetic markers (microsatellites, SNPs) and new statistical methods for their analysis, it has been shown that for commercial marine fish species, including herring and cod, population structuring can be evident even at small geographic scales (Hauser and Carvalho, 2008; Nielsen *et al.*, 2009), which is favourable for traceability. Indeed, while origin assignment of marine fish remains challenging compared to freshwater and anadromous species that typically display higher genetic differentiation, its feasibility has been proven in at least some cases (Manel *et al.*, 2005). Scientists of the Technical University of Denmark were able to assign Atlantic Cod (*Gadus morhua*) individuals to a North Sea, Baltic Sea or north-east Arctic Ocean population with almost 100% certainty, using microsatellite markers (Nielsen *et al.*, 2001). The same principle has been applied to generate evidence in court during a case where a fisherman claimed the wrong origin for his catch. The evidence provided the basis for a conviction (Personal Communication: Lars B. Erikson, Danish Directorate of Fisheries; Inspectorate of Fisheries; Einar E. Nielsen, Technical University of Denmark).

Approaches not based on DNA technologies for origin assignment are also valuable. Fatty acid analysis has been employed to distinguish between marine fish of wild and aquaculture origin (Seaborn *et al.*, 2000). It was also shown that trace element analysis on muscle tissue, applying forensic standards, can be used to distinguish groupings of pink ling (*Genypterus blacodes*) from different areas in New Zealand (Bremner, 2009). Such approaches are distinct from a genetic approach in that demographic relationships are revealed rather than reproductive relatedness: indeed, these approaches can complement each other.

15.4 Technology transfer: from the research laboratory to authorities, industry and policy makers

As reflected in a vast treasure of myths and legends, the oceans and seas have always been a vital and cherished source of food for humans. Seafood globally plays a fundamental role in the supply of nourishment, contributing in some parts of the world more than 50 percent of total animal protein intake (FAO, 2009). Moreover, beyond its role for sustaining livelihood, marine fisheries and aquaculture have nowadays tremendous socio-economic importance.

© Woodhead Publishing Limited, 2011

Reckless IUU fishing contributes greatly to the looming demise of marine fisheries in many parts of our world, and fraud along the supply chain distorts markets and impedes consumer protection. Globally, efforts are underway to restrain illegal fishing activities. Examples are the FAO International Plan of Action to Prevent, Deter and Eliminate Illegal, Unreported and Unregulated Fishing (IPOA IUU) (FAO, 2001), or the recently implemented IUU Regulation of the European Union (European Council, 2008), and modern control and monitoring technologies such as satellite-based vessel monitoring and detection (VMS/VDS). Traceability is also generally acknowledged as an important component fisheries control, and to assure product authenticity.

In this context, the prevailing under-utilisation of molecular technologies, in particular DNA-based technologies, is unexpected. As shown by many, yet sparse, examples (see above), such technologies are not only theoretically available for traceability purposes in the fisheries sector, but have already been used for fisheries control and enforcement, as well as by the industry. Moreover, while origin assignment of marine fish remains a challenge, species identification for traceability purposes is at a stage where it should become standard and routine internationally. The ease with which DNA-barcoding can be applied is exemplified in a test carried out by high-school students of the Manhattan Trinity School in collaboration with the University of Guelph, Ontario (Canada): 60 samples of seafood taken from New York sushi restaurants and seafood markets, were tested for their authenticity. DNA-barcoding revealed that approximately 25% of fish samples with identifiable DNA were mislabelled (Schwartz, 2008). In a similar way, it is anticipated that with appropriate investment and development, enterprises such as FishPopTrace will increase accessibility and reliability of origin assignment (Martinsohn and Ogden, 2009). The feasibility is shown by a collaboration between the Danish fisheries inspectorate and the Danish Technical University (DTU). Inspectors are equipped with tissue sampling kits and can send samples for genetic analysis to a laboratory of the DTU if suspicion about the identity or origin of fish landings arises (Personal communication: Lars B. Erikson, Danish Fisheries Inspectorate).

Yet, despite such largely anecdotal success, a coherent and consistent approach to fully integrate molecular technologies into a traceability framework supporting fisheries control authorities as well as the industry on an international level, is not yet in place. Since illegal fishing or fisheries fraud is barely impaired by borders, such apparent little uptake is of particular note. Similarly, it is appropriate to stress that developing countries are most exposed to the risks emerging from IUU. For example, along the West African coastline, total estimated catches were reported to be 40% higher than reported catches (MRAG and Development: DFID, 2009). Many countries, including those from the developing world, urgently require access to affordable valuable traceability tools for fish and fish products and, eventually, the use of forensic molecular tests as emphasised, for example, by the FAO (Personal communication: Michel Kuruc, FAO Fisheries and Aquaculture Department). However, such uptake places demands not only on the availability of appropriate

© Woodhead Publishing Limited, 2011

technologies, but crucially is constrained by complex socio-economic and political issues.

So what are the primary reasons for the insufficient transfer of advanced molecular technologies into traceability applications readily available for authorities and the industry, and how can existing obstacles be overcome? To a certain extent it is founded on the negligible uptake of such technologies into fisheries management, as the consequence of a prevailing gap between science and fisheries management (Waples *et al.*, 2008). A similar problem is apparent in wildlife forensics (Ogden, 2010).

A further considerable impediment is the current absence of a central data hub, where DNA and chemical data relevant to fish and fish product traceability is stored, professionally managed and easily accessible, similar to the European Bioinformatics Institutes (EBI) DNA databank (Brooksbank *et al.*, 2009). However, a great opportunity opened for species identification by the DNA-barcoding approach, since the reference sequences for currently more than 7500 fish species (June 2010) are compiled in a public database (http://www.boldsystems.org). In contrast, no comparable database is available that includes population baseline data for origin assignment, is unavailable, which is the basis for the recommendation of the ICES Working Group on Applied Genetics in Fisheries and Mariculture (WGAGFM) to create a meta-database cataloguing existing data in the field of fish and shellfish population genetics (Verspoor *et al.*, 2009).

Additional requirements have to be fulfilled to build a functional infrastructure ensuring the use of modern molecular technologies globally. Not only should data be easily accessible, but it must also be validated, data quality must be verified, and standardised formats must be applied. Secondly, a network of test laboratories, certified to carry out analysis for control and enforcement purposes, and applying protocols, validated through inter-laboratory trials, as well as forensic standards should be established. Such institutions should also be committed to share information and expertise. In many countries it will not even be necessary to create new laboratories, as facilities with appropriate expertise and capacity already exist. However, these typically operate independently, and respond to *ad hoc* queries. To ensure inter-operability and consistency, both on the national and international level, accreditation schemes for routine and reference laboratories should be put in place, as for instance observed in the GMO detection area (Zel *et al.*, 2006, 2008). Another crucial component of capacity building is training for inspectors and enforcement officers, including guidance in core sampling procedures and for laboratory personal (see Section 15.2.4). The existence of 'focal points', institutions, endowed with the necessary expertise and authority to produce technical guidelines and standards to be followed by laboratories, and to pursue targeted dissemination activities in order to inform stakeholders would be crucial.

Ideally such an approach should be accompanied by a sound cost-benefit analysis (CBA) to estimate cost-effectiveness and to provide a valuable reference and decision-support tool for stakeholders and policy analysts (Pearce

© Woodhead Publishing Limited, 2011

et al., 2006). However, the recent decline in costs, especially for DNA analysis, and the examples of applications, strongly indicates that the methods discussed here are cost effective. Moreover, advanced technologies feature an added value in that they are beneficial to fisheries management beyond traceability: they typically allow detection and monitoring of discrete fish assemblages, the dynamics of which underpin resilience and recovery (Waples *et al.*, 2008).

15.5 Conclusions

The state of art in genetics, chemistry and forensics, as discussed here, illustrates that molecular analytical technologies and forensics can support fisheries control and enforcement, as well as the verification of authenticity and origin of seafood for traceability from 'ocean to fork'. The transfer of such technologies into applications for control authorities and the industry worldwide would significantly benefit the consumer, as well as socio-economic development in the fisheries sector, and the ecological balance of our oceans. Ultimately such technologies could forcefully contribute to an escape route from the 'Tragedy of the commons'.

15.6 Acknowledgements

We would like to thank all members of the FishPopTrace consortium for continous intellectual input combined with unfaltering optimism, which greatly contributed to the completion of this chapter. We thank Petra Spaniol for valuable advice on EU legislative and policy aspects. FishPopTrace is funded by the European Community's Seventh Framework Programme under grant agreement no. KBBE-212399 in the area of 'Food, Agriculture and Fisheries, and Biotechnology'.

15.7 References

ADEY, E. A., BLACK, K. D., SAWYER, T., SHIMMIELD, T. M. and TRUEMAN, C. N. (2009) Scale microchemistry as a tool to investigate the origin of wild and farmed *Salmo salar*. *Marine Ecology Progress Series*, 390, 225–235.

AGNEW, D. J., PEARCE, J., PRAMOD, G., PEATMAN, T., WATSON, R., BEDDINGTON, J. R. and PITCHER, T. J. (2009) Estimating the worldwide extent of illegal fishing. *PLoS ONE*, 4.

ARNAUD, C. H. (2008) Moving from bench to bedside. *Chemical and Engineering News*, 86, 56–58.

AURSAND, M., STANDAL, I. B., PRAËL, A., MCEVOY, L., IRVINE, J. and AXELSON, D. E. (2009) 13C NMR pattern recognition techniques for the classification of Atlantic salmon (*Salmo salar* L.) according to their wild, farmed, and geographical origin. *Journal of Agricultural and Food Chemistry*, 57, 3444–3451.

BANKS, M. A. and EICHERT, W. (2000) WHICHRUN (version 3.2): A computer program for population assignment of individuals based on multilocus genotype data. *Journal of Heredity*, 91, 87–89.

© Woodhead Publishing Limited, 2011

BERTOJA, G., GIACCONE, V., CARRARO, L., MININNI, A. N. and CARDAZZO, B. (2009) A rapid and high-throughput real-time PCR assay for species identification: Application to stockfish sold in Italy. *European Food Research and Technology*, 229, 191–195.

BOWKER, J. (2006) *Definition and Use of the Standard Operating Procedure (SOP)*, U.S. Fish and Wildlife Services – Aquatic Animal Drug Approval Partnership, SOP No. GEN003.2, 1–2.

BRÉCARD, D., HLAIMI, B., LUCAS, S., PERRAUDEAU, Y. and SALLADARRÉ, F. (2009) Determinants of demand for green products: An application to eco-label demand for fish in Europe. *Ecological Economics*, 69, 115–125.

BREMNER, G. (2009) Marine fish of disputed origin: can trace elements provide a forensic solution? A case study from southern New Zealand. Ministry of Fisheries, PO Box 19261, Dunedin, New Zealand.

BROOKSBANK, C., CAMERON, G. and THORNTON, J. (2009) The European Bioinformatics Institute's data resources. *Nucleic Acids Research*, 38, D17–D25.

BRUMFIELD, R. T., BEERLI, P., NICKERSON, D. A. and EDWARDS, S. V. (2003) The utility of single nucleotide polymorphisms in inferences of population history. *Trends in Ecology and Evolution*, 18, 249.

BUTLER, J. M. (2005) *Forensic DNA typing: biology, technology, and genetics of STR markers*, Amsterdam ; Boston, Elsevier Academic Press.

CAMPANA, S. E. (2005) Otolith elemental composition as a natural marker of fish stocks. In Cadrin, S. X., Friedland, K. D. and Waldman, J. R. (Eds) *Stock Identification Methods: Applications in Fishery Science.* 1st edn. London, Elsevier Academic Press.

CARVALHO, G. R. and HAUSER, L. (1994) Molecular genetics and the stock concept in fisheries. *Reviews in Fish Biology and Fisheries*, 4, 326–350.

CORNUET, J. M., PIRY, S., LUIKART, G., ESTOUP, A. and SOLIGNAC, M. (1999) New methods employing multilocus genotypes to select or exclude populations as origins of individuals. *Genetics*, 153, 1989.

COSTA, F. O. and CARVALHO, G. R. (2007) The Barcode of Life Initiative: synopsis and prospective societal impacts of DNA barcoding of fish. *Genomics, Society and Policy*, 3, 29–40.

CUVELIERS, E. L., GEFFEN, A. J., GUELINCKX, J., RAEYMAEKERS, J. A. M., SKADAL, J., VOLCKAERT, F. A. M. and MAES, G. E. (2010) Microchemical variation in juvenile *Solea solea* otoliths as a powerful tool for studying connectivity in the North Sea. *Marine Ecology Progress Series*, 401, 211–220.

DEWOODY, J., NASON, J. D. and HIPKINS, V. D. (2006) Mitigating scoring errors in microsatellite data from wild populations. *Molecular Ecology Notes*, 6, 951.

DOOLEY, J. J., SAGE, H. D., CLARKE, M.-A. L., BROWN, H. M. and GARRETT, S. D. (2005) Fish species identification using PCR–RFLP analysis and lab-on-a-chip capillary electrophoresis: application to detect white fish species in food products and an interlaboratory study. *Journal of Agricultural and Food Chemistry*, 53, 3348–3357.

ELSDON, T. S., WELLS, B. K., CAMPANA, S. E., GILLANDERS, B. M., JONES, C. M., LIMBURG, K. E., SECOR, D. H., THORROLD, S. R. and WALTHER, B. D. (2008) Otolith chemistry to describe movements and life-history parameters of fishes – hypotheses, assumptions, limitations and inferences. *Oceanography and Marine Biology: An Annual Review*, 46, 297–330.

EUROPEAN COMMISSION (2001) Commission Regulation (EC) No 2065/2001 of 22 October 2001 laying down detailed rules for the application of Council Regulation (EC) No 104/2000 as regards informing consumers about fishery and aquaculture products. *Official Journal of the European Communities*, L 278, 6.

© Woodhead Publishing Limited, 2011

EUROPEAN COMMISSION (2006) New rules to combat illegal caviar trade. European Commission Press Release; IP/06/611; Brussels, 15 May 2006.

EUROPEAN COMMISSION (2007) Communication from the Commission to the European Parliament, the Council, the European Economic and Social Committee and the Committee of the Regions: On a new strategy for the Community to prevent, deter and eliminate Illegal, Unreported and Unregulated fishing. Brussels: European Commission.

EUROPEAN COMMISSION DG TRADE (2009) Fisheries Statistics. Date: 31/5/2010; URL: http://ec.europa.eu/trade/creating-opportunities/economic-sectors/fisheries/statistics/. Last updated: 23 January 2009.

EUROPEAN COUNCIL (2008) Council Regulation (EC) No 1005/2008 of 29 September 2008 establishing a Community system to prevent, deter and eliminate illegal, unreported and unregulated fishing. *Official Journal of the European Union*, L 286, 1.

EUROPEAN COUNCIL (2009) Council Regulation (EC) No 1224/2009 of 20 November 2009 establishing a Community control system for ensuring compliance with the rules of the common fisheries policy. *Official Journal of the European Union*, L 343, 1.

EUROPEAN PARLIAMENT AND COUNCIL (2002) Regulation (EC) No 178/2002 of the European Parliament and of the Council of 28 January 2002, laying down the general principles and requirements of food law, establishing the European Food Safety Authority and laying down procedures in matters of food safety. *Official Journal of the European Communities*, L 31/1, 1.

FAO (2001) *International Plan of Action to Prevent, Deter and Eliminate Illegal, Unreported and Unregulated Fishing (IPOA IUU)*. FAO, Rome.

FAO (2006) *FACT SHEET: The international fish trade and world fisheries.*

FAO (2009) *The State of World Fisheries and Aquaculture 2008*. Rome, FAO Fisheries Aquaculture, Department.

FISKE, E. (2008) Two more convicted in false labelling scheme. *FiS Worldnews*. Fish Information and Services.

FLOTHMANN, S., VON KISTOWSKI, K., DOLAN, E., LEE, E., MEERE, F. and ALBUM, G. (2010) Closing loopholes: getting illegal fishing under control. *Science*, 328 (5983), 1235–1236.

GEFFEN, A. J., JARVIS, K., THORPE, J. P., LEAH, R. T. and NASH, R. D. M. (2003) Spatial differences in the trace element concentrations of Irish Sea plaice *Pleuronectes platessa* and whiting *Merlangius merlangus* otoliths. *Journal of Sea Research*, 50, 245–254.

GILLANDERS, B. M. (2001) Trace metals in four structures of fish and their use for estimates of stock structure. *Fishery Bulletin*, 99, 410–419.

GRAHL-NIELSEN, O. (2005) Fatty acid profiles as natural marks for stock identification. In Cadrin, S. X., Friedland, K. D. and Waldman, J. R. (Eds.) *Stock Identification Methods: Applications in Fishery Science*. London, Elsevier Academic Press.

HANSEN, M. M. and HEMMER-HANSEN, J. (2007) Landscape genetics goes to sea. *Journal of Biology*, 6 (3), art. no. 6.

HARDIN, G. (1968) The tragedy of the commons. *Science*, 162, 1243–1248.

HAUSER, L. and CARVALHO, G. R. (2008) Paradigm shifts in marine fisheries genetics: ugly hypotheses slain by beautiful facts. *Fish and Fisheries*, 9, 333–362.

HIGGINS, R. M., DANILOWICZ, B. S., BALBUENA, J. A., DANIELSDOTTIR, A. K., GEFFEN, A. J., MEIJER, W. G., MODIN, J., MONTERO, F. E., PAMPOULIE, C., PERDIGUERO-ALONSO, D., SCHREIBER, A., STEFANSSON, M. and WILSON, B. (2010) Multi-disciplinary fingerprints reveal the harvest location of cod *Gadus morhua* in the northeast Atlantic. *Marine Ecology Progress Series*, 404, 197–206.

© Woodhead Publishing Limited, 2011

HUDSON, M. E. (2008) Sequencing breakthroughs for genomic ecology and evolutionary biology. *Molecular Ecology Resources*, 8, 3–17.

ISO 9000:2005, Quality management systems – Fundamentals and vocabulary.

JACQUET, J. L. and PAULY, D. (2008) Trade secrets: renaming and mislabeling of seafood. *Marine Policy*, 32, 309.

KOCHZIUS, M. (2008) Trends in fishery genetics. In Beamish, R. J. and Rothschild, B. J. (eds) *The Future of Fisheries Science in North America*, 1st edn. Springer, London, pp. 451–491.

KOCHZIUS, M., NÖLTE, M., WEBER, H., SILKENBEUMER, N., HJÖRLEIFSDOTTIR, S., HREGGVIDSSON, G. O., MARTEINSSON, V., KAPPEL, K., PLANES, S., TINTI, F., MAGOULAS, A., GARCIA VAZQUEZ, E., TURAN, C., HERVET, C., CAMPO FALGUERAS, D., ANTONIOU, A., LANDI, M. and BLOHM, D. (2008) DNA microarrays for identifying fishes. *Marine Biotechnology*, 10, 207.

KWETEGYEKA, J., MPANGO, G. and GRAHL-NIELSEN, O. (2008) Variation in fatty acid composition in muscle and heart tissues among species and populations of tropical fish in lakes Victoria and Kyoga. *Lipids*, 43, 1017–1029.

LIU, P., YEUNG, S. H. I., CRENSHAW, K. A., CROUSE, C. A., SCHERER, J. R. and MATHIES, R. A. (2008) Real-time forensic DNA analysis at a crime scene using a portable microchip analyzer. *Forensic Science International Genetics*, 2(4), 301–309.

LOGAN, C. A., ALTER, S. E., HAUPT, A. J., TOMALTY, K. and PALUMBI, S. R. (2008) An impediment to consumer choice: overfished species are sold as Pacific red snapper. *Biological Conservation*, 141, 1591–1599.

MACKENZIE, K. and ABAUNZA, P. (1998) Parasites as biological tags for stock discrimination of marine fish: A guide to procedures and methods. *Fisheries Research*, 38, 45.

MANEL, S., GAGGIOTTI, O. E. and WAPLES, R. S. (2005) Assignment methods: Matching biological questions with appropriate techniques. *Trends in Ecology and Evolution*, 20, 136.

MARKO, P. B., LEE, S. C., RICE, A. M., GRAMLING, J. M., FITZHENRY, T. M., MCALISTER, J. S., HARPER, G. R. and MORAN, A. L. (2004) Mislabelling of a depleted reef fish. *Nature*, 430, 309.

MARTINSOHN, J. T. and OGDEN, R. (2009) FishPopTrace-Developing SNP-based population genetic assignment methods to investigate illegal fishing. *Forensic Science International: Genetics Supplement Series*, 2, 294.

METZKER, M. L. (2010) Sequencing technologies – the next generation. *Nature reviews. Genetics*, 11, 31.

MILLER, D. D. and MARIANI, S. (2010) Smoke, mirrors, and mislabeled cod: poor transparency in the European seafood industry. *Frontiers in Ecology and the Environment*, doi:10.1890/090212, 14.

MRAG and DFID (2009) Policy Brief 8: Illegal, Unreported and Unregulated Fishing. *DFID Policy Brief*, 8, 1–4.

MURPHY, B. and MORRISON, R. D. (2007) *Introduction to environmental forensics*, Boston, MA, Academic Press.

NIELSEN, E. E. and HANSEN, M. M. (2008) Waking the dead: the value of population genetic analyses of historical samples. *Fish and Fisheries*, 9, 450–461.

NIELSEN, E. E., HANSEN, M. M., SCHMIDT, C., MELDRUP, D. and GRØNKJÆR, P. (2001) Population of origin of Atlantic cod. *Nature*, 413, 272.

NIELSEN, E. E., WRIGHT, P. J., HEMMER-HANSEN, J., POULSEN, N. A., GIBB, I. M. and MELDRUP, D. (2009) Microgeographical population structure of cod *Gadus morhua* in the North

© Woodhead Publishing Limited, 2011

Sea and west of Scotland: the role of sampling loci and individuals. *Marine Ecology Progress Series*, 376, 213–225.

OECD (ed.) (2005) *Why Fish Piracy Persists: The Economics of Illegal, Unreported and Unregulated Fishing*, Organisation for Economic Co-Operation and Development.

OGDEN, R. (2008) Fisheries forensics: the use of DNA tools for improving compliance, traceability and enforcement in the fishing industry. *Fish and Fisheries*, 9, 462.

OGDEN, R. (2010) Forensic science, genetics and wildlife biology: getting the right mix for a wildlife DNA forensics lab. *Forensic Science, Medicine, and Pathology*, 6(3), 172–179.

OGDEN, R., DAWNAY, N. and MCEWING, R. (2009) Wildlife DNA forensics – bridging the gap between conservation genetics and law enforcement. *Endangered Species Research*, 9, 179–195.

PEARCE, D., ATKINSON, G. and MOURATO, S. (EDS) (2006) *Cost-Benefit Analysis and the Environment: Recent Developments*, OECD.

PEPE, T., TROTTA, M., MARCO, I. D., ANASTASIO, A., BAUTISTA, J. M. and CORTESI, M. L. (2007) Fish species identification in surimi-based products. *Journal of Agricultural and Food Chemistry*, 55, 5.

PIRY, S., ALAPETITE, A., CORNUET, J. M., PAETKAU, D., BAUDOUIN, L. and ESTOUP, A. (2004) GENECLASS2: A software for genetic assignment and first-generation migrant detection. *Journal of Heredity*, 95, 536–539.

PRIMMER, C. R., KOSKINEN, M. T. and PIIRONEN, J. (2000) The one that did not get away: Individual assignment using microsatellite data detects a case of fishing competition fraud. *Proceedings of the Royal Society – Biological Sciences (Series B)*, 267, 1699.

PRITCHARD, J. K., STEPHENS, M. and DONNELLY, P. (2000) Inference of population structure using multilocus genotype data. *Genetics*, 155, 945.

RASMUSSEN, R. S. and MORRISSEY, M. T. (2008) DNA-based methods for the identification of commercial fish and seafood species. *Comprehensive Reviews in Food Science and Food Safety*, 7, 280.

RASMUSSEN, R. S. and MORRISSEY, M. T. (2009) Application of DNA-based methods to identify fish and seafood substitution on the commercial market. *Comprehensive Reviews in Food Science and Food Safety*, 8, 118–154.

REHBEIN, H. (2003) Identification of fish species by protein- and DNA-analysis. In Perez-Martin, R. and Soleto, C. G. (eds) *Authenticity of species in meat and seafood products*. Association 'International Congress on Authenticity of Species in Meat and Seafood Products'.

RENO, J., MARCUS, D., ROBINSON L., BRENNAN, L. and TRAVIS, J. (2000) Crime Scene Investigation: A Guide for Law Enforcement. US Department of Justice Technical Working Group on Crime Scene Investigation. Washington, DC.

SCHEFFER, M., CARPENTER, S. and DE YOUNG, B. (2005) Cascading effects of overfishing marine systems. *Trends in Ecology and Evolution*, 20, 579–581.

SCHWARTZ, J. (2008) Fish tale has DNA hook: Students find bad labels. *The New York Times*, 21 August.

SEABORN, G. T., JAHNCKE, M. L. and SMITH, T. I. J. (2000) Differentiation between cultured hybrid striped bass and wild striped bass and hybrid bass using fatty acid profiles. *North American Journal of Fisheries Management*, 20, 618–626.

SOBRINO, B., BRIÓN, M. and CARRACEDO, A. (2005) SNPs in forensic genetics: a review on SNP typing methodologies. *Forensic Science International*, 154, 181–194.

TAYLOR, W. J., PATEL, N. P. and JONES, J. L. (1994) Antibody-based methods for assessing

© Woodhead Publishing Limited, 2011

seafood authenticity. *Food and Agricultural Immunology*, 6(3), 305–314.

TELETCHEA, F. (2009) Molecular identification methods of fish species: reassessment and possible applications. *Reviews in Fish Biology and Fisheries*, 19, 265–293.

THRESHER, R. E. (1999) Elemental composition of otoliths as a stock delineator in fishes. *Fisheries Research*, 43, 165–204.

VERSPOOR, E., ARNAUDO, L. and MARTINSOHN, J. (2009) Establishment of a Meta-Database for Genetic Data on Fish and Shellfish Genetics covered under the ICES Remit – Progress and Prospects. In *ICES Report of the Working Group on the Application of Genetics in Fisheries and Mariculture (WGAGFM) 2009*. Sopot, International Council for the Exploration of the Sea.

VON DER HEYDEN, S., BARENDSE, J., SEEBREGTS, A. J. and MATTHEE, C. A. (2010) Misleading the masses: detection of mislabelled and substituted frozen fish products in South Africa. *ICES Journal of Marine Science*, 67, 176–185.

WAPLES, R. S. (1990) Conservation genetics of Pacific salmon: I. Temporal changes in allele frequency. *Conservation Biology*, 4, 144–156.

WAPLES, R. S. and GAGGIOTTI, O. E. (2006) What is a population? An empirical evaluation of some genetic methods for identifying the number of gene pools and their degree of connectivity. *Molecular Ecology*, 15, 1419–1439.

WAPLES, R. S., PUNT, A. E. and COPE, J. M. (2008) Integrating genetic data into management of marine resources: how can we do it better? *Fish and Fisheries*, 9, 423–449.

WARD, R. D., HANNER, R. and HEBERT, P. D. N. (2009) The campaign to DNA barcode all fishes, FISH-BOL. *Journal of Fish Biology*, 74, 329–356.

WIRGIN, I., KOVACH, A. I., MACEDA, L., ROY, N. K., WALDMAN, J. R. and BERLINSKY, D. L. (2007) Stock identification of Atlantic cod in US waters using microsatellite and single nucleotide polymorphism DNA analyses. *Transactions of the American Fisheries Society*, 136, 375–391.

WONG, E. H. K. and HANNER, R. H. (2008) DNA barcoding detects market substitution in North American seafood. *Food Research International*, 41, 828–837.

YANCY, H. F., ZEMLAK, T. S., MASON, J. A., WASHINGTON, J. D., TENGE, B. J., NGUYEN, N. L. T., BARNETT, J. D., SAVARY, W. E., HILL, W. E., MOORE, M. M., FRY, F. S., RANDOLPH, S. C., ROGERS, P. L. and HEBERT, P. D. N. (2008) Potential use of DNA barcodes in regulatory science: Applications of the regulatory fish encyclopedia. *Journal of Food Protection*, 71, 210–217.

ZEL, J., CANKAR, K., RAVNIKAR, M., CAMLOH, M. and GRUDEN, K. (2006) Accreditation of GMO detection laboratories: Improving the reliability of GMO detection. *Accreditation and Quality Assurance*, 10, 531–536.

ZEL, J., MAZZARA, M., SAVINI, C., CORDEIL, S., CAMLOH, M., STEBIH, D., CANKAR, K., GRUDEN, K., MORISSET, D. and VAN DEN EEDE, G. (2008) Method validation and quality management in the flexible scope of accreditation: an example of laboratories testing for genetically modified organisms. *Food Analytical Methods*, 1, 61–72.

© Woodhead Publishing Limited, 2011

Part IV

Consumer views and future trends

© Woodhead Publishing Limited, 2011

16

Communicating food and food chain integrity to consumers: lessons from European research

W. Verbeke, Ghent University, Belgium

Abstract: Effective and efficient communication is crucial for food chains active in today's global food market. This chapter highlights the importance of communication and related challenges based on consumer data from several EU-funded research projects dealing with traditional foods, beef, pork and seafood. Apart from intrinsic food qualities relating to safety and healthiness, emerging ethical issues are also recognised. Consumers are open to receiving additional and easily accessible food quality reassurance. This can be done through labelling or on-label indications that are backed up and guaranteed by trustworthy traceability systems that underscore the supply chains' integrity.

Key words: attitude, communication, consumer, food, labelling, traceability.

16.1 Introduction

Food quality, safety and healthiness are a continuous concern for food producers and their end consumers. One should be careful about being too optimistic and remain vigilant with respect to the wholesomeness of our food and its provisioning, but largely speaking and notwithstanding a few exceptions, most European consumers have been saved from major pan-European food safety incidents and crises during the last decade. The BSE (bovine spongiform encephalopathy) crisis during the second half of the 1990s and the political, institutional and private managerial reactions to this crisis constituted a landmark for European food quality and safety policy. Traceability systems were

© Woodhead Publishing Limited, 2011

established and product labelling gained momentum, in terms of quality, origin and nutrition labelling.

Transparency and communication were key components in strategies to re-establish perceived integrity of food and food supply chain. Labelling, in particular, became a valuable potential alternative and complement to private banding as a product differentiation strategy. Indications on food labels may indeed represent some value to consumers because they may be perceived as signalling a particular quality level.

Labels inform consumers during their food choice decision-making process and can contribute to reducing uncertainty and aligning food choices with true preferences (Verbeke, 2005). Nevertheless, food labelling was sometimes criticised for potentially overloading consumers with information they may neither really need nor use. Such an information overload entails the risk of adding to the existing uncertainty, which may trigger indifference or disinterest among consumers. Therefore, finding the right balance in terms of information provisioning and its formatting has become a key challenge for food policy makers and stakeholders involved in European food chains.

Apart from intrinsic food quality and safety, ethical food qualities (e.g. animal welfare, environmental impacts, child labour, benefit-sharing with poor countries and fair trade) have also emerged during the last decade as decisive factors in food choice decisions for a growing number of consumers. Food chains are increasingly expected to deliver these qualities as well, and to organise their supplies, processes and communications accordingly. Safety and ethical attributes are examples of so-called credence attributes, i.e. attributes that consumers can neither experience upon usage or consumption of the food, nor can they effectively verify them upon purchase, unless these qualities are appropriately communicated.

A crucial role is therefore reserved for the organisations and sources involved in the certification and communication of these qualities. In most cases, the sources of communication are the food chains themselves including producers, processing industries and retailers, whose real and perceived integrity has become vital to maintaining credibility, building corporate image and strengthening competitive position.

The contribution of this chapter is threefold. First, it will present the case of traditional foods based on EU (European Union)-funded research performed as a part of the project Truefood ('Traditional United Europe Food'). This case will demonstrate that a good understanding of how consumers define and perceive a particular food category is crucial for developing effective communication to European food consumers. Second, based on data collected during the EU-funded project ProSafeBeef ('Advancing Beef Safety and Quality through Research and Innovation'), this chapter will illustrate how European consumers think and feel about beef safety and novel beef processing technologies aimed at improving safety. The insights obtained from this study are topical to illustrate European consumers' attitudes towards beef one decade after the BSE crisis, which triggered numerous initiatives to re-establish consumer trust in beef and

© Woodhead Publishing Limited, 2011

perceived integrity of the beef supply chain. Third, consumer interest in traceability and traceability information will be discussed using insights from the EU-funded project SEAFOODplus ('Health promoting safe seafood of high eating quality in a consumer-driven fork-to-farm concept'). Finally, the chapter will highlight emerging ethical issues that are expected to increasingly shape the integrity of food and food chains in the near future. For this purpose, insights from the EU-funded project Q-PorkChains ('Improving the quality of pork and pork quality for the consumer') and other studies will be presented.

16.2 Definition and perception of traditional foods

Food chain experts have defined traditional foods along four dimensions including locality, authenticity, availability over a long period of time, and linkage with gastronomic heritage (Guerrero *et al.*, 2009). This definition was adopted as the initial working definition within the Truefood project where it was further researched from the specific perspective of food consumers.

Following a set of focus group interviews with consumers in Belgium, France, Italy, Norway, Poland and Spain, a first and all-embracing consumer-based definition of traditional foods was proposed by Guerrero *et al.* (2009). The diversity of consumer reactions towards the different elements comprised in this definition suggested that traditional foods may be appealing to different consumers in different ways depending on their cultural background and situational influences. In order to validate the different elements of the consumer-based definition of traditional foods, and to verify whether a concise consensus definition is realistic, the second phase of the study consisted of a quantitative cross-sectional consumer survey. Representative samples of about 800 consumers in Belgium, France, Italy, Norway, Poland and Spain (total sample size 4828) completed a web-based survey in autumn 2007. As a result of this quantitative phase, the following definition of traditional foods was agreed:

> 'A traditional food product is a product frequently consumed or associated to specific celebrations and/or seasons, transmitted from one generation to another, made in a specific way according to the gastronomic heritage, naturally processed, and distinguished and known because of its sensory properties and associated to a certain local area, region or country' (Vanhonacker *et al.*, 2010a).

Some elements of this definition were widely supported by European consumers from each of the countries involved in the study. These elements included the association with a long existence or history, a high degree of consumer familiarity, the presence of specific sensory properties, seasonal availability and authenticity. The latter aspect relating to authenticity has a clear link with the concept of integrity in traditional food chains, since it was related in particular to an authentic origin, the use of authentic recipes, the combination of authentic raw materials and authentic ways of processing. Other elements,

© Woodhead Publishing Limited, 2011

like associations with special occasions, locality, and natural processing were generally weaker and not equally shared across cultures. Hence, depending on the target audience, such elements may require more or less emphasis in communications about traditional foods.

This consumer study also revealed that traditional foods have a very favourable image among European consumers, which is mainly stemming from the products' specific taste, high and consistent quality, healthiness, safety and nutritional value (Vanhonacker *et al.*, 2010a, 2010b). Interestingly, perceptions relating to the low availability of traditional foods (a flavour of exclusivity) and their time-consuming preparation also contributed positively to the image of these foods. Hence, from a consumer perspective, traditional foods are a broad and multifaceted concept that is difficult to cover with a concise and short definition. The insights from this study are quite challenging from a policy and communications perspective. They suggest namely that the integrity of traditional foods and traditional food supply chains may be judged by consumers against a wide and flexible range of criteria as specified in the definition, most of which have to be met consistently in order for the food and its supply chain to be viewed as wholesome and integer, respectively.

16.3 Attitudes to beef safety and processing

The integrity of food supply chains has been criticised during the second half of the 1990s following several consecutive problems with meat safety in particular, such as growth hormone abuse, the presence of residues, risks associated with BSE and Creutzfeldt-Jacob Disease, and dioxin contaminations in animal feed and meat supply chains. The concept of traceability in the food chain has been introduced as part of the EU General Food Law as a direct consequence of these problems, most specifically with the establishment of Commission Regulation 2000/13/EC that defined the concept and spelled out its general provisions.

Research concerning consumer beliefs and perceptions towards the safety of beef has been undertaken through qualitative exploratory focus group discussions that were performed in 2008 with beef consumers in France, Germany, Spain and the United Kingdom (Van Wezemael *et al.*, 2010; Verbeke *et al.*, 2010). This exploratory phase was followed by a quantitative experimental study beginning 2010. In these studies, European consumers defined beef safety mostly in relation to their personal health and well-being. Consumers claimed to personally face difficulties and uncertainty when evaluating beef safety at the point of purchase. Nevertheless, a relatively high degree of trust was reported towards beef 'that is available' in the market, as well as confidence towards independent institutions involved in quality control and certification.

Cues perceived to signal safe beef were local, regional or national origin, labels and brands. Furthermore, 'natural' beef was perceived as safe beef, though consumers could not clearly define their opinion on what 'natural' beef is. Mostly, naturalness was seen as not enforcing the animals during their

© Woodhead Publishing Limited, 2011

lifespan, natural feeding and little or no processing of the meat after slaughter. The responsibility for beef safety was laid mainly with actors at the beginning of the chain, such as livestock farmers and veterinarians, and to a lesser extent with actors involved in the processing of beef.

The quantitative research findings suggested further that interventions to improve the safety of beef are better accepted when applied in the early stages of the beef supply chain, i.e. during primary production and through the feeding of the cattle at the farm level, rather than in later stages such as on the slaughter line or during beef processing and packaging.

Processing of beef emerged as a particularly sensitive issue among consumers, who turned out to be very sceptical when they were exposed to information about beef processing technologies; even on simple marinating technologies which have shown to be quite successful in the market (de Barcellos *et al.*, 2010). Technologies that can make beef healthier (more nutritious) and that contribute to guaranteeing beef eating quality were better perceived than technologies to improve beef safety, which were largely rejected.

Criticism stemmed mainly from a desire to restrict processing as much as possible and from fear of excessive manipulations and interventions that infringe the integrity and naturalness of the food product in the shelf and on the plate. Especially invasive technologies (e.g. injecting marinade) and restructuring of meat (e.g. with enzymes) were rejected. The insights from these consumer studies indicate that the perceived integrity of the beef chain relates mainly to beliefs and perceptions of naturalness, a preference for safeguarding beef safety from the early stages of the production process rather than curing during later stages, the avoidance of invasive technologies and excessive manipulation, and minimal processing in the supply chain.

16.4 Interest in seafood traceability and labelling

Detailed rules as regards the labelling and consumer information for fishery and aquaculture products in the EU are specified in Council Regulation 104/2000 and Commission Regulation 2065/2001. These regulations include the mandatory provisioning of information relating to the farmed or wild production method of seafood. During a consumer study of 2004 (Olsen *et al.*, 2007; Pieniak *et al.*, 2007), which included a sample of 4786 consumers from Belgium, Denmark, the Netherlands, Poland and Spain, European fish consumers were asked about their interest in different information cues on seafood labels. These consumers reported the strongest interest in a safety guarantee and a quality mark for fish, with almost 60% of the sample scoring five or more on a seven-point scale.

Mandatory indications such as country of origin (catch or harvest), as well as indications about the farmed or wild origin of the fish scored generally much lower in terms of consumer interest. Also consumers' direct interest in traceability information *per se* was very low. The study also included a consumer

© Woodhead Publishing Limited, 2011

segmentation based on claimed use of and trust in information sources related to seafood (Pieniak *et al.*, 2007). The segmentation yielded three distinct consumer segments, which were referred to as 'Enthusiasts' (consumers who reported a high degree of trust and a high claimed use of information sources), 'Confidents' (consumers who reported a high degree of trust but a low claimed use of information sources) and 'Sceptics' (consumers who reported both a low trust and low use of seafood information sources).

Confident consumers reported a moderate to low score on direct interest in traceability, most likely because they do not perceive a strong need for traceability information owing to their already confident attitude. In contrast, Enthusiasts scored highest on direct interest in traceability, whereas Sceptics scored the lowest. These findings suggest that consumers' needs and expectations with respect to seafood information in general, and traceability information in particular, are not uniform. Likewise, different consumer segments may expect different communication about the integrity of (sea)food chains. Last but not least, this study also revealed that no segment could be identified with a low level of trust and a high claimed usage of information sources. Clearly, some baseline trust is required before consumers even think about examining and using information sources. This again stresses the relevance of real and perceived integrity of food chains.

Interestingly, similar findings have been reported from a national study in Belgium dealing with the case of fresh beef (Verbeke and Ward, 2006). Information cues that relate directly to product quality and with which consumers are familiar (e.g. expiry date and type of product) were much more attended to than traceability information cues on beef labels. Also beef labels that guarantee quality and safety scored relatively well in terms of attracting consumers' attention.

The conclusions from both the seafood and beef traceability studies with consumers are twofold. First, consumers will use those information cues they are most familiar with. Second, consumer interest is much stronger in direct indications of safety and quality instead of traceability information, alone. However, meeting quality and safety interests requires an effective traceability system as the backbone for providing quality and safety guarantees. Hence, although traceability information may have little apparent value for consumers directly and therefore might have not to be communicated in great detail to end-users, its primary role is clearly within the chain where it functions as a vital tool for safeguarding food chain integrity and allowing related communication with customers.

16.5 Emergence of ethical issues

Real and perceived integrity of food chains is no longer determined only by the intrinsic quality, including safety and healthiness of the foods it procures. More intangible qualities that relate in many cases to ethical issues such as

© Woodhead Publishing Limited, 2011

sustainability, environmental impact, carbon footprint, and animal welfare are emerging. Sustainability pertains to the chain's potential for balancing and achieving social, economic and ecological goals, sometimes expressed as the 3 Ps of People, Profit and Planet. Carbon footprints, energy-efficiency and food miles (e.g. local versus airplane cargo-shipped food), fair trade (e.g. coffee) and animal friendliness (e.g. free-range eggs and poultry) have become prominent in debates about the integrity of food and food chains and have even managed to make their way to product labels.

Many studies show that consumers increasingly value ethical aspects in food products; they are increasingly open to receiving information about the impact of production processes on agro-ecological and environmental conditions. Nevertheless, market shares of food products with ethical qualities remain relatively small in most cases, but these shares are growing. Potential explanations for the differences between individuals' pro-ethics attitudes as citizens and their marketplace behaviour as consumers have been found in many different factors such as low perceived consumer effectiveness (i.e. low beliefs that one's behaviour can effectively contribute to solving ethical problems), low perceived availability of products with ethical quality, social norms or influences from the social environment and peers, and obviously also price differences and related price perceptions and limited willingness-to-pay (Vermeir and Verbeke, 2006, 2008).

The citizen-consumer duality was investigated as part of a 2008 pork consumer survey including 1931 citizens from Belgium, Denmark, Germany and Poland (Krystallis *et al.*, 2009; Verbeke *et al.*, 2010). Apart from a majority segment with weak attitudes towards pig production systems, three small but distinct segments were identified: one with stronger preferences for small-scale pig farming, one with stronger attitudes relating to environmental issues associated with pig production, and one with stronger attitudes towards animal welfare in pig production (Krystallis *et al.*, 2009). Interestingly, only weak relationships could be identified between the segments based on citizen attitudes towards pig farming and segments based on consumer behaviour, namely pork choices and pork consumption frequency. The identified citizen attitudes towards pig production (more so than their interest in specific pork attributes) can be expected to shape individuals' opinions and expectations with respect to the integrity of pork supply chains on these emerging dimensions that clearly extend beyond intrinsic product quality and safety.

16.6 Conclusions

Communication is crucial throughout food chains in order for food producers and related stakeholders in food supply chains to reassure citizens and consumers of the wholesomeness of their foods, to gain public support for their activities, and to maintain a favourable competitive position vis-à-vis other global suppliers. Communicating effectively and efficiently requires a good

© Woodhead Publishing Limited, 2011

understanding of how consumers define and perceive the concerned foods, and how they feel and what they believe relative to the production processes involved. People are not all alike. Therefore, the existence of multiple citizen and consumer segments with different interests, as well as the citizen-consumer duality and the factors that determine the gap between attitudes and behaviour have to be taken into account. In general, European consumers' perceptions of food safety were found to be quite positive, and a relatively high degree of trust and confidence in the food that is available on the European market has been reported. Nevertheless, consumers are open to receiving more easily accessible information and additional reassurance with respect to the intrinsic as well extrinsic quality of the food they consume. Easily recognisable labels signalling desired qualities backed up by trustworthy traceability systems are expected to continue playing an important role in future communications about food and food chain integrity.

16.7 Acknowledgements

Insights and conclusions reported in this chapter are mainly based on research performed with financing by the European Commission (EC) under its sixth Framework Programme (FP6). Parts of the work presented were performed within the integrated research projects Truefood (FOOD-CT-2006-016264), ProSafeBeef (FOOD-CT-2006-36241), SEAFOODplus (FOOD-CT-2004-506359) and Q-PorkChains (FOOD-CT-2007-036245). Financing by the European Commission is gratefully acknowledged. The information in this document reflects the author's view and the EC is not liable for any use that may be made of the information contained herein.

16.8 References

DE BARCELLOS MD, KÜGLER JO, GRUNERT KG, VAN WEZEMAEL L, PÉREZ-CUETO FJA, UELAND O and VERBEKE W (2010) 'European consumers' acceptance of beef processing technologies: A focus group study', *Innovative Food Science and Emerging Technologies*, 11, 721–732.

GUERRERO L, GUÀRDIA MD, XICOLA J, VERBEKE W, VANHONACKER F, ZAKOWSKA S, SAJDAKOWSKA M, SULMONT-ROSSÉ C, ISSANCHOU S, CONTEL M, SCALVEDI LM, GRANLI BS and HERSLETH M (2009) 'Consumer-driven definition of traditional food products and innovation in traditional foods. A qualitative cross-cultural study', *Appetite*, 52, 345–354.

KRYSTALLIS A, DE BARCELLOS MD, KÜGLER J, VERBEKE W and GRUNERT KG (2009) 'Investigating European citizen attitudes towards pig production systems', *Livestock Science*, 126, 46–56.

OLSEN SO, SCHOLDERER J, BRUNSØ K and VERBEKE W (2007) 'Exploring the relationship between convenience and fish consumption: a cross-cultural study', *Appetite*, 49, 84–91.

© Woodhead Publishing Limited, 2011

PIENIAK Z, VERBEKE W, SCHOLDERER J, BRUNSØ K and OLSEN SO (2007) 'European consumers' use of and trust in information sources about fish', *Food Quality and Preference*, 18, 1050–1063.

VAN WEZEMAEL L, VERBEKE W, KÜGLER J, DE BARCELLOS MD and GRUNERT KG (2010) 'European consumers and beef safety: perceptions, expectations and uncertainty reduction strategies', *Food Control,* 21, 835–844.

VANHONACKER F, VERBEKE W, GUERRERO L, CLARET A, CONTEL M, SCALVEDI L, ZAKOWSKA-BIEMANS S, GUTKOWSKA K, SULMOT-ROSSÉ C, RAUDE J, GRANLI BS and HERSLETH M (2010a) 'How European consumers define the concept of traditional food: Evidence from a survey in six countries', *Agribusiness*, 26, 453–476.

VANHONACKER F, LENGARD V, HERSLETH M and VERBEKE W (2010b) 'Profile and projected image of traditional food consumers in Europe', *British Food Journal*, 112, 871–886.

VERBEKE W (2005) 'Agriculture and the food industry in the information age', *European Review of Agricultural Economics*, 32, 347–368.

VERBEKE W and WARD RW (2006) 'Consumer interest in information cues denoting quality, traceability and origin: an application of ordered probit models to beef labels', *Food Quality and Preference*, 17, 453–467.

VERBEKE W, PÉREZ-CUETO FJA, DE BARCELLOS MD, KRYSTALLIS A and GRUNERT KG (2010) 'European citizen and consumer attitudes and preferences regarding beef and pork', *Meat Science,* 84, 284–292.

VERMEIR I and VERBEKE W (2006) 'Sustainable food consumption: exploring the "attitude behavioural intention" gap', *Journal of Agricultural and Environmental Ethics*, 19, 169–194.

VERMEIR I and VERBEKE W (2008) 'Sustainable food consumption among young adults in Belgium: Theory of planned behaviour and the role of confidence and values', *Ecological Economics*, 64, 542–553.

© Woodhead Publishing Limited, 2011

17

The role of traceability in restoring consumer trust in food chains

M. Garcia Martinez and F. M. Brofman Epelbaum, University of Kent, UK

Abstract: This chapter presents an overview of the recent literature on consumer interest in food labels and the role of traceability information in restoring consumer trust in food chains. Consumers consider food traceability to be an important tool for improving food safety. Although consumers show little understanding of the mechanism underpinning food traceability, they appreciate its potential to improve food safety as well as to promote consumer protection. It is important to develop effective communication strategies to inform consumers about the benefits traceability systems can provide in addressing their concerns if consumer trust in food safety monitoring systems is to be developed and maintained.

Key words: traceability, trust, consumer confidence.

17.1 Food safety regulation, trust and traceability

The number of food scares and controversies in recent years has left consumers confused about exactly what is and is not safe and uncertain as to whether or not the government is acting in their best interests (Davies, 2001). This has resulted in increasing public distrust and reduced consumer confidence in the safety of food, the food industry and the government's ability to adequately regulate, manage and communicate food risks (Cantley, 2004; Caduff and Bernauer, 2006; Halkier and Holm, 2006). Consequently, government oversight of food safety has undergone profound reform in many countries with the establishment of independent food safety agencies. These reforms have

© Woodhead Publishing Limited, 2011

responded to increasing demands for a more transparent and inclusive process of food safety governance (Vos, 2000; Flynn *et al.*, 2004; Ansell and Vogel, 2006).

The outbreak of the *bovine spongiform encephalopathy* (BSE) crisis in 1996 in the United Kingdom (UK) and later in the European Union (EU) clearly revealed significant dysfunctions both in industry practices and their supervision by governments, undermining general public trust in food safety governance (Borraz, 2007; Vos, 2000). The institutional failures put pressure on the European Commission to expand its scope of influence in regulating food safety across the EU (Caduff and Bernauer, 2006). It also forced the Commission to present a more coherent approach to food safety based on principles of separation of risk assessment and risk management; i.e. the separation of the responsibility for legislation, scientific advice and inspection, which leads to greater transparency and information throughout the decision-making process and inspection (Vos, 2000).

The private sector, for its part, has responded to consumer demand for effective food safety controls with the development of a technologically sophisticated and organisationally complex range of approaches to improving product safety and quality standards (Garcia Martinez *et al.*, 2007). These are becoming a prominent driving force for agri-food systems in many countries (Henson and Hooker, 2001; Henson and Reardon, 2005; Fulponi, 2006; Havinga, 2006). As supply chains become more global, so do private standards. For global retailers with a multitude of suppliers scattered globally, private schemes like GlobalGAP[1] have become a key means to manage procurement costs and to ensure the integrity of their supply chains.

17.2 Initiatives to restore consumer confidence

Producers and regulatory institutions in the EU have attempted to restore consumer confidence in policy making and industry practices by introducing food and food ingredient traceability systems. Under the framework of the General Food Law, Regulation (EC) 178/2002 has set new standards to ensure stricter implementation of traceability in food and feed chains by enforcing mandatory traceability of food and food ingredients as of 1 January 2005. The regulatory requirements necessitate all food and feed business operators to be able to identify the source of all foods and ingredients and to provide the basis for further monitoring throughout supply chains. Traceability systems and subsequent quality and origin labelling schemes are expected to increase transparency throughout the food chain and to result in the development and maintenance of consumer trust in food and food producers (Van Rijswijk *et al.*,

1. GlobalGAP is a private-sector body that sets voluntary standards for the certification of production processes of agricultural (including aquaculture) products around the globe.

© Woodhead Publishing Limited, 2011

2008). Quality signalling can transform credence attributes[2] into search attributes and strengthen consumer trust, allowing the reduction of consumer perceived risk towards food quality and safety, and of information asymmetry between consumers and producers (Mojduszka and Caswell, 2000). Research shows that labelling of meat safety and quality attributes can positively influence consumer perceptions of meat products (Bredahl, 2004; Loureiro and Umberger, 2007).

Consumers across Europe, however, have divergent associations, perceptions and expectations regarding traceability (Giraud and Halawany, 2006; van Rijswijk and Frewer, 2008; Van Rijswijk *et al.*, 2008). Cultural values influence consumer food decision making (Briley *et al.*, 2000; Hoogland *et al.*, 2005), so expectations and attitudes towards food safety issues and risk management may vary depending on consumers' cultural backgrounds (Houghton *et al.*, 2008; Van Kleef *et al.*, 2006). Giraud and Halawany (2006) show that consumers do not fully understand traceability systems; however, they are able to indicate what benefits are expected from these systems. Product origin is the first thing consumers have in mind when asked to define traceability. Consumer research has also linked consumer perceptions of traceability to safety issues like the BSE crises (Latouche *et al.*, 1998) and dioxin contamination of the food chain (Verbeke, 2001) as well as the acceptance of GMO foods and ingredients (Miles *et al.*, 2005). van Rijswijk *et al.* (2008) study on consumer perceptions of traceability in Europe shows similarities across countries regarding the benefits consumers associate with traceability (i.e., health, quality, safety and control). Cross-country differences were also identified. German consumers are more sensitive about the processing conditions, while French consumers care more about origin and give a higher importance to past product experience such as taste. Spanish consumers perceive a high quality product as a sign of trust and Italian consumers are more sensitive to safety conditions. In other words, consumers' background would highly affect their attitude towards traceability-related attributes.

In the United States (US), consumer interest in food traceability has increased after recent food safety incidents such as the discovery of BSE in 2003 and the case of *E.coli* contaminated spinach in 2006 (by the time the outbreak was over, 204 people had become ill across 26 States and Canada, 104 had been hospitalised, 31 had developed the serious complication of hemolytic-uremic syndrome (HUS), and 3 had died). The California leafy-green industry required traceability in its marketing agreement less than a year after the discovery of *E. coli* in 2006 (Calvin, 2007). In March 2009, the President's Food Safety Working Group recommended a new national traceback and response system to deliver food safety alerts to consumers.[3] However, progress in the

2. Credence characteristics of a product (i.e., food safety, animal welfare, organic production) cannot be observed or inferred by direct inspection, on consumption or even after consumption, whereas search characteristics (i.e., appearance or size) can be known to consumers prior to purchase (Darby and Karni, 1973).
3. http://www.foodsafetyworkinggroup.gov/ContentKeyFindings/HomeKeyFindings.htm

© Woodhead Publishing Limited, 2011

implementation of traceability systems has been slow; for instance the adoption of the National Animal Identification System (NAIS) has encountered numerous obstacles since it was first proposed and currently the system is ineffective as a result of low participation (Anderson, 2010). The lack of interest by some industries and government may stem from the lack of knowledge about the effects of traceability on consumers' welfare and on firms' profits (Pouliot, 2010).

Traceability systems have evolved differently across sectors and countries depending upon direct market pressures. In Australia, a previously voluntary system for national livestock identification (NILS) became mandatory in 2005 in an attempt to maintain competitive advantage in export markets (Synergies, 2004). In Canada, in 2002 after the verification of some BSE cases, the Canadian Cattle Identification Agency introduced a mandatory national cattle identification system for a short time, only ten months, aimed at facilitating the retracting of meat products in the event of food safety problems (Carlberg, 2010). Likewise, in major exporting countries like Argentina and Uruguay, traceable systems have become mandatory in order to comply with export international rules (Mathews and Vandeveer, 2007).

Most traceability systems involve a combination of mandatory and voluntary standards or quality assurances. In most cases, the mandatory government-driven programme covers the traceability of animals from origin to slaughterhouses. Further tracing from processor to consumer is usually governed by industry groups, and is usually independently audited and associated with production standards of interest to consumers, such as animal welfare, animal feed and environmental management (Trautman *et al.*, 2008).

17.3 Translation of techniques into labels

Consumer interest in traceability information cannot be taken for granted (Hobbs *et al.*, 2005; Verbeke and Ward, 2006). Consumers have little technical understanding of what traceability is (Giraud and Amblard, 2003) and are not particularly interested in cues directly related to traceability and product identification, despite uncertainty following a number of meat safety crises (Giraud and Halawany, 2006; Verbeke and Ward, 2006; Verbeke *et al.*, 2002). Direct indications of traceability (e.g. bar codes and licence numbers) and the provision of technical information associated with it are unlikely to increase consumer confidence given the low quality or safety inference potential of such information cues (Verbeke *et al.*, 2007). Conversely, quality labels and safety guarantees are much more valued by consumers. Hence, the presentation of simpler information on traceability systems accompanied by labels (quality and safety related) would increase the probability of traceability being valued by consumers (Verbeke *et al.*, 2007; Van Rijswijk *et al.*, 2008). Indeed traceability could be used to back up product claims such as origin and quality labelling (Verbeke and Ward, 2006) for which some consumers show higher willingness

© Woodhead Publishing Limited, 2011

to pay (Skuras and Vakrou, 2000; Krystallis and Chryssochoidis, 2005; Batte *et al.*, 2007). Traceability provides the history of a product and is a precaution against fraud (i.e., labelling conventional products as organic) and a quality signal to consumers (Smyth and Phillips, 2002).

Evidence supporting a price premium and the intrinsic added value of traceability for consumers is uncertain (Trautman *et al.*, 2008). Consumers tend to associate traceability with higher product prices (Giraud and Halawany, 2006) and often state that they would be willing to pay more for safer food; however, actual buying decisions show that it is economic convenience that most affects purchasing decisions rather than safety or traceability information on the label of a product (Meuwissen *et al.*, 2003). Loureiro and Umberger (2007) calculated that US consumers were willing to pay a price premium of up to 38% for 'US certified steak' and even up to 58% for 'US certified hamburger'. Hobbs *et al.* (2005) suggest that Canadian consumers consider traceability to be an important system to guarantee food safety, especially if associated with other quality assurances, but results show that traceability does not reduce the information asymmetry between producer and consumer with respect to quality attributes. Dickinson and Bailey (2005) found that consumers in the US, Canada, Britain and Japan were ready to pay a nontrivial price for red meat traceability. The willingness to pay (WTP) rises further for traceability provided characteristics (e.g. additional meat safety and humane animal treatment guarantees). Conversely, despite the importance of food safety to Spanish consumers, the majority (73%) are not willing to pay a price premium for traceable beef (Angulo and Gil, 2007). Traceability alone plays a very small role in Spanish consumer choices compared to products labelled with a Protected Designation of Origin (PDO) label linked to a particular region with a reputation for food safety or food quality.

17.4 Communicating traceability information to consumers

The potential scope of economic benefits of traceability systems would not be realised unless consumers become aware of and are influenced by a traceability system. However, communication efforts aimed at increasing consumer awareness and meaning of beef traceability have failed to evoke active information search by consumers (Verbeke *et al.*, 2002) reinforcing already established evidence that consumers are not really interested in traceability cues on beef labels (Verbeke and Ward, 2006).

Traceability is needed as the regulatory and logistic backbone for providing guarantees related to origin and quality and its role in restoring consumer confidence in food chains would be through the general benefits consumers perceive associated with food purchases in general, such as origin and quality labels (Van Rijswijk *et al.*, 2008). Hence, communicating the importance of traceability to consumers should be in terms of its benefits in relation to health, safety, quality control, origin, organic production, animal welfare, carbon

© Woodhead Publishing Limited, 2011

footprint, child labour, benefits for local businesses and fair trade; in other words issues relating to food chain integrity about which consumers are concerned or interested in (Gregory, 2000). Research highlights the importance of cultural context regarding the impact of potential risk communication strategies (van Dijk *et al.*, 2008). Appropriate market segmentation and targeting are key success factors for food labelling and information provision programmes (Verbeke, 2005). A generic approach through the provision of vast amounts of information to the general public stands the risk of information overload leading to confusion and lack of interest among the majority of consumers (Petty and Cacioppo, 1996). Hence, specific cultural characteristics may determine the need to adapt risk communication strategies to cross-country variations. This finding is particularly relevant within the EU context where the establishment of the European Food Safety Authority (EFSA) would lead to a standardised pan-European approach to risk communication. van Dijk *et al.* (2008) cross-country study shows that while communication of uncertainty had a positive impact in Germany, the same information had a negative impact in the UK and Norway. UK consumers have been found to be more sceptical about the efficiency of risk assessment practices compared to Germany and Greece (Van Kleef *et al.*, 2009). Cultural differences were also found in the perceived quality of food risk management associated with different hazards (van Dijk *et al.*, 2008). The evaluation of risk management of mycotoxins was rated the highest in all countries, while the risk management of pesticide residues and GM potatoes differed among countries.

17.5 Conclusions

Consumers consider food traceability to be an important tool for improving food safety. Although consumers show little understanding of the mechanism underpinning food traceability, they appreciate its potential to improve food safety as well as to promote consumer protection. It is important therefore to develop effective communication strategies to inform consumers about the benefits traceability systems can provide in addressing their concerns if consumer trust in food safety monitoring systems is to be developed and maintained (Kher *et al.*, 2010).

17.6 References

ANDERSON, D. P. (2010) The US animal identification experience. *Journal of Agricultural and Applied Economics*, 42, 543–550.

ANGULO, A. M. and GIL, J. M. (2007) Risk perception and consumer willingness to pay for certified beef in Spain. *Food Quality and Preference*, 18, 1106–1117.

ANSELL, C. and VOGEL, D. (2006) The Contested Governance of European Food Safety Regulation. In: Ansell, C. and Vogel, D. (eds) *What's the Beef: The Contested*

© Woodhead Publishing Limited, 2011

Governance of European Food Safety Regulation. Cambridge, MA: MIT Press.

BATTE, M. T., HOOKER, N., HAAB, T. C. and BEAVERSON, J. (2007) Putting their money where their mouths are: consumer willigness to pay for multi-ingredient, process organic food products. *Food Policy*, 32, 145–159.

BORRAZ, O. (2007) Governing standards: the rise of standardization processes in France and in the EU. *Governance: An International Journal of Policy, Administration, and Institutions*, 20, 57–84.

BREDAHL, L. (2004) Cue utilisation and quality perception with regard to branded beef. *Food Quality and Preference*, 15, 65–75.

BRILEY, D. A., MORRIS, M. W. and SIMONSON, I. (2000) Reasons as carriers of culture: Dynamic versus dispositional models of cultural influence on decision making. *Journal of Consumer Research*, 27, 157–178.

CADUFF, L. and BERNAUER, T. (2006) Managing risk and regulation in European Food Safety Governance. *Review of Policy Research*, 23, 153–168.

CALVIN, L. (2007) Outbreak linked to spinach forces reassessment of food safety practices. *Amber Waves*. Economic Research Service/USDA.

CANTLEY, M. (2004) How should public policy respond to the challenges of modern biotechnology? *Current Opinion in Biotechnology*, 15, 258–263.

CARLBERG, J. G. (2010) Development and Implementation of a Mandatory Animal Identification System: The Canadian Experience. *Journal of Agricultural and Applied Economics*, 42, 559–570.

DARBY, M. and KARNI, E. (1973) Free competition and the optimal amount of fraud. *Journal of Law and Economics*, 16, 67–88.

DAVIES, S. (2001) Food choice in Europe – the consumer perspective. In: Frewer, L. J., Risvik, E. and Schifferstein, H. (eds) *Food, People & Society: A European Perspective of consumers food choice*. Berlin: Springer.

DICKINSON, D. L. and BAILEY, D. (2005) Experimental evidence on willingness to pay for red meat traceability in the United States, Canada, the United Kingdom and Japan. *Journal of Agricultural and Applied Economics*, 37, 537–548.

FLYNN, A., CARSON, L., LEE, R., MARSDEN, T. and THANKAPPAN, S. (2004) The Food Standards Agency: Making a Difference? Cardiff: The Centre for Business Relationships, Accountability, Sustainability and Society (BRASS), Cardiff University.

FULPONI, L. (2006) Private voluntary standards in the food system: the perspective of major food retailers in OECD countries. *Food Policy*, 31, 1–13.

GARCIA MARTINEZ, M., FEARNE, A., CASWELL, J. A. and HENSON, S. (2007) Co-regulation as a possible model for food safety governance: opportunities for public-private partnerships. *Food Policy*, 32, 299–314.

GIRAUD, G. and AMBLARD, C. (2003) What does traceability mean for beef meat consumers? *Food Science*, 23, 40–64.

GIRAUD, G. and HALAWANY, R. (2006) Consumers' perception of food traceability in Europe. In: Paper presented at the 98th EAAE seminar: Marketing Dynamics within the Global Trading System: New Perspectives, 29 June–2 July 2006, Chania, Crete, Greece.

GREGORY, N. G. (2000) Consumer concerns about food. *Outlook on Agriculture*, 29, 251–257.

HALKIER, B. and HOLM, L. (2006) Shifting responsibilities for food safety in Europe: an introduction. *Appetite*, 47, 127–133.

HAVINGA, T. (2006) Private regulations of food safety by supermarkets. *Law and Policy*, 28, 515–533.

© Woodhead Publishing Limited, 2011

HENSON, S. and HOOKER, N. H. (2001) Private sector management of food safety: public regulation and the role of private controls. *International Food and Agribusiness Management Review*, 4, 7–17.

HENSON, S. and REARDON, T. (2005) Private agri-food standards: Implications for food policy and the agri-food system. *Food Policy*, 30, 241–253.

HOBBS, J. E., BAILEY, D. V., DICKINSON, D. L. and HAGHIRI, M. (2005) Traceability in the Canadian red meat sector: do consumers care? *Canadian Journal of Agricultural Economics*, 53, 47–65.

HOOGLAND, C. T., DE BOER, J. and BOERSEMA, J. J. (2005) Transparency of the meat chain in light of food culture and history. *Appetite*, 45, 15–23.

HOUGHTON, J. R., ROWE, G., FREWER, L. J., VAN KLEEF, E., CHRYSSOCHOIDIS, G., KEHAGIA, O., KORZEN-BOHR, S., LASSEN, J., PFENNING, U. and STRADA, A. (2008) The quality of food risk management in Europe: Perspectives and priorities. *Food Policy*, 33, 13–26.

KHER, S. V., FREWER, L. J., DE JONGE, J., WENTHOLT, M. and HOWELL DAVIES, O. (2010) Experts' perspectives on the implementation of traceability in Europe. *British Food Journal*, 112, 261–274.

KRYSTALLIS, A. and CHRYSSOCHOIDIS, G. (2005) Consumers' willingness to pay for organic food: factors that affect it and variation per organic product type. *British Food Journal*, 107, 320–343.

LATOUCHE, K., RAINELLI, P. and VERMERSCH, D. (1998) Food Safety issues and the BSE scare: some lessons from the French case. *Food Policy*, 23, 347–356.

LOUREIRO, M. L. and UMBERGER, W. J. (2007) A choice experiment model for beef: what US consumer responses tell us about relative preferences for food safety, country-of-origin labeling and traceability. *Food Policy*, 32, 496–514.

MATHEWS, K. H. and VANDEVEER, M. (2007) *Beef Production, Markets, and Trade in Argentina and Uruguay: An Overview*. Economic Research Service, US Department of Agriculture.

MEUWISSEN, P. M. M., VELTHUIS, A. G. J., HOGEVEEN, H. and HUIRNE, R. B. M. (2003) Traceability and certification in meat supply chains. *Journal of Agribusiness*, 21, 167–181.

MILES, S., UELAND, O. and FREWER, L. J. (2005) Public attitudes towards genetically-modified food. *British Food Journal*, 107, 246–262.

MOJDUSZKA, E. M. and CASWELL, J. A. (2000) A test of nutritional quality signaling in food markets prior to implementation of mandatory labelling. *American Journal of Agricultural Economics*, 82, 298–309.

PETTY, R. E. and CACIOPPO, J. T. (1996) *Communication and persuasion: central and peripheral routes to attitude change*, New York: Springer.

POULIOT, S. (2010) Welfare effects of mandatory traceability when firms are heterogeneous. In: Paper presented at the Agricultural and Applied Economics Association 2010 AAEA, CAES & WAEA Annual Meeting, 25–27 July 2010, Dever, Colorado.

SKURAS, D. and VAKROU, A. (2000) Consumer willingness to pay for origin labelled wine: a Greek case study. *British Food Journal*, 104, 898–912.

SMYTH, S. and PHILLIPS, P. W. B. (2002) Product differentiation alternatives: identify, preservation, segregation and traceability. *AgBioForum*, 5, 30–42.

SYNERGIES (2004) The implications for the Queensland beef industry from NLIS implementation. A report to the Department of Primary Industries & Fisheries.

TRAUTMAN, D., GODDARD, E. and NILSSON, T. (2008) *Traceability – a literature review*. Edmonton, Canada: Department of Rural Economy, Faculty of Agricultural, Life and Environmental Sciences, University of Alberta.

VAN DIJK, H., HOUGHTON, J. R., VAN KLEEF, E., VAN DER LANS, I., ROWE, G. and FREWER, L. J.

© Woodhead Publishing Limited, 2011

(2008) Consumer response to communication about food risk management. *Appetite*, 50, 340–352.

VAN KLEEF, E., FREWER, L. J., CHRYSSOCHOIDIS, G. M., HOUGHTON, J. R., KORZEN-BOHR, S., KRYSTALLIS, T., LASSEN, J., PFENNING, U. and ROWE, G. (2006) Perceptions of food risk management among key stakeholders: results from a cross-European study. *Appetite*, 47, 46–63.

VAN KLEEF, E., UELAND, Ø., THEODORIDIS, G., ROWE, G., PFENNING, U., HOUGHTON, J., VAN DIJK, H., CHRYSSOCHOIDIS, G. and FREWER, L. J. (2009) Food risk management quality: consumer evaluations of past and emerging food safety incidents. *Health, Risk and Society*, 11, 127–163.

VAN RIJSWIJK, W. and FREWER, L. J. (2008) Consumer perceptions of food quality and safety and their relation to traceability. *British Food Journal*, 110, 1034–1046.

VAN RIJSWIJK, W., FREWER, L. J., MENOZZI, D. and FAIOLI, G. (2008) Consumer Perceptions of Traceability: A Cross-National Comparison of the Associated Benefits. *Food Quality and Preference*, 19, 452–464.

VERBEKE, W. (2001) Beliefs, attitudes and behaviour towards fresh meat revisited after the Belgian dioxin crisis. *Food Quality and Preference*, 12, 489–498.

VERBEKE, W. (2005) Agriculture and the Food Industry in the Information Age. *European Review of Agricultural Economics*, 32, 347–368.

VERBEKE, W. and WARD, R. W. (2006) Consumer interest in information cues denoting quality, traceability and origin: an application of ordered probit models to beef labels. *Food Quality and Preference*, 17, 453–467.

VERBEKE, W., WARD, R. W. and AVERMAETE, T. (2002) Evolution of publicity measures relating to the EU beef labeling system in Belgium. *Food Policy*, 27, 339–353.

VERBEKE, W., FREWER, L. J., SCHOLDERER, J. and DE BRABANDER, H. F. (2007) Why Consumers Behave as They Do With Respect to Food Safety and Risk Information. *Analytica Chimica Acta*, 586, 2–7.

VOS, E. (2000) EU Food safety regulation in the aftermath of the BSE crisis. *Journal of Consumer Policy*, 23.

© Woodhead Publishing Limited, 2011

18

Future trends in food chain integrity

J. Hoorfar, Technical University of Denmark, Denmark, R. Prugger, Tecnoalimenti S.C.p.A, Italy, F. Butler, University College Dublin, Ireland and K. Jordan, Teagasc, Ireland

Abstract: Food chain integrity is multi-disciplinary, covering all aspects of the food chain from producers to consumers. Our present understanding of food chain integrity is based on microbial and chemical food safety, authenticity of origin, fraud and quality. This will inevitably be expanded, as future consumers will demand food that is not only safe, healthy and tasty, but also, is sustainable in nature, minimizes the carbon footprint, is considerate of animal welfare, free of child labour and favours trade with developing countries. More research is needed to convert these subjective perceptions into quantitative and objective criteria in order to produce foods of high integrity.

Key words: pathogens, chemicals, contaminants, tracing, authenticity, risk assessment, food chain, tracking, identification, e-platforms, small and medium enterprises, food industry, agriculture, distribution.

18.1 Globalization impacts on the food supply chain

Food has never been as safe as it is today, yet it has also never been at greater risk. Not only can the rapid transport of ingredients and products across the globe spread hazards quickly, contaminated food can affect many more people because of large production units. In addition, the globalization of news (TV, internet, etc.) can instantly spread the 'bad' press, which can have detrimental economic consequences, even for large food producers.

The globalization of food supply and increasingly long and complex supply chains has been a key driver in the establishment of the Global Food Safety

© Woodhead Publishing Limited, 2011

Initiative (GFSI) by global manufacturers. The GFSI has set requirements for food safety schemes through a benchmarking process in order to improve cost efficiency throughout the food supply chain. Its origin was in 2000, when The Consumer Goods Forum was founded in Europe by major global retailers and food manufacturers (e.g. Wal-Mart, Carrefour, Tesco, Metro Group, Kraft, etc.) in order to seek a solution that would reduce certification redundancy. The forum has harmonized standards to ensure competing programmes deliver the same basic requirements.

Under the auspices of the GFSI, several interchangeable schemes exist, such as the Food Safety Management Systems (ISO 22000), Synergy 22000, and the International Food Standard. Each of the schemes promotes documentation that guides manufacturers to implement and manage an effective food quality and safety program. These systems support upstream and downstream supply chains that often span the globe.

18.2 Broader understanding of food integrity

The term integrity (besides its meaning in ethics) typically refers to being whole, entire, or undiminished. A definition of integrity for the food supply chain might be the requirement that the system performs its intended function in an unimpaired manner free from deliberate or inadvertent manipulation (Anon. 1995). This analogy is borrowed from computer networks such as the World Wide Web.

The wide range and scope of the chapters presented in this book show that 'food chain integrity' is multi-disciplinary, covering all aspects of the food chain from producers to consumers. The current definition includes microbial and chemical food safety, authenticity of origin, fraud and quality. However, the definition will certainly be expanded, as future consumers will demand food that is not only safe, healthy and tasty, but also, amongst others, produced in a way that minimizes the impact on the environment, is sustainable in nature, is certified in origin and variety, minimizes the carbon footprint, is considerate of animal welfare, free of child labour, favours trade with developing countries, and so on. This trend is already visible in the industrialized world.

Food producers thus face the tough challenge of living up to these demands, and at the same time producing food that is cheap, safe, good quality and supplies the fast-growing population on the earth. In this regard, the technological solutions described in the book chapters, most of it financed by the European Union, can provide valuable tools for ensuring food chain integrity.

18.3 Closing the EU gaps in traceability

In its effort to assure food chain integrity, the European Union has taken steps to increase traceability by funding a number of research projects. Despite this, many knowledge gaps remain. In the General Food Law (EC Regulation no.

© Woodhead Publishing Limited, 2011

178/2002), product tracing is defined as 'The ability to trace and follow a food, feed, food-producing animal or substance intended to be, or expected to be incorporated into a food or feed, through all stages of production, processing and distribution' meaning 'any stage, including import'.

The legal framework for product tracing in the EU requires food and feed businesses, as of 1 January 2005, to identify one step backwards from whom and one step forwards to whom a food or food-producing animal or any substance intended to be incorporated into a food or feed has been supplied. The EU legislation does not, however, require whole food chain tracing or internal product tracing, that is, the matching up of all inputs to outputs, which is a feature of international and commercial standards (Campden BRI 2009). Guaranteeing the authenticity, geographical origin, traceability and safety of foods requires new diagnostic tools and implementation of new information systems.

18.4 Future traceability solutions

A food product is a combination of many components that all need to be controlled along the entire food chain. Each of them is potentially the subject of voluntary or involuntary contamination, tampering, cheating or spoilage, and thus could represent a hazard to consumers. Supply chains embracing global markets of finished products require extensive handling of commodities, ingredients, components and finished food products by a large number of people before reaching the final consumer. Another factor causing an increased hazard is the increasing popularity of ready-to-eat foods which undergo minimal processing therefore maintaining the required level of freshness and taste without reducing the complexity of the ingredients' supply chain. In all these cases, the potentially increased hazard can only be reduced to a minimum if the entire food chain is controlled. Independent control by each food chain player is not sufficient and not cost-effective: the entire food chain needs to be goverened with advanced tools.

Control of production along the food chain can be exerted if each element of the full product chain can be identified. Identification is therefore a necessary component that allows differentiation of a single element from the others and so influences the potential need for intervention to re-establish the desired safety or quality level after an incident.

Currently, the concept of 'lot' is used as the identification unit. However, the 'lot', as a minimum group of homogeneously manufactured products, is a relative concept that changes with different food chain players and with different product chains. The larger the 'lot' is, the less complex are the controls and the more cost-intensive is a product recall. A new concept of 'lot' is being discussed which could base its definition on a homogeneous risk level. The new challenge would be to reduce the size of a lot, but this needs a substantial re-thinking of current product handling and traceability tools.

© Woodhead Publishing Limited, 2011

Traceability is one of the most powerful instruments for guaranteeing food chain integrity as well as reducing the withdrawal cost of a product. The speed of the response and the reliability of information are key elements in any solution to traceability. Real-time detection of critical contaminants at different levels along the food chain is an essential component of traceability systems. All these elements lead to moving from manually operated and paper-based traceability to automated ICT-enabled and empowered traceability using mathematical modelling and source level inferences.

Basic traceability is a consumer requirement for which companies are not always willing to pay. Extended traceability that incorporates the ability to trace other components such as quality, safety and non-tangible attributes (e.g. origin, methods of production, food miles) could become a business opportunity for the food chain companies while fulfilling social requirements. The ability to collect relevant information from the food chain and use it to ensure product quality in real-time provides tangible benefits to the food industry.

The incorporation of many components into food chain traceability systems would highlight a development of the food chain infrastructures from basic traceability towards food chain e-platforms applying collaborative supply chain management principles, which is the eventual requirement for lasting food chain integrity.

The future directions for traceability are:

- Adoption by industry of the technical solutions already developed by the research community (see Chapters 1, 3, 8 and 14). Particular challenges remain for food SMEs where the cost of adoption of the new technologies may be prohibitive.
- Increasing regulation, in part reflecting the technical solutions now available. It is recognized that the traceability requirements of EU 178/2002 were a first attempt to introduce regulatory requirements for traceability within the European Union. One key future consideration is how feasible it is to adopt internal traceability as a regulatory requirement across all sectors of the food industry.
- Harmonization of approaches and protocols for traceability systems. One of the key messages from the many studies presented in this book is that for full chain integrity, information must flow with the product as it passes through different ownership along the food chain. For this to work successfully, harmonization of approaches and specifically the protocols for information storage and retrieval are essential.
- Environmental and sustainability issues including the drive to reduce 'food miles' and carbon footprint will influence food chains, perhaps by reducing the length of chains. To fully quantify food miles and carbon footprint for food products, accurate information is required on the actual total food/feed chain associated with the product.
- Further increase in interest by consumers in food traceability issues will drive

© Woodhead Publishing Limited, 2011

the further development of web- and mobile phone-based applications used by consumers to access traceability information.
- Increased focus on full feed/food chain analysis will help drive the adoption of objective risk ranking approaches to hazards in the food chain, which in turn will require the further development and adoption of real-time sensors for detection of hazards. This will need to be coupled with better statistical process control techniques to adequately manage and control hazards in the food chain.
- Mathematical modelling will not only be for tracing forward in the food chain, but also tracking back to identify sources of contamination, and therefore facilitate more targeted actions and more rapid responses (see Chapter 2).

18.5 Future food safety solutions

Over the decades, food safety issues change. During the 1960s the focus was on chemicals, the 1980s saw the emergence of listeria and VTEC as foodborne pathogens, while the 1990s was focused on BSE and other food safety scares. This was the backdrop to the emergence of 'traceability' in the food chain in the late 1990s and early 21st century.

Future trends that will facilitate food safety include:

- Functional genomics: the developments in genomics, proteomics, transcriptomics, etc., have generated a very large amount of data. To extract the maximum information from this data, expertise in bioinformatics and functional genomics is needed.
- Microbial quantification: with advances in detection technologies, particularly molecular methods, leading to possibilities of improved typing and quantification, and improvements in statistically based sampling methods, better data will be available for mathematical modelling (Hoorfar *et al.*, 2011).
- Nanotechnology and its potential uses in smart packaging, sensors and traceability is currently in its infancy from an applications point of view. Future trends will likely see wider applications of such technology.
- Risk informed decision making: a systems approach to risk-based assessment, agent-based modelling, decision-making tools, better communication and risk-informed decision making will facilitate traceability.

A major risk with these approaches that must be guarded against is 'information overload'. Too much information, possibly even some conflicting data, could lead to even greater difficulty in decision making, resulting in 'decision paralysis'. Whatever the advances in traceability, detection methods, modelling or information technology, it is clear that they must be 'user friendly' so that they have a broad application and result in improved food safety. However, despite these advances, humans will still make errors – deliberate or otherwise; reduction of these errors is paramount.

© Woodhead Publishing Limited, 2011

18.6 References

ANON. (1995) National Institutes of Standards and Technology. An introduction to computer security. The NIST Handbook special publication 800-12. Available at http://csrc.nist.gov/publications/nistpubs/800-12/handbook.pdf

CAMPDEN BRI (2009) *Traceability in the food and feed chains: General principles and basic system requirements*. Guideline nr. 60. Gloucestershire, UK: Campden BRI.

HOORFAR J, CHRISTENSEN BB, PAGOTTO F, RUDI K, BHUNIA A, GRIFFITH M. (2011) Future trends in rapid methods. In: Hoorfar J (ed.), *Rapid Detection, Characterization and Enumeration of Foodborne Pathogens*. American Society for Microbiology, Washington, DC.

© Woodhead Publishing Limited, 2011

Part V

Appendix: Project abstracts

© Woodhead Publishing Limited, 2011

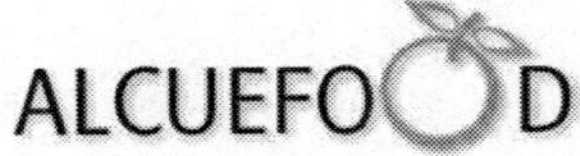

ALCUE-FOOD

From European fork to Latin American farm: an innovative networking platform for EU-LAC partnerships in food quality and safety R&D

www.inta.gov.ar/alcuefood/

Further to the Rio summit, the bi-regional scientific and technological (S&T) dialogue between Latin American and Caribbean (LAC) countries and the EU's Members States (ALCUE) successfully highlighted opportunities in areas of co-operation and provided a concrete outline of issues to be addressed in the agrifood area. Based on the recommendations of this dialogue, this proposal focused on the implementation of a 'networking platform' to promote and enhance S&T co-operation in 'Food Quality and Safety'. Considering the role of agriculture and agrifood in the Southern Cone Countries, the importance of agricultural products trade to the EU, the presence of recognized research groups and the bilateral EU agreements in S&T, the networking platform was implemented and focused its activities in this area. Nevertheless, the dissemination of outputs was addressed to all LAC countries. The **ALCUE-FOOD** project proposed a 'European fork to Latin American farm' approach, to:

- meet EU safety and quality requirements
- develop safer food and feed chain production
- provide high-quality and health-enhancing foods.

Promoting in LAC countries a 'total food chain approach' for EU exported products will subsequently contribute to improving the welfare of Latin American populations and must be considered as an important impact in terms of co-operation and mutual benefit. The main expected results were:

- implementation of R&D bi-regional groups of scientists
- organization of an innovative and global information system
- proposal of new relevant R&D topics based on analysis of social demand and productive sector needs
- synergy between the R&D national policies and those of the European Commission.

The implementation of the networking platform necessarily promoted dialogue between and within the two regions, with implications for national and regional policies. It contributed to strengthening the food quality and safety priority and internationalizing the European Research Area.

© Woodhead Publishing Limited, 2011

ALCUE-FOOD partners

Argentina
Secretariat for Science, Technology and Productive Innovation
Instituto Nacional de Tecnología Agropecuaria
Coordinadora de las Industrias de Productos Alimenticios

Belgium
Universiteit Gent

Brazil
Brazilian Agricultural Research Corporation
Chambre de Commerce France-Brésil

Chile
Fundación Chile
Organización Internacional de Asociaciones de Consumidores

France
Centre de Coopération Internationale en Recherche Agronomique pour le Développement

Portugal
Instituto de Biologia Experimental e Tecnológica

Uruguay
Asociación de las Universidades del Grupo de Montevideo

© Woodhead Publishing Limited, 2011

BIOTOX

Development of cost-effective tools for risk management and traceability systems for marine biotoxins in seafood

The general objectives of the **BIOTOX** project were:

- development and validation of a confirmatory analytical method for the detection of lipophilic toxins in shellfish, and of robust, rapid and cost-effective assays that cover all lipophilic shellfish toxins mentioned in the EU legislation and to enable a phasing out of all animal tests currently applied in this field
- development of advanced early warning tools for the detection of shellfish toxins in seawater and algae
- development of decontamination methods for toxic shellfish
- implementation of the developed methods in adequate hazard analysis and critical control point (HACCP) control and verification procedures in Europe for the monitoring, depuration and traceability of biotoxins, and advice on the harmonization of these procedures in European member states.

With respect to the last objective:

A review of 'European Shellfish producing countries and EU Food Safety legislation on monitoring and control of bivalves' has been produced and an overview of industry practice on shellfish toxin control by industry case studies has been given. Recommendations were made to national and community authorities. These authorities should take the responsibility in mentoring Food Business Operators (FBOs) to help fill the gaps in scientific and technological knowledge where this is needed, and to make official control data more accessible on the web.

© Woodhead Publishing Limited, 2011

BIOTOX partners

Belgium
Scientific Institute of Public Health

France
Institut Français De Recherche Pour L'Exploitation De La Mer

Ireland
Oyster Creek Seafoods Limited
National University of Ireland
Marine Institute

Italy
University of Modena

The Netherlands
Rikilt Institute of Food Safety
Institute for Marine Resources and Ecosystem Studies

Norway
Biosense Laboratories AS
The Norwegian School of Veterinary Science
National Veterinary Institute

© Woodhead Publishing Limited, 2011

BIOTRACER

Improved biotraceability of unintended micro-organisms and their substances in food and feed chains

www.biotracer.org

The **BIOTRACER** consortium included Europe's largest food/feed industries, several SMEs, relevant international co-operation (INCO) countries, experts in predictive microbiology, database developers, software companies, risk assessors, risk managers, system biologists, food and molecular microbiologists, legislative officers, standardization and validation members and food retailers. The strategic objective was to improve traceability of micro-organisms (biotraceability) contaminating the food/feed chain (including bottled water), whether accidental or deliberate. Using a total food chain approach, and integrating novel genomic and metabolimic data combined with a better understanding of the physiology of the micro-organisms, and advances in predictive food-based microbiological models, progress has been made towards reducing food safety risks.

The outputs of **BIOTRACER** include:

- improved detection methods for *Salmonella*, *Clostridium botulinum* BoNT genes, norovirus
- development of modelling methodology to include source level inference
- domain models for *Salmonella* in the pork chain and *S. aureus* in the milk chain
- better understanding of the physiology of *Listeria, Salmonella, S. aureus* and *Campylobacter* in the relevant food chains
- a greater understanding of contamination (*Salmonella* and mycotoxins) in the feed chain.

The results of the project have been published in more than 100 scientific journal articles and many interactions with industry and regulatory authorities. All of this will help ensure a more reliable and rapid response to a contamination event in the food chain.

BIOTRACER partners

Austria
University of Veterinary Medicine Vienna

Belgium
Nutrition Sciences N.V.

© Woodhead Publishing Limited, 2011

Brazil
Brazilian Agricultural Research Corporation

Czech Republic
Institute of Chemical Technology Prague

Denmark
National Food Institute – Technical University of Denmark
Faculty of Life Sciences – University of Copenhagen
ilochip A/S
HUGIN Expert Ltd
Danish Meat Research Institute
State Serum Institute

Finland
University of Helsinki

France
France Agency for Food Safety
Centre National Interprofessionnel de l'Économie Laitière

Germany
Free University of Berlin
University of Regensburg
Technical University of Munich
Federal Institute for Risk Assessment

Greece
Agricultural University of Athens

Indonesia
Gadjah Mada University

Ireland
Glantreo Ltd
TEAGASC

Italy
Italian National Institute of Health
COOP Italia

Lithuania
Lithuanian Veterinary Academy
Joint Stock Company Agaras

Netherlands
Agrotechnology and Food Innovations B.V.
Check-Points Ltd
RIKILT Institute of Food Safety
National Institute of Public Health and Environment
VION Fresh Meat West

Norway
Norwegian Food Research Institute

Romania
'Dunărea de Jos' University of Galaţi

Russian Federation
A.N. Bakh Institute of Biochemistry

Slovakia
Food Research Institute

Slovenia
University of Ljubljana

South Africa
Microchem
University of Johannesburg

Spain
Valencian Institute for Agricultural Research

Sweden
Lund University
Prevas Ltd
National Veterinary Institute
Swedish Feed

Switzerland
University of Bern
SmartGene Services Ltd
NESTEC, Nestlé Research Centre

United Kingdom
University of Manchester
London School of Hygiene and Tropical Medicine
Institute of Food Research

© Woodhead Publishing Limited, 2011

CHILL-ON

Developing and integrating novel technologies to improve safety, transparency and quality assurance of the chilled/frozen food supply chain

www.chill-on.com

The consumption of chilled and frozen food products within Europe is growing above average (growth rates >10%) due to changes in society whereas food safety and control are of great concern for consumers. With 11 million tons consumed per year, the European Union is the second largest market for chilled and frozen food products. The reinforcement of confidence in chilled food products, especially regarding supply-chain aspects, is of high priority for all involved in food manufacturing, trade, logistics and distribution, especially as Directive 178/2002/EC on General Food Law required traceability to be established at all stages of the food chain from January 2005. This Directive forces the introduction of community controls of the treatment of agricultural products and foodstuffs. More than 70% of food products are traded internationally which makes the food supply chain very complex with regard to quality and safety management. Before the project started, there were no integrated concepts available for the chilled and frozen food chain which allowed both the complete and continuous monitoring of food safety and quality (together with environmental concerns) and the traceability of the entire supply chain. The pro-active and integrated approach of the **CHILL-ON** project, implemented in-depth research for each step of the chilled/frozen supply chain. This included reviewing state-of-the-art technologies, development of new technological concepts, their validation and potential scale up with focus on monitoring managing and tracing the entire supply chain by a newly developed system called eCHILL-ON. As fish is in third place in the European food market and is one of the most sensitive goods with regard to food poisoning, the chilled and frozen fish supply chain was selected to be the test case in this IP. The proposed integrated **CHILL-ON** concept will result in the 2020s generation of the frozen/chilled food supply chain.

© Woodhead Publishing Limited, 2011

CHILL-ON partners

Brazil
Fundação Universidade Federal do Rio Grande
Seara Cargill S.A.
Companhia Minuano de Alimentos

Chile
Fundación Chile

China
Chinese Agricultural University
Beijing Beishui Food Industry Co. Ltd

Germany
ttz Bremerhaven, Verein zur Förderung des Technologie-Transfers an der Hochschule Bremerhaven e.V.
Rheinische Friedrich-Wilhelms-Universität Bonn
Q-Bioanalytic GmbH

Greece
Research Laboratories Ltd

Iceland
Matís ohf.
University of Iceland

Israel
Afcon Software and Electronics Ltd
Motorola Israel Ltd
OSM-DAN Ltd
Technion – Israel Institute of Technology

Italy
ActValue Consulting & Solutions S.R.L.
University of Parma

The Netherlands
Chainfood B.V.

Spain
Instituto Technológico del Embalaje, Transporte y Logística

Switzerland
Freshpoint Quality Assurance Ltd

United Kingdom
Wessex Institute of Technology
University of Kent
Research Relay Ltd
Traceall Ltd

© Woodhead Publishing Limited, 2011

CO-EXTRA

GM and non-GM supply chains: their CO-EXistence and TRAceability

Co-Extra has been studying biological containment methods and pollen flow dissemination, in particular over long distance on fragmented landscape. **Co-Extra** noted that companies are using a contractual threshold (between 0.1 and 0.01%) largely below the 0.9% European labelling threshold. It can be explained by the sampling and measurement uncertainties as well as a safety margin kept by companies for further admixtures. Several tools to detect unapproved genetically modified organisms (GMOs) have been produced. They range from a day-to-day cost-effective method to highly refined and costly methods depending on the aim of detection. Apparatus and chemistries used in quantitative real-time polymerase chains reactions (QRT-PCRs) show that all technologies provide equally accurate results. The most cost- and time-effective can thus be chosen. The taxa reference systems are still inappropriate in several cases, including some already validated. Harmonization of taxa identification and quantification is still required and core collections of taxa should be established. Several sampling plans were assessed for both fields and shipments. Information structure, content and flow management for ensuring reliable and cost-effective documentary traceability has been studied all along supply chains.

Co-Extra surveyed GMO-related legal regimes and practices that exist in and beyond the EU with an important work on liability regimes applicable in supply chains. **Co-Extra** developed decision-support modules integrating tools, methods, models and guidelines needed to deal with the coexistence and traceability issues. Stakeholders have been involved in the project. Surveys and analyses of consumers' attitudes and opinions were made. Results were presented to different EC Directorate General and European Network of GMO Laboratories. A book on coexistence and traceability of GM and non-GM supply chains is currently in preparation.

COEXTRA partners

European Commission
Joint Research Centre

Argentina
Instituto Nacional de Tecnología Agropecuaria

Austria
European Centre of Tort and Insurance Law

Belgium
Centre Wallon de Recherches Agronomiques

© Woodhead Publishing Limited, 2011

The Institute for Agricultural and Fisheries Research – Technology & Food Unit
Eppendorf Array Technologies SA
Flanders Interuniversity Institute for Biotechnology
Hogeschool Gent – Department of Biotechnological Sciences
Scientific Institute of Public Health

Brazil
Instituto de Tecnologia do Paraná
Faculdades Integradas do Brasil

Bulgaria
AgroBioInstitute

Denmark
Danish Institute of Food and Resource Economics

France
ADRIANT
ARVALIS – Institut du Végétal
Centre National de la Recherche Scientifique
Centre Technique Interprofessionnel des Oléagineux Métropolitains
Groupe d'Intérêt Public – Groupe d'Etude et de contrôle des Variétés et des Semences
Institut National de la Recherche Agronomique
INRA Transfert
Service Commun des Laboratoires du MINEFI – Laboratoire de Strasbourg

Germany
Bundesinstitut für Risikkobewertung
Julius Kuehn Institute
Fraunhofer Gesellschaft zur Forderung der Angewandten Forschung – Institute for Process Engineering & Packaging
GeneScan Analytics GmbH
Genius – Biotech GmBH
Max Planck Society for the Advancement of Science
Weihenstephan-Triesdorf University of Applied Sciences

Greece
National Hellenic Research Foundation

Italy
Centro Ricerche Produzioni Animali SpA
Istituto Superiore di Sànita
University of Parma

The Netherlands
RIKILT Institute of Food Safety
Schuttelaar and Partners

Norway
Nofima Mat
National Veterinary Institute

Poland
Warsaw Agricultural University

Russia
Center Bioengineering Russian Academy of Sciences

Slovenia
Agricultural Institute of Slovenia
Institut 'Jozef Stefan'
National Institute of Biology

Spain
Centro de Investigación en Economía y Desarrollo Agroalimentarios
Consejo Superior de Investigaciones Científicas

Switzerland
Agro-ProjektManagement
Biolytix AG
Delley Semences et Plantes
Forschungsinstitut für Biologischen Landbau
Swiss Federal Institute of Technology Zurich

United Kingdom
LGC Ltd
Lumora Ltd
Central Science Laboratory Defra
National Institute of Agricultural Botany
University of Reading

© Woodhead Publishing Limited, 2011

GTIS CAP

Geotraceability Integrated System for the Common Agricultural Policy (GTIS CAP)

http://gtis-cap.cra.wallonie.be/

The aim of **GTIS CAP** was to define and to validate an integrated information system that would serve both the European and national administrative bodies in charge of the Common Agricultural Policy and the producers of vegetal produce for consumers and for livestock. Some aims of **GTIS CAP** were to complete the Integrated Administration and Control System/Land Parcel Information System (IACS/LPIS) data with other data obtained from remote sensing and to define simple geotraceability indicators aimed at the management, control and monitoring of the CAP, which could also be used in integrated agricultural management systems. **GTIS CAP** aimed to facilitate online access to this geo-traceability data and ensure greater efficiency in checking and conformity of good agricultural practices, in line with CAP recommendations. The **GTIS CAP** project comprised four different types of work:

1. The evaluation of new functionalities necessary for the integration of geo-traceability into existing tools and for it to become a CAP management, follow-up and monitoring tool which will also be useful for producers.
2. Definition of norms and standards that will enable the interoperability of IACS/LPIS geo-referenced data and functionalities for CAP management, follow up and monitoring.
3. Definition of simple and easy-to-use geo-indicators provided by satellite images.
4. Implementation and validation of the integrated system in two test regions in order to produce after experimentation by producers groups a recommendation for European and national administrative bodies.

© Woodhead Publishing Limited, 2011

GTIS-CAP partners

Belgium
Centre Wallon de Recherches Agronomiques
Centre Interprofessionnel pour la Gestion en Agriculture
Université de Liège

France
ACTA Informatique
CDER
Spot Image

© Woodhead Publishing Limited, 2011

OTAG

Operational management and geodecisional prototype to track and trace agricultural production

www.otag-project.org

Specific Support Actions (SSAs) were designed to complement the implementation of the 6th Framework Programme and were possibly to be used to help in preparations for future Community research and technological development policy activities including monitoring and assessment activities related to food safety and security. They involved conferences, seminars, improvement studies and prototype implementation analyses, working groups, expert groups and user groups, operational support and dissemination, information and communication activities, or a combination of these, as appropriate. The aim of **OTAG** was to find a sustainable and easily understood method to track and trace emerging risks in beef production in the context of Southern Cone Countries and EU policies.

OTAG focused on improving:

- innovative economically viable mechanisms, methods and geotechnologies, for recording reliable and accurate data on beef origin and primary production as well as the environmental conditions of pasture territories and
- putting in place an operational geodecisional system under control conditions, to track and trace the mobility, provenance and state of beef cattle using emerging geospatial and geocommunication technologies.

This SSA project was strengthened by European and Canadian knowledge, existing operational systems and precision geoinformatics tools as well as the interaction of experts and users groups from Southern Cone Countries, Canada and EU. This SSA took advantage of previous EU projects on related topics as well as the actual state-of-the-art to demonstrate by a combination of operational prototype, dissemination, information and communication activities how the management of emerging risks in beef production in the context of Southern Cone Countries and the EU policies can be efficiently supported.

© Woodhead Publishing Limited, 2011

OTAG partners

Argentina
Programa Cooperativo para el Desarrollo Agro-alimentario y Agro-industrial del Cono Sur

Brazil
Brazilian Agricultural Research Corporation

Canada
Université Laval

France
CIRAD-Agricultural Research for Development
CEMAGREF-Agricultural and Environmental Engineering Research

© Woodhead Publishing Limited, 2011

PathogenCombat
for safe food

PATHOGENCOMBAT

Control and prevention of emerging and future pathogens at cellular and molecular level throughout the food chain

www.pathogencombat.com

Food safety is of fundamental importance to the European consumer, food industry and economy. The impact on trade and competitiveness is very substantial. Despite significant investment the incidence of food derived disease still increases in the EU. **PathogenCombat** attacked this pan-European problem through a holistic, multidisciplinary approach towards threats from new/ emerging pathogens in the entire food chain. A number of advanced platforms have been explored to investigate the survival and virulence expression of pathogens in feed and food, and on contact surfaces in the food chain including the intestinal tract of farm animals. The platforms, several of which were used for the first time in food safety studies, comprised bioimaging, laser tweezers, functional mammalian cell models, functional genomics and microarrays. New/ emerging foodborne bacteria, yeast, filamentous fungi and viruses were targeted in the contexts of milk and dairy products, beef and lamb, poultry, pigs and their meat products. The key deliverables were:

- food with none or acceptably low levels of pathogens
- new methods to detect and predict the occurrence of viable pathogens and their virulence in the food chain and at the time of consumption
- control methods to eliminate biofilms based upon hygienic design and processing
- novel processing technologies to inactivate pathogens
- new probiotic and protective cultures to eliminate pathogens
- pathogen control optimisation throughout the food chain based upon knowledge of pathogen behaviour
- cost effective food safety management systems
- unified awareness of food safety for the European consumers, food industries and regulatory agencies.

A distinctive aspect of this project was the facilitation of proactive, preventive, more coherent, effective and flexible responses to new/emerging pathogens. The deliverables will be widely exploited by SMEs to produce safe food and improve the health and well-being of Europeans and the competitiveness of the food industry.

© Woodhead Publishing Limited, 2011

PATHOGENCOMBAT partners

Australia
The Australian Food Safety Centre of Excellence

Belgium
University of Ghent

Cyprus
Pittas Dairy Industries Ltd

Czech Republic
Veterinary Research Institute

Denmark
Bactoforce A/S
Danisco A/S
Technical University of Denmark
University of Copenhagen

France
Agence Francaise de Securite Sanitaire des Aliments
Institute National de la Recherche Agronomique
Institut Technique Français des Fromages

Germany
Bergpracht-Milchwerk Gmbh & Co.
Biomax Informatics AG
CONGEN Biotechnologie GmbH
Max Rubner-Institut – Federal Research Institute of Nutrition and Food
Forschungzentrum Karlsruhe GmbH
Geflügelspezialitäten Ziegler
University of Stuttgart
Vermicon AG

Greece
Agricultural University of Athens
EBTE Consultants

Hungary
Corvinus University of Budapest

Ireland
Cocker Consulting Ltd

Italy
University of Udine/Torino
Alma Mater Studiorum – Universitá di Bologna
Anidral
Granarolo

The Netherlands
Wageningen University

Slovenia
University of Ljubljana
University of Maribor

Spain
Colear Castilla Sociédad Cooperative
Cooperativa Avicola y Ganadera de Burgos
Jamones Segovia, S.A.
Martinez Loriente, S.A.
Universidad de Burgos

Sweden
Lund University
Swedish University of Agriculture

United Kingdom
The Manchester Metropolitan University
Scottish Agricultural College

© Woodhead Publishing Limited, 2011

PROSAFEBEEF

Improving the safety of beef and beef products for the consumer in production and processing

www.prosafebeef.com

The EU has a strategically important global position in terms of beef. In 2003 the EU-25 region produced 8.06 million tons and consumed 8.32 million tons of beef and beef products, estimated at €75 billion. However, Common Agricultural Policy (CAP) reform, increased globalization, reduced commodity prices and an increasingly sophisticated, health-conscious consumer are motivating the beef-chain to produce beef and beef products that are convenient, traceable, nutritious and of consistent eating quality.

Alongside these considerations and of paramount importance given the serious impact of beef health scares, today's consumer demands assurances about the safety of beef and beef products. The overall objective of **ProSafeBeef** is to reduce microbiological and chemical contaminants in beef and beef products and to enhance quality, choice and diversity in the beef-chain in order to boost consumer trust and invigorate the industry.

ProSafeBeef will achieve this by deployment of four main instruments concerned with:

1. Development and application of quantitative risk assessment models to monitor, trace and reduce microbial/chemical contamination in the beef chain
2. Develop novel control and intervention strategies for microbial pathogens at key points in the beef chain to maximize the safety of beef and beef products
3. Satisfy consumer demand for diversity and choice and invigorate and add value to the beef chain by developing innovative processing techniques and products and
4. Engage with SMEs, expert collaborators from third countries and international co-operation (INCO) partners with a vested interest in beef export to assist in delivery of the ProSafeBeef project.

These activities will permit **ProSafeBeef** to deliver to the beef industry an innovative toolbox of novel methodologies and practices that will contribute to the imperative need for a viable European beef chain, operating at a competitive and sustainable level for its stakeholders while at the same time delivering safe, high-quality product to its many consumers.

© Woodhead Publishing Limited, 2011

PROSAFEBEEF partners

Australia
Co-operative Research Centre, Cattle and Beef Quality Australia

Austria
University of Veterinary Medicine Austria
International Atomic Energy Agency Austria
Josef Strobel und Sohne

Belgium
Ghent University

Brazil
Universidad Federale de Sao Paulo
Universidade de Sao Paulo
Microbioticos (SME)

Canada
Canadian Research Institute, University of Guelph

Denmark
Aarhus School of Business, University of Aarhus
Danish Meat Research Institute

France
Institut National de la Recherche Agronomique
Association pour le Développement de l'Institut de la Viande
Union Nationale des Cooperatives d'Elevage et d'Insemination Animale

Germany
Institute of Farm Animal Biology Germany
Greifenfleisch GmbH

Greece
Agricultural University of Athens
Aristotle University of Thessaloniki
Apostolos Papadopoulos & Sia OE
Biofarma Peloponissos AE

Ireland
Teagasc – The Agriculture and Food Development Authority
University College Dublin
University College Cork

The Netherlands
RIKILT, Institute of Food Safety

New Zealand
Institute of Environmental Science and Research Ltd

Norway
Nofima Mat AS, The Norwegian Food Research Institute
Furuseth Slakteri AS
Prima Jæren AS

Poland
Agricultural University of Poznan, Institute of Meat Technology
National Veterinary Research Instituteoland

Serbia & Montenegro
University of Novi Sad

Spain
L'Organizacion de Consumidores y Usuarios
Institute of Agro-Food Research and Technology

United Kingdom
Aberystwyth University
University of Bristol
Queen's University Belfast
British Nutrition Foundation
Ensors Abattoir Limited
Celtic Pride Limited

USA
USDA, Western Regional Research Centre

© Woodhead Publishing Limited, 2011

SEAFOODPLUS

Health-promoting, safe seafood of high eating quality in a consumer driven fork-to-farm concept

www.seafoodplus.org

The strategic objective of the **SEAFOODplus** Integrated Programme was to reduce health problems and to increase well-being among European consumers by applying the benefits obtained through consumption of health promoting and safe seafood products of high eating quality. The relevance of seafood in the diet to diminish the increasing incidences of, for example, cardiovascular, cancer and inflammatory diseases was assessed by performing dietary intervention and epidemiological studies. Other focus areas were the health of young population, weight reduction, prevention of osteoporosis and alleviation of postpartum depression observed in women after childbirth. Seafood's importance for consumer well-being and behaviour was assessed to understand determinants of consumers' seafood consumption and to adapt seafood products to consumer demands. The impact of health-related communication strategies on consumer seafood decision-making also needed assessment. The objectives of the seafood safety component were to make seafood safe for the consumer, by identifying risk factors, avoiding risks caused by viral and bacterial contamination, biogenic amines in seafood and to undertake risk-benefit analysis. The total value chain was addressed by developing consumer driven tailor-made, functional seafood products to improve health and to ensure nutritional quality and safety by full utilization of raw materials from aquaculture and from traditional fisheries. The aquaculture component studied the effects of dietary modulation, husbandry, fish physiology, genetics and pre-slaughter conditions. The challenge was to find a compromise between the trends towards intensive rearing and consumer demand for healthy, high quality seafood that is ethically acceptable, and has low impact on the environment. Validated traceability systems were assessed within **SEAFOODplus** to make it possible to apply a total chain approach from the live fish to the consumer product, and to trace back any feature from fork-to-farm.

SEAFOODPLUS partners

Belgium
Association Européenne des Producteurs de Mollusques

Canada
AquaNet (UBC)

Denmark
BioMar AS
Danish Institute for Fisheries Research
Danish Veterinary and Food Administration

© Woodhead Publishing Limited, 2011

International Organisation for the Development of Fisheries in Eastern and Central Europe
Marinova aps
Royal Greenland Seafood A/S
Statens Serum Institut
Aarhus School of Business

Germany
Federal Research Centre for Nutrition and Food
Friedrich-Schiller-University of Jena

Finland
University of Helsinki

France
Coopérative de Traitement des Produits de la Pêche
Ecole Nationale d'Ingénieurs des Techniques des Industries Agricole et Alimentaires
Institut Français de Recherche pour l'Exploitation de la Mer
Institut National de la Recherche Agronomique
Muséum National d'Histoire Naturelle
Université de Bretagne Sud – GIS PROGEBIO
University of La Rochelle
University of West Brittany

Hungary
Double Delta Kereskedelmi Termelo es Kutatasfejlestesi Beteti Tarsasag

Iceland
Icelandic Fisheries Laboratories
Landspitali-University Hospital & University of Iceland
Primex ehf

Ireland
Marine Institute
TEAGASC
University College Cork
University College Dublin

Italy
Istituto Superiore di Sanità

The Netherlands
Academisch Ziekenhuis bij de Universiteit van Amsterdam
Animal Science Group of Wageningen UR
Ghent University
Institute for Marine Resources and Ecosystem Studies
Maastricht University
National Institute of Public Health and the Environment
Plant Research International
TNO Nutrition and Food Research
Wageningen Centre for Food Sciences
Wageningen University
Zeeland Vis BV

Norway
AGA Halibut AS
Fjord Seafood ASA
Matforsk – Norwegian Food Research Institute
Moere Research
Norwegian Institute of Fisheries and Aquaculture Research
Norwegian University of Science and Technology
SINTEF Fisheries and Aquaculture
Trace Tracker Marine AS
University of Tromsø

Poland
Fish_Farm Tomasz Król
Institute of Ichthyobiology and Aquaculture of the Polish Academy of Sciences

Portugal
National Research Institute on Agriculture and Fisheries

Spain
Albacora, S.A.
AZTI Fundazioa
Consejo Superior De Investigaciones Científicas
GEASA, Gestión Empresarial Alavesa, S.A.
Salica Industria Alimentaria, S.A.
Universidad de Santiago de Compostela
Universitat de Barcelona
University of Coruna
University of Navarra

Sweden
Chalmers University of Technology
Göteborg University

United Kingdom
Centre for Environment Fisheries and Aquaculture
EWOS
Institute of Food Research
Johnson Seafarms Ltd
Unilever UK Central Resources Ltd
University of Glasgow
University of St Andrews

© Woodhead Publishing Limited, 2011

sigma
Σchain

SIGMAΣCHAIN

Developing a stakeholders' guide on the vulnerability of food and feed chains to dangerous agents and substances

www.sigmachain.eu

SigmaΣchain was a Specific Targeted Research Project (STREP) with 11 partners from the European Union, Norway and Brazil. With a duration of 36 months (starting on 1 April 2006) and EU funding of €2.93 million the project achieved a critical mass comprising centres of expertise from universities, research institutes and industry.

The project coordinated by University College Dublin, Ireland, aimed to develop methodologies to optimize the traceability process with respect to chain vulnerability to contamination.

Current chain traceability systems were evaluated, including methodologies for the identification of contaminants. Chain vulnerability varies depending on chain type, so case studies were conducted on four 'high vulnerability' products, representing three major categories of chains.

The chains are

1. drinking water (i.e. *a rapid contamination chain*)
2. milk powder (i.e. *a batch mixing chain*)
3. poultry meat and farmed salmon (i.e. *long geographic chains*).

Each of these chains were mapped and their vulnerability to contamination assessed. A Risk Model was developed to provide quantitative risk assessments of chain vulnerability. A generic framework was constructed for the assessment of chain vulnerability and the prioritization of chain contamination risk. This framework was validated using the case studies and wider inputs from stakeholders.

The outputs from these tasks were synthesized into a *Stakeholders' Guide* to food and feed chain vulnerability to contamination. The Guide is in book format, supported by software that enables the stakeholder to input specific chain data for a product and produce associated chain maps and assessments of contamination risks. It will also enable the stakeholder to examine risk minimization options (namely, corrective measures) such as enhanced security of the most vulnerable links.

© Woodhead Publishing Limited, 2011

SIGMAΣCHAIN partners

Brazil
Instituto de Tecnologia de Alimentos

France
Institut National de la Recherche Agronomique

Germany
Max Rubner-Institut
Syncom

Ireland
University College Dublin
AgriTech Solutions
Global Trust Certification

The Netherlands
TNO Quality of Life TNO
Wageningen University

Norway
SINTEF

Poland
Poznan University of Life Sciences

© Woodhead Publishing Limited, 2011

SIGMΣA

Sustainable introduction of GMOs into European Agriculture

http://ec.europa.eu/research/fp6/ssp/sigmea_en.htm
http://www.international.inra.fr/research/some_examples/results_of_the_european_sigmea_project

This project studied temporal and spatial gene flow across Europe in order to determine the measures needed for the co-existence of GM and non-GM production systems. Crops selected for the study were maize, oilseed rape, beet and wheat in both seed and crop production systems. The partnership exploited information from recent and current national research programmes as well as generating new research in the project. The results included:

- novel and improved methods for detecting GM plants/products, methods (including models) for monitoring and measuring seed and pollen mediated gene flow and predicting levels of impacts on whole farm and regional scales
- measures needed to segregate GM from other cropping systems, at the farm and regional scale, including changes in agronomic practices, new investments in equipment and machinery, zoning of crop production, the practicality and cost of these measures, including testing, authentication and stewardship of crops
- the socio-economic impacts of introducing GM crops into certain regions including increased costs on both the non-GM and GM sectors, the relationships between farmers adopting different technologies, constraints on farmers, farm saved seed implications, liability for purity of crops, conflicts and disputes between farmers. The results will inform all sectors of European agriculture and help with decision making at local, regional, national and European level on the management of farming systems in relation to GM and non-GM cropping.

© Woodhead Publishing Limited, 2011

SIGMΣA partners

Commission of the European Communities
Directorate General Joint Research Centre

Belgium
Arcadia International
Eigen Vermogen Van Het Instituut Voor Landbouw-En Visserijonderzoek
Université Catholique De Louvain

Czech Republic
Czech University of Agriculture in Prague
University of South Bohemia in Ceske Budejovice

Denmark
Danmarks Tekniske Universitet
Aarhus Universitet
Kobenhavns Universitet

France
Arvalis-Institut du Vegetal
Institut National de la Recherche Agronomique
Universite De Paris-Sud XI
Centre Technique Interprofessionnel des Oleagineux Metropolitains

Germany
University of Applied Sciences of Weihenstephan
Technische Universitaet Muenchen
Technical University of Braunschweig
Bundesamt für Verbraucherschutz und Lebensmittelsicherheit
Universitaet Hohenheim
Universitaet Bremen
Fraunhofer Gesellschaft zur Foerderung der Angewandten Forschung E.V.
Federal Biological Research Centre for Agriculture and Forestry

Italy
Consiglio Per La Ricerca E Sperimentazione In Agricoltura
Polytechnics University of Marche

The Netherlands
Plant Research International B.V.

Poland
Plant Breeding and Acclimatization Institute

Slovenia
Jozef Stefan Institute

Spain
Institut De Recerca i Tecnologia Agroalimentaries
Consejo Superior de Investigaciones Cientificas
Instituto Nacional de Investigacion y Tecnologia Agraria y Alimentaria

Switzerland
Forschungsinstitut Fuer Biologischen Landbau
Swiss Federal Institute of Technology
Universitaet Bern

United Kingdom
NIAB
Silsoe Research Institute
Institute of Grassland and Environmental Research
Rothamsted Research Ltd
University of Sheffield
University of Reading
Adas Consulting Ltd
Central Science Laboratory, DEFRA
Cambridge Environmental Research Consultants Ltd
Scottish Crop Research Institute

© Woodhead Publishing Limited, 2011

TRACE

Tracing food commodities in Europe

www.trace.eu.org

Trace developed generic and commodity specific traceability systems to enable the objective verification of the origin of food and animal feed. Natural tracers such as trace elements, ratios of heavy (geo) and light (bio) isotopes and genetic markers were measured to determine geographical, species and production origin. Geo and bioclimatic mapping was undertaken to extrapolate geographical origin and to reduce the need for commodity specific databases. Post genomic technology created rapid and sensitive methods for species identification. Profiling methods were used to identify markers, to characterize food products and facilitate cost effective screening methods. The project focused on meat, cereals, honey, olive oil and mineral water. Particular attention was paid to commodities labelled as 'Designated Origin' and 'Organic'. The European Virtual Institute for Chemometrics developed novel specifications obtained from analytical data, which characterized these foods. **Trace** developed and drafted standardized XML request-response schemes for coding and electronic information exchange and established 'Good Traceability Practice'. An electronic information platform incorporating verifiable specifications and thresholds was developed. This permits objective verification of origin and enables rapid and cost effective product withdrawal. The new systems were demonstrated and benchmarked in the five chosen chains, and the cost/benefit determined. Consumer scientists investigated perceptions and attitudes towards the 'ability-to-trace food' through a pan-European consumer study. A multi-tiered interactive information system was developed for communication both within the project and for external use by industry, consumers and regulatory bodies. Research workers, control agencies and industry have been educated in the new methods, procedures and systems via short-term secondments and workshops.

TRACE partners

European Commission
DG Joint Research Centre of the European Commission

Argentina
Consejo Nacional de Investigaciones Cientificas y Tecnicas
National University of La Plata
Universidad Nacional de Córdoba

Austria
Austrian Institute of Technology GmbH
WPA Beratende Ingenieure GmbH

© Woodhead Publishing Limited, 2011

Belgium
Walloon Agricultural Research Centre
Vrije Universiteit Brussels

China
Institute of Quality Standards and Testing Technology for Agricultural Products

Czech Republic
Institute of Chemical Technology Prague

France
Eurofins Scientific Analytics
Institut National de la Recherche Agronomique
Ecole Nationale d'Ingenieurs des Techniques Agricoles de Clermont-Ferrand
Famille Michaud Apiculteur

Germany
Bundesinstitut für Risikobewertung
LGL Bayern
Bavarian State Collection for Palaeontology and Geology
Isolab GmbH
Hydroisotop GmbH
Qiagen GmbH
Bundesanstalt für Geowissenschaften und Rohstoffe

Greece
EKPIZO
Agricultural University of Athens
The Hellenic Research House

Iceland
Maritech

Ireland
National University of Ireland
TEAGASC

Italy
Fondazione Edmund Mach
Institute of Chemical Metodologies of CNR
Catholic University of S. Cuore
Universita di Genova
University of Parma

The Netherlands
RIKILT Institute of Food Safety
University of Utrecht
Radboud University Nijmegen
Geochem Research BV
Wageningen University

Norway
The Norwegian Institute of Food, Fisheries and Aquaculture Research
TraceTracker Innovation
SINTEF Fisheries & Aquaculture Ltd

Poland
University of Silesia

Slovenia
National Institute of Chemistry

Spain
Universitat Rovira I Virgili
Universidad Politecnica de Madrid
Agua Insalus
FoodReg

Switzerland
Biolytix AG
Geschäftsstelle BATS

United Kingdom
The Food and Environment Research Agency
Institute of Food Research
Kenneth Pye Associates
University of East Anglia

© Woodhead Publishing Limited, 2011

TRACEBACK

Integrated system for reliable traceability of food supply chains

www.traceback-ip.eu

Assuring total traceability of food and feed along the whole chain from production to consumption is a cornerstone of EU policy on the quality and safety of food. This is a complex procedure involving identification, detection and processing of a vast amount of information. Profit margins of food producers and processors are already very tight, so they require a tracking mechanism that is not only reliable and easy to use, but does not entail a major cost burden. With a concerted effort and input from expert institutions, modern technology could provide such a system.

The integrated project **TRACEBACK** developed a system based on a service-oriented infrastructure model, including biosensors and microdevices tools, to build a new traceability architecture for the entire food supply chain extended to safety and quality. A key driver is the development of an information management system compatible with the technology level of all food chain players that is, without the burden of changing the technology currently operating in the field. The system is further supplemented by the aim to keep costs to a minimum.

Throughout Europe, 27 partners active in food research and the food industry, including food enterprises, teamed up with two from Egypt and Turkey to develop and test the innovative system in two representative and dominant food chains: from feed until dairy and in tomato.

TRACEBACK will have the following long-term socio-economic impacts:

- turn traceability from a burden for industry into an opportunity for the food chain
- underpin the safety and quality of the entire food chain
- help enterprises to reduce the cost of traceability
- give SMEs new market opportunities in high-tech and food products
- enable faster reaction to outbreaks of food contamination across the entire food chain, minimizing the consequent economic and health risks.

© Woodhead Publishing Limited, 2011

TRACEBACK partners

Egypt
NSCE Ltd

Finland
MTT Agrifood Research Finland

France
KBS

Germany
Centiv GmbH
Institut für Agrar- und Stadtökologische Projekte an der Humboldt – Universität zu Berlin
Max Planck Society

Ireland
Teagasc Agriculture & Food Development Authority

Italy
Tecnoalimenti S.C.p.A. (the Co-ordinator)
CEMAT – Combined European Management And Transportation S.p.A.
Consiglio Nazionale delle Ricerche, Istituto di Chimica del Riconoscimento Molecolare
Engineering Ingegneria Informatica S.p.A.
Federalimentare
Parmalat S.p.A
Scuola Superiore ISUFI, eBusiness Management Section, University of Lecce
Selex Communications S.p.A.
Technobiochip Scarl
University of Parma

Poland
ILIM – Institute of Logistics and Warehousing
Regionalna Wielkopolska Izba Rolno-Przemyslowa

Spain
AINIA – Asociaciòn de Investigaciòn de la Industria Agroalimentaria
Atos Origin Sociedad Anonima Espanola
CONSUM Sociedad Cooperativa Valenciana
SGS ICS Ibérica, S.A.

Sweden
Swedish University of Agricultural Sciences Faculty of Alnarp

Turkey
Akdeniz University, Economic Research Center on Mediterranean Countries

United Kingdom
City University London – Centre for HCI Design
University of Kent

© Woodhead Publishing Limited, 2011

Index

© Woodhead Publishing Limited, 2011

© Woodhead Publishing Limited, 2011

© Woodhead Publishing Limited, 2011

© Woodhead Publishing Limited, 2011

© Woodhead Publishing Limited, 2011

© Woodhead Publishing Limited, 2011

© Woodhead Publishing Limited, 2011

© Woodhead Publishing Limited, 2011

© Woodhead Publishing Limited, 2011

© Woodhead Publishing Limited, 2011

© Woodhead Publishing Limited, 2011